Engineering Trouble

China Studies

PUBLISHED FOR THE INSTITUTE FOR CHINESE STUDIES, UNIVERSITY OF OXFORD

Series Editor

Rana Mitter, *University of Oxford*

VOLUME 52

The titles published in this series are listed at *brill.com/chs*

Engineering Trouble

US–Chinese Experiences of Professional Discontent, 1905–1945

By

Thorben Pelzer

BRILL

LEIDEN | BOSTON

Funding information: DFG (SFB-1199, 2020), DAAD (IPID4all, 2018)

Cover image credit: Recreation of historical maps showing the Salaqi Irrigation Project, courtesy of Yi Meng Wu, Studio Wu 無 Design, Berlin

The Library of Congress Cataloging-in-Publication Data is available online at https://catalog.loc.gov
LC record available at https://lccn.loc.gov/2023016430

Typeface for the Latin, Greek, and Cyrillic scripts: "Brill". See and download: brill.com/brill-typeface.

ISSN 1570-1344
ISBN 978-90-04-54947-0 (hardback)
ISBN 978-90-04-54955-5 (e-book)

This book is printed on acid-free paper and produced in a sustainable manner.

Contents

Acknowledgments IX
List of Figures and Tables XI
List of Abbreviations XIII
Glossary of Recurring Civil Engineers XIV
Note on Transcriptions XVI

Introduction
Open Roads in China 1
1 Historical Background 2
2 The Segment of US-Trained Civil Engineers 7
3 "Specialized but Equally Ordinary Men" 11
4 Engineering and Its Discontents 17
5 Chapter Overview 21

1 Nationalism and the Cosmopolitical
Education Overseas, 1905–1918 24
1 Early Ways Into the United States 24
2 Engineering Education 31
3 Engineering Practice 34
4 The Cosmopolitan Dimension 36
5 The National Dimension 39
6 Bringing the Profession Home 45
7 Conclusion 50

2 Financial Constraint
The Grand Canal Board, 1918–1922 53
1 A Transnational Venture 53
2 Recruiting a Team 56
3 Surveying an Unstable Environment 63
4 The Money Goes Astray 67
5 Solidarity—The Yellow River Bridge Controversy 70
6 Frustration—Huai River Improvement Schemes 74
7 Conclusion 78

3 Political Dependency
Wang Jingchun at the Chinese Eastern Railway, 1919–1924 81
1 The Nationalist Paradigm 81
2 The Efficiency Paradigm 84

3 Wang Jingchun's Ascent to Power 88
4 The Chinese Eastern—"A Railroad Born in Sin" 92
5 The Wang–Ostroumov Dyad 100
6 The End of Expert Management 107
7 Conclusion 110

4 **The Visible College**
The Early Association of Chinese & American Engineers, 1919–1927 113
1 Formation and Organizational Makeup 113
2 Functions of the Association and Its *Journal* 120
3 Ethics and Socialization 125
4 Career Opportunities 134
5 Network for Collaboration 136
6 Transnational Friendship 138
7 Conclusion 142

5 **Cohesion and Exclusion**
The Relief Commission Paves the Provinces, 1926–1934 145
1 "Good Roads" and Labor Relief 146
2 Contested Authority in Yunnan 150
3 Guizhou—The Strong State 156
4 Guizhou—Mass Mobilization 158
5 Guizhou—Beneficiaries and Burden Bearers 165
6 Continuities between Xi'an and Lanzhou 168
7 Conclusion 172

6 **Demise without Exhaustion**
The Withdrawal of US Engineers, 1928–1941 174
1 Early Troubles of the ACAE 174
2 Support for the Command Economy 175
3 The Resurrection of the ACAE 182
4 Crisis of Repute—The Salaqi Irrigation Project 185
5 The Militarization of Civil Engineering 191
6 The Plight of US Engineers 196
7 Conclusion 201

7 **Wartime Engineering**
Ling Hongxun under Pressure, 1932–1945 205
1 A Career Start of Ups and Downs 205
2 Abortive Western Expansions 209
3 From Guangzhou to Hankou: Cutting Time 213

4 Lessons Learnt and Lessons Dealt 216
5 Drawbacks of the Strong State 219
6 The Gansu–Xinjiang Railway 224
7 Conclusion 228

8 **Conclusion** 230
1 Summary 230
2 Findings 232
3 Implications 235
4 Outlook 236
5 Coda: Loose Ends 237

Bibliography 241
Index 268

Acknowledgments

This book is the hard labor of many years. As everyone who has written a monograph is well aware, carrying out a research project is like climbing a tall mountain. But real-world mountaineers have an advantage: before setting out, they are already aware of the distance that lies between them and the mountaintop. For the researcher, the extent of the remaining ascent is always uncertain. The intellectual journey only reveals its length at the very end, when the fog lifts and the mountain peak reappears on the horizon.

Thanks to generous funding, I did not have to remain a backpacker on this trip. The German Academic Exchange Service (DAAD) financed my research in London and the US East Coast, the German Research Foundation (DFG) my travel to the West Coast and the Midwest, and the Leipzig University International Relations Office my stay in China. The private resources I invested into the project came from my employment at Leipzig University, where I taught at the Institute for East Asian Studies and researched at the Research Center Global Dynamics. For both employments, without which I would not have been in the financial position to consider writing this monograph, I need to thank Elisabeth Kaske, my restless and ambitious boss.

On my journey, I had stimulating discussions with Peter Bol, Eugenia Lean, Peter Perdue, Victor Seow, Ying Jia Tan, Emma Teng, Yuan Yi, and You Zhanhong. In Leipzig, I especially enjoyed the recurring comments from Hailian Chen, Steffi Marung, and Katja Naumann. Furthermore, I thank Martina Keilbach for coordinating the dissertation procedure out of which this book emerged. Among the many more individuals who have helped me on my journey, I cannot name all of them here. Sometimes, just in passing, they may have recommended a particular book or referred me to a specific term. All these cues, no matter how minor, were vital for the study and highly appreciated.

The book also profits from data tediously entered by my team at the Leipzig Research Center. While our dataset, known as the Chinese Engineers Relational Database, is a separate publication not directly related to this book, it has enabled additional analyses from which my argument benefited. It, too, was sponsored by the DFG and co-financed through the Federal Ministry of Education and Research (BMBF). I would like to thank Tsz-kit Chan, Oliver Dieckmann, Ariane Kolden, Chin Fung Ng, Felix Opper, Christian Staudte, and Mirjam Stretz for their careful assistance and smooth cooperation with this database.

I use this opportunity to thank my earliest mentors. The people who discovered and supported me were exclusively strong women. From Susanne Wiese, my schoolteacher, to Christine Moll-Murata, who fostered me during my undergraduate times, to Carmen Meinert, who encouraged me during my graduate studies: without people like these, I may never have realized, let alone been able to pursue, my love for the humanities.

Many more friends and colleagues have entertained me through the years. My friends never abandoned me, even when Covid regulations prohibited us from long-distance commuting. Merle and Xin deciphered the worst archival handwriting when I had already given up. Phelicitas, San, and Steven always cared.

Figures and Tables

Figures

1 Native places of civil engineering graduates, 1905–1918 27
2 Wang Jingchun as a student, c. 1908 29
3 Yang Baoling as a student, c. 1911 29
4 Xue Zhuobin, c. 1917 33
5 Tan Zhen, c. 1917 33
6 Ling Hongxun as an intern of the American Bridge Company, 1916 35
7 Caricature "Opportunity of the Returned Students," 1911 48
8 Map of the Grand Canal surveys, 1918–1920 52
9 Surveyors of the Grand Canal Improvement Board, between 1919 and 1921 64
10 A gauge, installed along the Yellow River banks near the Tianjin–Pukou Railroad Bridge, 1918 65
11 The Chinese Eastern Railway and its environs, 1919 80
12 Portrait of Wang Jingchun, c. 1921 97
13 Graph of the ACAE network main component, 1919–1941 112
14 Group photo of the annual ACAE fall meeting, 1921 121
15 Chronological development of *Journal* coverage, 1920–1941 125
16 Map of ANRC and CIFRC Projects, 1920–1933 144
17 Caricature "Transforming Soldiers into Workers," 1922 149
18 Peng Binglu as a student, c. 1920 152
19 Road inspection at Dabanqiao fields, 1927 155
20 Middle school students digging earth in Guizhou, 1926 159
21 Governor Zhou Xicheng lifting a Chinese hoe, 1926 162
22 New road in Guiyang, 1927 163
23 Bad grade near Zunyi, Guizhou, 1927 167
24 Manual cutting of rocks at Tingkou, West Shaanxi, 1933 170
25 Yang Baoling, c. 1927 178
26 The Salaqi irrigation project's state in 1934 187
27 Digging the Salaqi main canal intake, May 1932 189
28 Civil engineering degrees, 1905–1937 198
29 Map of Ling Hongxun's engagements 204
30 Ling locating a bridge site for the Gansu–Haizhou Railway, 1932 211
31 Ling completing the Guangzhou–Hankou Railway, April 28, 1936 215

Tables

1 US schools most popular with Chinese civil engineering students 26
2 National ratio of the ACAE members 118
3 Timeline of ACAE executives, 1920–1928 119
4 Comparison of the most prominent positions held 135
5 Timeline of ACAE executives after its "revival," 1929–1941 186

Rights and Permissions

Abbreviations

ACAE	Association of Chinese & American Engineers
AIC	American International Corporation
ANRC	American National Red Cross
ASCE	American Society of Civil Engineers
CER	Chinese Eastern Railway
CES	China Engineering Society
CIE	Chinese Institute of Engineering
CIFRC	China International Famine Relief Commission
CERD	Chinese Engineers Relational Database
CSA	Chinese Students Alliance
CPB	Central Planning Board
GMD	Guomindang (Nationalist Party)
IATB	Inter-Allied Technical Board
MAC	Military Affairs Commission
NDPC	National Defense Planning Commission
NEC	National Economic Commission
NGRA	National Good Roads Association
NRA	National Revolutionary Army
NRC	National Resources Commission
TSA	Triple System of Administration

Glossary of Recurring Civil Engineers

Name	Characters	Lifetime	Prominent function	Chapters
Freeman, John Ripley		1855–1932	Hydraulic engineer at the Grand Canal Board	2, 4, 6
Huang, Rongde	黃榮德	d. 1932	Highway engineer at the CIFRC	4, 5, 6
Ling, Hongxun	淩鴻勛	1894–1981	Railway, highway, and bridge engineer at the Ministries for Communications and Railways	1, 4, 6, 7
Peng, Binglu	彭炳祿	b. 1891	Highway engineer at the CIFRC	4, 5
Tan, Zhen	譚真	1899–1976	Hydraulic engineer at the Grand Canal Board and the CIFRC	1, 2, 4, 6
Todd, Oliver Julian		1899–1973	Hydraulic engineer at the Grand Canal Board, highway and hydraulic engineer at the CIFRC	2, 4, 5, 6
Waddell, John Alexander Low		1854–1938	Bridge engineer, consultant of the Nanjing government	2, 4, 6, 7
Wang, Jingchun	王景春	1882–1956	Railroad administrator at the Ministry of Communications	1, 3, 4, 6
Xue, Zhuobin	薛卓斌	1895–1991	Hydraulic engineer at the CIFRC	1, 2, 4, 6

(cont.)

Name	Characters	Lifetime	Prominent function	Chapters
Yang, Baoling	楊豹靈	1887–1966	Hydraulic engineer at various government river conservancies	1, 2, 4, 6
Zhang, Jichun	張季春	1902–1946	Highway engineer at the CIFRC	4, 5, 6

Note on Transcriptions

Chinese transcriptions follow the rules of the Hanyu Pinyin system. Although now an academic standard, the system was unknown to the historical individuals researched in this study.

To allow for cross-referencing with earlier material and to pay respect to the spellings which the historical actors preferred, original transcriptions are appended at the first occurrence of a person's name. To establish consistency and to avoid irregular exceptions on a case-by-case basis, this rule pertains to all people, including individuals whose original spellings are known to a foreign audience, e.g., Jiang Jieshi (Chiang Kai-shek).

The principle of consistency extends to all proper nouns, including the English rendition of railway names, for which in the past an inconsistent mix of terminal stations (e.g., "Peking–Kalgan Railway") and direct transcriptions (e.g., "Longhai Railway") was used. In this study, all railway lines are referred to by their terminals, approximating the meaning of their Chinese names (e.g., *Long Hai* 隴海 becomes "Gansu–Haizhou").

For reasons of consistency and clarity, the term *weiyuanhui* 委員會, when used to mark the rank of an institution inside the Nationalist government, is consistently rendered as "commission," even though one of the commissions, the National Economic Commission, historically referred to itself as the "National Economic Council."

Throughout the book, Chinese characters follow personal names and transcribed nouns at their first occurrence only. For Chinese and Japanese sources, characters are omitted from the footnotes but can be found in the bibliography.

The book includes an analysis of data from the Chinese Engineering Relational Database (CERD). In these cases, footnotes include search parameters with which the findings can be reproduced. The notation is explained in: Pelzer, Thorben. *Chinese Engineers Relational Database* (CERD) *Design Manual*. Working Paper Series SFB 1199 21. Leipzig: Leipzig University Press, 2020. Alternatively, an R script is also available.

Introduction

Open Roads in China

In 1915 San Francisco, the Panama-Pacific International Exposition opened its doors. Situated opposite Alcatraz, a temporary "Jewel City" celebrated the technical progress of human civilization. Curious visitors witnessed electrical light installations, automobiles raced the exhibition grounds, and a telephone line transmitted the sound of the Pacific Ocean to New York City. The expo's name came in celebration of the opening of the Panama Canal, a prestigious engineering feat seen as a triumph of humankind over nature. The exhibition's location was a triumph in its own right: opened in 1869, the Transcontinental Railway had prevailed over nature and fueled the rapid development of the recently annexed region. After the devastating earthquake of 1906, engineers had rebuilt San Francisco with record speed and efficiency.

Apart from more than 18 million domestic visitors, three Chinese engineers had come to join the celebration of technical achievements.[1] Independent from one another, they had traveled across the Pacific Ocean at a time when most Chinese had no access to electricity, never sat in a car, and only started seeing telegraphy stations sprout up three decades ago.[2] The three Chinese men at the expo marked three different generations of civil engineers. Zhan Tianyou 詹天佑 (Jeme Tien-yow, 1861–1919) was a guest of honor. He had originally come to the States in the 1870s and held the first US engineering degree awarded to a Chinese citizen. In his homeland, he had risen to fame as the first domestic chief railway engineer. The second engineer on site, Wang Jingchun 王景春 (Ching-chun Wang, 1882–1956), had graduated in civil engineering at Yale and had completed his PhD in railway management in Illinois a few years prior. He had recently started his career at the Ministry of Communications and was in San Francisco on a diplomatic mission. Finally, Ling Hongxun 凌鴻勛 (Hung-hsun Ling, 1894–1981) stopped over to visit the exposition as a young tourist. He had set sail to the States to work as an intern at US engineering companies and to take evening classes at Columbia University. As the three individually roamed the expedition grounds, they may have been unaware of one another's presence. However, at different phases of their career, their occupational background ultimately connected them as a group.

---

1 For an overview of the expo and its Chinese participation, see Yǒngzé Wáng, *Túshuō 1915 Bānámǎ sàihuì: Guāngyào shìbóshǐ de Zhōngguó piānzhāng* (Shanghai: Shànghǎi yuǎndōng chūbǎnshè, 2010).

2 Erik Baark, *Lightning Wires: The Telegraph and China's Technological Modernization, 1860–1890* (Westport, London: Greenwood Press, 1997), 158–63.

DOI:10.1163/9789004549555_002

Through their US training, they were coming to share professional upbringing, knowledge, and experiences. They were coming to understand themselves as a particular segment of civil engineers.

What did it mean to be a civil engineer in China during that time? How did this meaning come to be? What developments challenged it and forced it to evolve? At a time when the concept, territory, and political entity of "China" was under reconstruction and uncertainty, the engineering profession was under construction as well. On the one hand, domestically, a reimagined engineering occupation had only recently initialized the process of professionalization under the impression of import and utilization of modern technologies. On the other hand, internationally, conflicts and reorientations of the engineering profession unfolded in a dynamic, global(ized) environment. Civil engineering in China was, at the same time, a domestic endeavor that promised national salvation through technology and a transcultural practice inextricably linked with international finance and politics. Engineers, especially those that had received their education abroad, navigated through this dual context. What they found did not always meet their expectations and ideals. In this book, I argue that their experiences of contradictions, disillusionments, and discontent kept the profession in a constant flux of searching for its meaning and its place in society.

1 Historical Background

The age and origins of civil engineering in China depend on definition. According to Needham's classic, sympathetic approach, civil engineering had been a thousands of years old practice, and China one of the most prolific and influential regions of the craft.[3] Alternatively, engineering emerged with the reception of Western thought through Jesuit converts of the 17th century.[4] In yet another version, the move from a trial-and-error approach to a cohesive body of knowledge differentiated pre-modern artisans from modern

3 Joseph Needham and Ling Wang, *Physics and Physical Technology: Part II: Mechanical Engineering,* Science and Civilisation in China 4.2 (Cambridge: Cambridge University Press, 1965), 9–10; Joseph Needham, Ling Wang and Gwei-djen Lu, *Physics and Physical Technology: Part III: Civil Engineering and Nautics,* Science and Civilisation in China 4.3 (Cambridge: Cambridge University Press, 1971), 1. Also Francesca Bray, "Technics and Civilization in Late Imperial China: An Essay in the Cultural History of Technology," *Osiris* 13 (1998): 32–33.

4 Iwo Amelung, "Weights and Forces: The Introduction of Western Mechanics into Late Qing China," in *New Terms for New Ideas: Western Knowledge & Lexical Change in Late Imperial China*, ed. Michael Lackner, Iwo Amelung and Joachim Kurtz (Leiden, Boston, Cologne: Brill, 2001), 198–201.

engineers.[5] If one takes the scientific method as a prerequisite of engineering, one could follow reformers like Liang Qichao 梁启超 (1873–1929) and Lu Xun 鲁迅 (1881–1936), who could not find scientific thinking in Chinese history,[6] and, in consequence, strike out engineering from pre-modern China. Even if technological and scientific progress in China was retroactively "forgotten,"[7] a new generation of civil engineers in the early 20th century shared the vision of belonging to a new development. They believed themselves to be part of a radically new profession that may have shared some of its work with historical artisan crafts, but that followed scientific rules, operated under a modern efficiency paradigm, and organized itself in exclusive professional communities.

There is less ambition of (dis)placing civil engineering back in time for Western history. According to the common narrative, civil engineering as a self-conscious concept gradually emerged with the chartering of the British Institution of Civil Engineers in 1828, followed by the founding of the American Society of Civil Engineers in 1852. The process was slow and gradual: historians of European and American civil engineering agree that for the remainder of the 19th century, there was no cohesive paradigm, standardized certification, united cause, or common identity between the practitioners.[8] A "professional

---

5 Xiucheng Le, "The Struggle for Autonomy from the State: Professionalization of Engineering in China" (PhD, Cornell University, 1989), 279–80; Zhèng Fáng, *Jìndài gōngchéngshī qúntǐ de 'mínjiān lǐngxiù': Zhōngguó gōngchéngshī xuéhuì yánjiū (1912–1950)* (Beijing: Jīngjì rìbào chūbǎnshè, 2014), 11–14.

6 Iwo Amelung, "Die ‚vier großen Erfindungen': Selbstzweifel und Selbstbestätigung in der chinesischen Wissenschafts- und Technikgeschichtsschreibung," in *Selbstbehauptungsdiskurse in Asien: China–Japan–Korea*, ed. Iwo Amelung, Matthias Koch and Joachim Kurtz (Munich: Iudicium, 2003), 245–47. See also Fáng, *Jìndài gōngchéngshī qúntǐ de 'mínjiān lǐngxiù'*, 14–16.

7 Benjamin Elman, *On their Own Terms: Science in China, 1550–1900* (Cambridge, Massachusetts, London: Harvard University Press, 2005), 386–87. Also Bócōng Lǐ, "Zhōngguó jìnxiàndài gōngchéng shǐ yánjiū de ruògān wèntí," *Kēxuéjìshù zhéxué yánjiū* 30, no. 6 (2013): 63–64; Bócōng Lǐ, "Lüèlùn Zhōngguó jìnxiàndài gōngchéng shǐ de kāiduān," *Shāndōng kējì dàxué xuébào* 17, no. 2 (2015): 4–8.

8 Konrad Jarausch, *The Unfree Professions: German Lawyers, Teachers, and Engineers, 1900–1950* (New York, Oxford: Oxford University Press, 1990), 17–18, 89; Edwin Layton Jr., *The Revolt of the Engineers: Social Responsibility and the American Engineering Profession* (Baltimore, London: The John Hopkins University Press, [1971] 1986), 4; David Noble, *America by Design: Science, Technology, and the Rise of Corporate Capitalism* (Oxford, New York, Toronto: Oxford University Press, [1977] 1979), 39–40; Frank Uekötter, "Techniker an der Macht: Der Ingenieur-Politiker im 20. Jahrhundert," *Historische Zeitschrift* 306, no. 2 (2018): 398; Dirk van Laak, *Weiße Elefanten: Anspruch und Scheitern technischer Großprojekte im 20. Jahrhundert* (Stuttgart: Deutsche Verlags-Anstalt Stuttgart, 1999), 38–39.

spirit" only developed during the first two decades of the 20th century.[9] Therefore, even if one follows Liang Qichao's pessimistic view about a late start of scientifically grounded civil engineering in China, domestic practitioners did not have much professional development to catch up. Rather, they found themselves to be participants in an ongoing global professionalization process.

Political actors inside the Qing Empire first acknowledged the need for technological and infrastructural development in the 1860s. In this sense, modernization in China did not start later than in Russia and Japan, but it happened under political disorientation and disarray and was ultimately less successful.[10] The late Qing Self-Strengthening Movement (*ziqiang yundong* 自强運動, 1861–1895) brought with it the emergence of technology as a discourse.[11] At the onset of this reform program, scholars at the *Tongwen guan* 同文館 in Beijing, the Jiangnan shipyard in Shanghai, and the Fuzhou navy yard school started translating Western technical texts that would serve as a basis for workshop training.[12] A decade later, the government ran the Chinese Education Mission (CEM), during which it sent 120 young boys to the US to socialize and receive a modern college education. The project failed when the Qing state found its subjects to have become too "Westernized." The government ordered its students to return to China before all but two of them had completed their studies. Still, of 106 CEM returnees, 14 ended up working in civil engineering, and 39 in the broader engineering sector.[13] Meanwhile in China, the first higher education

9 Layton Jr., *The Revolt of the Engineers*, 44–46. Also William Akin, *Technocracy and the American Dream: The Technocrat Movement, 1900–1941* (Berkeley, Los Angeles, London: University of California Press, 1977), 20; Michael Davis, *Thinking Like an Engineer* (Oxford, New York: Oxford University Press, 1998), 45.

10 Cyril Black, Marion Levy Jr., and Gilbert Rozman, "Conclusions: China's Modernization in Historical Perspective," in *The Modernization of China*, ed. Gilbert Rozman et al. (New York: The Free Press, 1981), 488–91.

11 Francesca Bray, "Technology," in *The Making of the Human Sciences in China: Historical and Conceptual Foundations*, ed. Howard Chiang (Leiden, Boston: Brill, 2019), 35–36; Hailian Chen, "Technology for Re-Engineering the Qing Empire: The Concept of 'Arts' and the Emergence of Modern Technical Education in China, 1840–1895," *ICON: Journal of the International Committee for the History of Technology* 26, no. 1 (2021); Zhèng Fáng, "Yángwù yùndòng yǔ Zhōngguó jìndài gōngchéngxué de fāduān," *Lánzhōu xuékān* 10 (2011); Per Sörbom, "The Reception of Western Technology in China and Japan," in *Technological Development in China, India and Japan: Cross-Cultural Perspectives*, ed. Erik Baark and Andrew Jamison (New York: Palgrave Macmillan, 1986), 38–46.

12 Benjamin Elman, "Naval Warfare and the Refraction of China's Self-Strengthening Reforms into Scientific and Technological Failure, 1865–1895," *Modern Asian Studies* 38, no. 2 (1999): 291–314; Fáng, *Jìndài gōngchéngshī qúntǐ de 'mínjiān lǐngxiù'*, 20–27.

13 Thomas La Fargue, *China's First Hundred* (Pullman: The State College of Washington Press, 1942), 173–76; Fáng, *Jìndài gōngchéngshī qúntǐ de 'mínjiān lǐngxiù'*, 29–33.

institutions opened in the late 1890s, including Beiyang College in Tianjin, Nanyang College in Shanghai, and the Imperial College in Beijing.[14] During this time, reform politicians like Zhang Zhidong 張之洞 (1837–1909), whose dictum "Chinese learning as substance, Western learning for application" (*Zhong ti Xi yong* 中體西用) became the motto of the era, favored an easily accessible, but often subpar, education in Japan over resuming to send students to the remote and expensive US.[15]

Numbers of Chinese students in the US only started to rise again in 1905. The year 1905 marked a significant turning point for China through educational reform, institutional development, and change in political climate.[16] Foremost, the Qing court abolished the traditional examination system that year. Thereby, it reappraised Western thought and made way for an urban, foreign-trained strand of intellectuals within the state hierarchy. As the status group of scholar-officials, which defined itself through imperial degrees, lost its certified legitimization, those who could afford it sought qualification abroad. Secondly, across the Pacific Ocean, the 1905 Chinese merchant boycott of American merchandise resulted in US politicians exploring ways to better their country's image toward the Chinese.[17] The 1905 boycott movement led to President Theodore Roosevelt (1858–1919) agreeing to fund scholarships for incoming Chinese students out of excess sums the Qing government had been paying as 1901 Boxer war reparations.[18] Thirdly, 1905 was the year when Chinese students in Japan rioted and returned to the Mainland,[19] ending a craze of seeking education in Japan and shifting attention to the States. In summary,

14 Ruth Hayhoe, *China's Universities 1895–1995: A Century of Cultural Conflict* (New York, London: Garland Publishing, 1996), 3, 36.

15 Stacey Bieler, *"Patriots" or "Traitors"? A History of American Educated Chinese Students* (Armonk: M. E. Sharpe, 2004), 24–51; Paula Harrell, *Sowing the Seeds of Change: Chinese Students, Japanese Teachers, 1895–1905* (Stanford: Stanford University Press, 1992), 69. For mediocracy of Japanese education, see also Fu-ch'ing Huang, *Chinese Students in Japan in the Late Ch'ing Period,* East Asian Cultural Studies Series 22 (Tokyo: The Centre for East Asian Cultural Studies, 1982), Translated by Katherine P. K. Whitaker, 57. Also wage differences in Fumihiko Ishida, "Rigaku/kōgaku wo senkō shita chūgokujin no ryūnichi gakusei shi," *Gijutsu-shi kyōiku gakkai shi* 6, no. 2 (2005); Fumihiko Ishida, "Rikōgaku wo senkō shita chūgokujin ryūnichi gakusei no shakaiteki katsudō," *Gijutsu-shi kyōiku gakkai shi* 8, no. 2 (2006).

16 Black, Levy Jr. and Rozman, "Conclusions," 488–91.

17 Madeline Hsu, *The Good Immigrants: How the Yellow Peril Became the Model Minority* (Princeton, Oxford: Princeton University Press, 2015), 36–45; Guanhua Wang, *In Search of Justice: The 1905–1906 Chinese Anti-American Boycott* (Cambridge, London: Harvard University Press, 2001), 192–98.

18 Bieler, *"Patriots" or "Traitors"?*, 40–51.

19 Harrell, *Sowing the Seeds of Change*, 164.

the events of that year laid the foundation that mobilized young Chinese to seek engineering training in the US through new public funding options and through the outlook of returning to China as a sought-after group of experts that could help reconstruct the country.

A history of professionalization that focuses on the interaction between Chinese and American actors requires some words on power inequality between the two states and its people. Antiimperialist critique of foreign presence in China closely correlated with the popularization of Marxism in the 1920s. It has continued to influence contemporary historiography.[20] While the US government was eager to promote its "Open Door" policy as a relationship that promised more bilateral equality than European and Japanese interventions, later president Woodrow Wilson (1856–1924) nonetheless asked for Western standards to be "imposed" on the country.[21] Likewise, China-stationed railway engineer William Barclay Parsons (1859–1832) asked for an American "commercial invasion" of China,[22] and his colleague in Manchuria, Benjamin Oliver Johnson (b. 1878), dry-wittedly called himself an "imperialist."[23] There cannot be one simple characterization of Sino-American relations without distorting the complexity of the era.[24] Prominently, the critical historian Williams characterized the "Open Door" policy as a form of "imperial *anti*colonialism" (added emphasis) that followed practical, economic motives.[25] Generally speaking, the Chinese civil engineers of the era did not see US involvement in China as critical. Nor did foreign privileges, for example, quotas on foreign railway staff, cause much irritation for the young engineers who expected to start their careers in junior positions. If anything, Chinese civil engineers welcomed the attested imperialist opportunities of the railroad and actively applied them to their domestic interior and frontier.[26] Rather than criticizing foreign presence in China, the new generation of engineers relied on their

20 Arthur Waldron, *From War to Nationalism: China's Turning Point, 1924–1925* (Cambridge: Cambridge University Press, [1995] 2002), 246–54.

21 Woodrow Wilson, "Democracy and Efficiency," *The Atlantic Monthly* 87, no. 521 (1901): 297.

22 William Parsons, "The American Invasion of China: A Personal Survey of Commercial Conditions in the Interior of China," *McClure's Magazine* 14, no. 6 (1900): 503, 510.

23 Benjamin Johnson, "New Light on Railroading in Sibiria," *The Journal of the Worcester Polytechnic Institute* 25, no. 2 (1922): 82.

24 Roger Dingman, "1917–1922," in *American–East Asian Relations: A Survey*, ed. Ernest May and James Thomson Jr. (Cambridge: Harvard University Press, 1972), 217.

25 William Williams, *The Tragedy of American Diplomacy*, 2nd ed. (New York: Dell, [1959] 1972), 47–57.

26 For imperialism and railroads, see Daniel Headrick, *The Tools of Empire: Technology and European Imperialism in the Nineteenth Century* (New York, Oxford: Oxford University Press, 1981), 54; Ronald Robinson, "Introduction: Railway Imperialism," in *Railway Imperialism*, ed. Clarence Davis and Kenneth Wilburn Jr. (New York: Greenwood Press, 1991),

international contacts and foreign education, which they proudly displayed and utilized as markers of distinction and as credentials of expertise.

The Nationalist and Communist governments that assumed control over China after 1928 and 1949 excoriated the earlier Beijing years. Both successive regimes had an interest in portraying the Beijing Republic as a "dark age of chaos."[27] But despite political instability, as Eastman points out, the state implemented "modern education, social change, industrialization, modern banking, improved sanitation, and electrical systems," while "an educated elite that had been broadly exposed to both Western knowledge and values" created the momentum from which the Nationalist government later profited.[28] The book takes seriously the many progressive individuals, including the US-trained civil engineers, who sincerely intended to reform and reconstruct China and included "the most impressive" officials to serve the country in the 20th century,[29] but whose scholarship has often been neglected and belittled under the premise of an allegedly "anarchic" and/or (semi-)colonial period in history.

2 The Segment of US-Trained Civil Engineers

Civil engineers concern themselves with, to quote the title of an unpublished 1926 book manuscript by China-based civil engineer Oliver Julian Todd (1899–1973), "opening roads."[30] "Roads," here, are meant in a material and in a metaphorical sense. The same holds true for the Chinese expression "transport and communications" (*jiaotong* 交通), an umbrella term with much ambivalence. Republican authors used the term to refer to railways, highways, and waterways, to radio and other broadcasting technologies, and to the exchange of people and ideas made possible thereby. It hints at the potential of civil engineering to link regional and national spaces, and to reconfigure and enhance these connections through professional practice.

In an era where professional differentiation and the development of different *habitus* subsets within the engineering profession were in their infancy, practitioners of civil engineering (*tumu gongcheng* 土木工程) simultaneously built bridges, roads, and canals without deep specialization. Early professionals saw

---

2. For frontier colonialism of Chinese engineers, see my chapter in: Elisabeth Kaske and Elisabeth Köll, *Age of Discovery*, forthcoming.

27 Waldron, *From War to Nationalism*, 11, 264, 273; Frank Dikötter, *The Age of Openness: China before Mao* (Hong Kong: Hong Kong University Press, 2008), 8–17.

28 Lloyd Eastman, *The Abortive Revolution: China under Nationalist Rule, 1927–1937* (Cambridge, London: Harvard University Press, 1990), 271–72.

29 Waldron, *From War to Nationalism*, 30.

30 Oliver Todd, "The Open Road in China," OJTP 17.03 (unpublished manuscript, 1926).

themselves as broad "infrastructural" engineers and understood practitioners of different sub-disciplines of civil engineering, namely highway, railway, and hydraulic engineering, as belonging to a collective group.[31] Hydraulic engineers paved highways, and bridge engineers constructed railways (chapters 5, 7). Excluded from this broad understanding of civil engineering were disciplines of engineering that, as the lack of fluctuation between their practitioners reveals, had already been distinct categories during the Republic. These included chemical engineering (*huaxue gongcheng* 化學工程), mechanical engineering (*jixie gongcheng* 機械工程),[32] and metallurgical engineering (*yejin gongcheng* 冶金工程), as well as the only distantly related disciplines of architecture (*jianzhu xue* 建築學) and military engineering (*binggong jixie* 兵工機械 or *junshi gongcheng* 軍事工程).[33]

If one takes the 1870s CEM students as an abortive generation of overseas students, the generation explored in this book qualifies as the first proper, regular one. Their CEM seniors remained influential, but the post-1905 generation was the first to study engineering and other techno-sciences in the US on a broad basis. I define this generation as students who commenced their studies after the Qing court debased the traditional literati class by abolishing the imperial examinations and who completed their studies by 1918, when the end of the First World War heralded the start of a new phase of Sino-American engineering opportunities. In that year, in the wake of the "golden age" of Republican economy,[34] the civil presidency of Xu Shichang 徐世昌 (1855–1939) led the Beijing state into relative stability. US investors renewed their interest in financing Chinese infrastructure projects, and work on the first grand-scale cooperative undertaking, the Grand Canal, commenced (chapter 2). In addition, in 1918, the founding of the US-based China Engineering Society marked the beginning of a new era in which overseas students started discussing their field in a more specialized environment. I discuss the work of some junior

31 They all reported "civil engineering" (*tumu*) as their field of study in the 1941 government directory of engineers: Zīyuán wěiyuánhuì, *Zhōngguó gōngchéng rénmínglù* (Changsha: Shāngwù yìnshūguǎn, 1941).

32 For the emergence of mechanical engineering as distinct from civil engineering, see Monte Calvert, *The Mechanical Engineer in America, 1830–1910: Professional Cultures in Conflict* (Baltimore: The John Hopkins University Press, 1967); Noble, *America by Design*, 36.

33 For China and military 'engineering,' see Qin Zhu, "Engineering Ethics Studies in China: Dialogue between Traditionalism and Modernism," *Engineering Studies* 2, no. 2 (2010): 89–90. See also Davis, *Thinking Like an Engineer*, 9. For architecture, see Davis, *Thinking Like an Engineer*, 9–11; Eliot Freidson, *Professionalism: The Third Logic* (Cambridge, Malden: Polity, [2001] 2007), 157n5.

34 Marie-Claire Bergère, *The Golden Age of Chinese Bourgeoisie: 1911–1937* (Cambridge: Cambridge University Press, [1986] 1989), 67–70.

engineers that graduated after 1918 to highlight that networks and work ethos passed over to later Sino-American cohorts (chapter 5). However, the group of civil engineers who graduated between 1905 and 1918 formed the influential core of the Sino-American civil engineering group (chapters 4, 6) and its representatives are consequently the focus of research.

The Chinese Engineers Relational Database (CERD), a public dataset under my direction that holds over 17,400 engineering biographies from the Republican period,[35] helps to locate and contextualize the US-educated civil engineers within the broader engineering cosmos. CERD draws its biographical data, including each individual's native place, tertiary education, and work assignments, primarily from a directory that the National Resources Commission (*ziyuan weiyuanhui* 資源委員會, NRC) published in 1941. For the compilation, the NRC relied on 80,000 questionnaires it had collected from professional and nonprofessional technical personnel between 1936 and the year of publication, and from which it extracted biographies of professional engineers.[36] Although the directory includes retired and deceased engineers, the directory's purpose of identifying and managing technical experts for the reconstruction efforts of the state means the database has a bias toward professionals still active in China in the late 1930s. To complement the data, the database further integrates biographical data from membership lists of a number of professional organizations.[37] Taken together, the dataset includes sufficient data about the early group of engineers of importance for this book. In total, the dataset includes 1,747 engineers confirmed to have graduated between 1905 and 1918.[38]

According to CERD, about 34% of Republican-era engineers received a degree in civil engineering.[39] The proportion did not change remarkably over time. When limiting the period to the years between 1905 and 1918, 703 individuals, or 40% of a total of 1,747 engineering graduates, studied the subject.[40] Limiting the study to civil engineering and to the specific period significantly decreases the number of people to research, but the number is still too large for generalized statements. Professions are not a cohesive, homogeneous

35 Thorben Pelzer, Elisabeth Kaske, and Hailian Chen, "Chinese Engineers Relational Database (CERD)," Zenodo, https://home.uni-leipzig.de/cerd/.

36 Zīyuán wěiyuánhuì, *Zhōngguó gōngchéng rénmínglù*, 1.

37 For a complete list of primary sources, see Pelzer et al., CERD.

38 Pelzer et al., CERD, [{"t":"10"}, {"f:10":"<=1900"}]; Pelzer et al., CERD, [{"t":"10"}, {"linked_to:940": [{"t":"58"}, {"f:9":"<=1918"}, {"f:9":">=1905"}]}].

39 Pelzer et al., CERD, [{"t":"10"}, {"linked_to:940":[{"t":"58"}, {"f:997":"9483"}]}].

40 Pelzer et al., CERD, [{"t":"10"}, {"linked_to:940":[{"t":"58"}, {"f:9":">=1905"}, {"f:9":"<=1918"}, {"f:997":"9483"}]}].

community but rather contain "many identities, many values, and many interests."[41] To study the civil engineering profession meaningfully, it must be broken down into smaller segments. These segments promote their ideals, methods, and techniques within the greater profession, and form close formal and informal circles of colleagueship.[42] Of the above 703 graduates, the 66 individuals who studied in the US—not in China, Europe, or Japan—mark such an influential, privileged segment. The US background was a marker that united the returned students.[43] Their reputable training added authority, competence, and legitimacy.[44] Certification through a prestigious US college helped US-returned engineers to set themselves apart from their colleagues.

Vital to the study of this segment are the American engineers that cooperated with and influenced the young Chinese professionals. Their transnational collaboration and engagement transcended strictly colonial relationships. While the US-trained Chinese engineers were a remarkable and influential group on their own, only the inclusion of these American voices makes the segment a truly transnational enterprise. As reflected in calling the segment "Sino-American," the exchange and collaboration between Chinese and American actors became a pivot of identification for the cohort. Together, their transpacific voices form an overall picture.

Concentrating on this Sino-American segment helps keep the transnational ties and the cosmopolitan dimension of the profession in mind. As will become clear, professionalization was not a teleological history of national pride. Especially in the early 20th century, in the eyes of the border-crossing elite, patriotism and internationalism did not contradict each other.[45] Likewise, I argue, the professionalization of civil engineering remained an ongoing transnational process. Professionals of different nationalities shared a sense

---

41 Rue Bucher and Anselm Strauss, "Professions in Process," *American Journal of Sociology* 66, no. 4 (1961): 325–26.

42 Bucher and Strauss, "Professions in Process," 326–30.

43 Dikötter, *The Age of Openness*, 56.

44 Magali Larson, "In the Matter of Experts and Professionals, or How Impossible It Is to Leave Nothing Unsaid," in Torstendahl; Burrage, *The Formation of Professions*, 30; John Jackson, "Professions and Professionalization: Editorial Introduction," in Jackson, *Professions and Professionalization*, 4–5.

45 See Edward Fung, *In Search of Chinese Democracy: Civil Opposition in Nationalist China, 1929–1949* (Cambridge, New York: Cambridge University Press, 2000), 10–12; Liang Pan, "National Internationalism in Japan and China," in *Internationalisms: A Twentieth-Century History*, ed. Glenda Sluga and Patricia Clavin (Cambridge: Cambridge University Press, 2017); Sheldon Pollock et al., "Cosmopolitanisms," *Public Culture* 12, no. 3 (2000): 578. Further discussion in chapters 1.4, 1.5.

of mutual belonging. The reconstruction of China was a pronounced aim, but while pursuing this goal, joint transcontinental experiences continued to shape and transform the profession.

3 "Specialized but Equally Ordinary Men"

In recent years, China historians have explored the emergence and development of various professions. Researchers have studied the professionalization of occupations and their roles in Republican Chinese nation-building, including journalism, law, and medicine,[46] philology,[47] geology,[48] chemistry,[49] and the military.[50] In contrast, histories of civil engineering in China have predominantly focused on institutions, such as professional associations, powerful commissions, or railway bureaus.[51] Unquestionably, institutional history provides an important angle to explain the dependencies of professionalization. The professionalization of engineering required a system of standardization, fieldwork, new colleges, and a bureaucratic structure that

46 Xiaoqun Xu, *Chinese Professionals and the Republican State: The Rise of Professional Associations in Shanghai, 1912–1937* (Cambridge: Cambridge University Press, 2001).

47 Benjamin Elman, *From Philosophy to Philology: Intellectual and Social Aspects of Change in Late Imperial China* (Cambridge, London: Harvard University Press, 1984), 98–99, 133–136.

48 Grace Shen, *Unearthing the Nation: Modern Geology and Nationalism in Republican China* (Chicago: University of Chicago Press, 2014).

49 James Reardon-Anderson, *The Study of Change: Chemistry in China, 1840–1949* (Cambridge: Cambridge University Press, 1991).

50 Nicolas Schillinger, *The Body and Military Masculinity in Late Qing and Early Republican China: The Art of Governing Soldiers* (Lanham, Boulder, New York, London: Lexington Books, 2016).

51 For associations, see, e.g., Fáng, *Jìndài gōngchéngshī qúntǐ de 'mínjiān lǐngxiù'*; Lin-chun Wu, "Foreign Engineers' Activities in China and the Process of China's Internationalization: The Case of the Engineering Society of China, 1901–1941," in *China's Development from a Global Perspective*, ed. María Elizalde and Jianlang Wang (Newcastle upon Tyne: Cambridge Scholars Publishing, 2017). For commissions, see, e.g., William Kirby, "Engineering China: Birth of the Developmental State, 1928–1937," in *Becoming Chinese: Passages to Modernity and Beyond*, ed. Wen-hsin Yeh (Berkeley, Los Angeles, London: University of California Press, 2000). For railway bureaus, e.g., Elisabeth Köll, *Railroads and the Transformation of China* (Cambridge, London: Harvard University Press, 2019). See also Sherman Cochran, *Encountering Chinese Networks: Western, Japanese, and Chinese Corporations in China, 1880–1937* (Berkeley, Los Angeles, London: University of California Press, 2000); Lin-chun Wu, *Měiguórén wèijìng de Zhōngguó mèng: Qǐyè, jìshù yǔ guānxì wǎng* (Xinbei: Linking Publishing, 2020).

could handle the proposed infrastructural projects.[52] Furthermore, the modernization of engineering in China is connected to matters of science, economics, and ethics.[53] The analysis of greater historical contexts explains broader structural developments very adequately. The drawback is that they miss the perspective of the individuals who experienced and shaped history by actively applying science and technology through their professional practice. When describing and analyzing the professional history of modern civil engineers, a study of the relationship between engineering and economic, political, and infrastructural dimensions does not suffice.

Institutional histories of engineering may have furthermore led to an oversimplified understanding of the role of engineers within the Nationalist state. The rise of powerful "technocratic" structures that administered works of engineering in the 1930s and 1940s may have led to the impression that engineers themselves had risen to power.[54] The biographical perspective, however, unveils that institutional empowerment of the field did not necessarily lead to the empowerment of civil engineers as professional actors. Neither were the technocratic institutions solely comprised of engineers, nor did all practicing engineers necessarily support the government that employed them.

As may have become clear by now, while the book shares its topical interest with the history of science and technology, it is primarily a history of technolog*ists*, not of technology. Researchers of professionalism and of engineering—historians and sociologists alike—have called to shift attention to the practitioners.[55] Using cues from historical anthropology, this study aims at

52 Elisabeth Köll, "The Making of the Civil Engineer in China: Knowledge Transfer, Institution Building, and the Rise of a Profession," in *Knowledge Acts in Modern China: Ideas, Institutions, and Identities*, ed. Robert Culp, Eddy U and Wen-hsin Yeh (Berkeley: Institute of East Asian Studies, 2016), 149.

53 Pierre-Étienne Will, "The Emergence of the Modern Civil Engineer in China, 1900–1940," in *China and the World—The World and China: Transcultural Perspectives on Modern China*, ed. Barbara Mittler and Natascha Gentz (Gossenberg: Ostasien Verlag, 2019), 3:93.

54 Cf., e.g., William Kirby, "Technocratic Organization and Technological Development in China: The Nationalist Experience and Legacy, 1928–1953," in *Science and Technology in Post-Mao China*, ed. Denis Simon and Merle Goldman (Leiden: Brill, 1989), 29–33, 42; Kirby, "Engineering China," 149–150. Further discussion in chapters 6.5, 7.6.

55 Lino Camprubí, *Engineers and the Making of the Francoist Regime* (Cambridge: The MIT Press, 2014), 164; Davis, *Thinking Like an Engineer*, 172, 180; Larson, "In the Matter of Experts and Professionals, or How Impossible It Is to Leave Nothing Unsaid," 32; Monika Renneberg and Mark Walker, "Scientists, Engineers and National Socialism," in *Science, Technology and National Socialism*, ed. Monika Renneberg and Mark Walker (Cambridge: Cambridge University Press, 1994), 4; Grace Shen, "Murky Waters: Thoughts on Desire, Utility, and the 'Sea of Modern Science'," *Isis* 98, no. 3 (2007): 586, 591; Uekötter, "Techniker an der Macht," 397.

putting the human element in the spotlight. As proponents of the "sociotechnical" approach argue, the cognitive ability to imagine makes human actors central in bestowing meanings onto technologies.[56] The resulting biographical studies are primarily not about facts but about the "individual's sifting of the facts."[57] The individual interpretation and subjective meaning of events and technologies can help make sense of how actors understood their profession, their position within it, and the profession's role within a larger society.

Outside the context of China, histories that deal with civil engineers as individuals usually build on the sociology of professions and professionalization. The study of professions has always had a comparably weak stand within sociology. Founding fathers Weber and Marx instead emphasized the study of bureaucracy and class, and as interest in these fields declined, studies about gender, race, and ethnicity eclipsed them in popularity. Among the few grand figures who wrote about professions were Durkheim and Parsons, who described their functions and attributes.[58] Since the late 1970s, various sociologists have deemed their approach naïve,[59] exaggerated in its universality,[60] and prone to the fallacy of reproducing historical claims as fixed and natural.[61]

56 Sheila Jasanoff, "Future Imperfect: Science, Technology, and the Imaginations of Modernity," in *Dreamscapes of Modernity: Sociotechnical Imaginaries and the Fabrication of Power*, ed. Sheila Jasanoff and Sang-hyun Kim (Chicago: University of Chicago Press, 2015), 16–19.

57 Malcolm Johnson, "Professional Careers and Biographies," in Dingwall; Lewis, *The Sociology of the Professions*, 252.

58 Emile Durkheim, *Professional Ethics and Civil Morals* (London, New York: Routledge, [1957] 2018); Talcott Parsons, "The Professions and Social Structure," *Social Forces* 17, no. 4 (1939).

59 Dietrich Rueschemeyer, "Professional Autonomy and the Social Control of Expertise," in Dingwall; Lewis, *The Sociology of the Professions*, 48; Thomas Brante, "Professional Types as a Strategy of Analysis," in Burrage; Torstendahl, *Professions in Theory and History*, 75.

60 Brante, "Professional Types as a Strategy of Analysis," 75–76, 79–80. Rueschemeyer, "Professional Autonomy and the Social Control of Expertise," 47; Eliot Freidson, "The Theory of Professions: State of the Art," in Dingwall; Lewis, *The Sociology of the Professions*, 23–26; Rolf Torstendahl, "Introduction: Promotion and Strategies of Knowledge-based Groups," in Torstendahl; Burrage, *The Formation of Professions*, 7; Rolf Torstendahl, "Essential Properties, Strategic Aims and Historical Development: Three Approaches to Theories of Professionalism," in Burrage; Torstendahl, *Professions in Theory and History*, 44–45.

61 Paul Atkison, "The Reproduction of the Professional Community," in Dingwall; Lewis, *The Sociology of the Professions*, 227; Brante, "Professional Types as a Strategy of Analysis," 75; Michael Burrage, Konrad Jarausch, and Hannes Siegrist, "An Actor-Based Framework for the Study of the Professions," in Burrage; Torstendahl, *Professions in Theory and History*, 204–5; Robert Dingwall, "Accomplishing Profession," *The Sociological Review* 24, no. 2 (1976): 334–35. See also Jarausch, *The Unfree Professions*, 5–6.

Historians may fall into this trap by confusing claims of professionalization with the actual process.

As an alternative, proponents of the "Chicago school" argued that instead of trying to define professions, historians should understand professionalism as a claim, a symbolic label, or a discourse,[62] and study the change in paradigms surrounding this process.[63] They point out that a profession is an ideology that exists before its institutionalization.[64] Practitioners gradually agree on membership, standards, education, and code.[65] In short, engineers are those whom the engineering profession recognizes as such.[66] From this scholarly perspective, the professionals' historical statements, such as Thomas Tredgold's (1788–1829) proclamation that civil engineering was "the art of directing the great sources of power in Nature for the use and convenience of man,"[67] become normative definitions that created the profession rather than describe it.

The "death" of the older, functionalist approach does not mean contemporary studies have to discard earlier sociological findings. Rather, supposedly essential attributes of the profession, such as specialized, esoteric knowledge,[68] an explicit or implicit code of ethics,[69] professional traditions,[70] reference to

62 Atkison, "The Reproduction of the Professional Community," 227; Dingwall, "Accomplishing Profession," 334–35.

63 Brante, "Professional Types as a Strategy of Analysis," 78.

64 Dingwall, "Accomplishing Profession," 332; Burrage, Jarausch and Siegrist, "An Actor-Based Framework for the Study of the Professions," 209.

65 Davis, *Thinking Like an Engineer*, 49–50, 164–65.

66 Davis, *Thinking Like an Engineer*, 175–77.

67 Institution of Civil Engineers, *Transactions* 1 (London: John Weale, 1836), 37.

68 Jackson, "Professions and Professionalization," 6–7; Larson, "In the Matter of Experts and Professionals, or How Impossible It Is to Leave Nothing Unsaid," 43; Layton Jr., *The Revolt of the Engineers*, 137–39; Staffan Selander, "Associative Strategies in the Process of Professionalization: Professional Strategies and Scientification of Occupations," in Burrage; Torstendahl, *Professions in Theory and History*, 139.

69 Burrage, Jarausch and Siegrist, "An Actor-Based Framework for the Study of the Professions," 209; Randall Collins, "Changing Conceptions in the Sociology of the Professions," in Torstendahl; Burrage, *The Formation of Professions*, 13; Jackson, "Professions and Professionalization," 5–6; Hannes Siegrist, "Professionalization as a Process: Patterns, Progression and Discontinuity," in Burrage; Torstendahl, *Professions in Theory and History*, 181.

70 Siegrist, "Professionalization as a Process," 190.

practical experience,[71] and a universal concern,[72] can serve as a research aid for identifying historical claims and strategies of engineers—without taking a position as to their factuality.[73] When following the newer interactionist or "historical" approach,[74] it also no longer matters whether engineering is the late coming "profession that failed,"[75] or to what extent civil engineers in the Han Chinese context matched lists of attributes originally created in the analysis of British medical doctors and barristers.[76] The uncertainty of what professionalization and professionalism mean, criticized by skeptics of this approach,[77] is an opportunity that this book builds on. Rather than defining what a professional civil engineer is or ought to be, it looks at what historical actors believed themselves to be, at what they believed they should aim to become, and at how the discrepancy between perceived *status quo* and imagined ideal had them continuously redefine and relocate their own position in society.

Chinese histories of civil engineers regularly concentrate on protruding individuals.[78] While influential men undoubtedly achieved milestones in the history of engineering in China, they were unable to run their operations on their own. Regardless of their genius, "hero-inventors" required a network to successfully pursue their work and claim their role.[79] As Galbraith notes in his classic study on bureaucracy in the age of modern technology: the flights to the moon were engineered not by an "individual genius" but by "specialized but equally ordinary men."[80] The idea that innovations were "the work of a new

71 Svante Beckman, "Professionalization: Borderline Authority and Autonomy in Work," in Burrage; Torstendahl, *Professions in Theory and History*, 119; Brante, "Professional Types as a Strategy of Analysis," 83–84; Burrage, Jarausch and Siegrist, "An Actor-Based Framework for the Study of the Professions," 215–16.

72 Jackson, "Professions and Professionalization," 8.

73 Burrage, Jarausch and Siegrist, "An Actor-Based Framework for the Study of the Professions," 206.

74 Brante, "Professional Types as a Strategy of Analysis," 75.

75 See Brante, "Professional Types as a Strategy of Analysis," 88.

76 For gradual differences between professions but also inside professions, see, e.g., Brante, "Professional Types as a Strategy of Analysis," 81; Larson, "In the Matter of Experts and Professionals, or How Impossible It Is to Leave Nothing Unsaid," 45; Siegrist, "Professionalization as a Process," 190.

77 Freidson, "The Theory of Professions," 20–21, 35–36.

78 Francis Moon, *Social Networks in the History of Innovation and Invention,* History of Mechanism and Machine Science 22 (Dordrecht, Heidelberg, New York, London: Springer, 2014), 7–8. Thomas Carlyle popularized "great man theory" in the 19th century. See, e.g., Qǐdí Wú, ed., *Zhōngguó gōngchéngshī shǐ* (Shanghai: Tongji University Press, 2017).

79 Moon, *Social Networks in the History of Innovation and Invention*, 177–83.

80 John Galbraith, *The New Industrial State* (Boston: Houghton Mifflin, [1967] 1985), 73.

and quite remarkable race of men" were the product of "pure vanity." Rather, technological progress is a predictable and gradual group effort.[81]

This study follows the observation that civil engineering is a collective effort but strictly limits the studied group to professional practitioners. This sets the approach apart from studies that include all kinds of public advocates of the field, regardless of whether they actively worked in the promoted field or not.[82] The common marker of professional practice helps identify a set of actors with similar experiences but also leads to conceptual problems. A prosopography, i.e., the study of a historical group with common attributes, brings with it the difficulty of accounting for the individual prone to bog down inside an anonymous pool of group characteristics. Consequently, the challenge is letting go of the genius but not of the individual.

For this purpose, the case studies utilize student scorecards, hand- and typewritten correspondences, autobiographies, memoirs, diaries, essays, newspaper columns, and speeches, as well as internal and official project reports. As for the correspondences, apart from communication through government channels,[83] the study especially relies on vast letter collections preserved in the archives of US engineers who corresponded with Chinese engineers during that period.[84] The letters from young Chinese professionals present a rare, often informal, insight into their thoughts about the profession and the larger social and political issues they connected with it. Rather than describing historical facts, these sources convey a subjective and often normative element. For a study that emphasizes the inner world of thought of practicing engineers, the lack of objectivity in many of the sources benefits the study in that the sources constitute the subjective testimonies required to substantiate the claims.

---

81 Galbraith, *The New Industrial State*, 62–68.

82 See, e.g., Benjamin Elman, "Toward a History of Modern Science in Republican China," in *Science and Technology in Modern China, 1880s–1940s*, ed. Jing Tsu and Benjamin Elman (Leiden, Boston: Brill, 2014), 15; Yìshùn Lǐ, "Jiǎnlùn *Xīnmín cóngbào* zhōng de kēxué rènzhī," *Húnán kējìdàxué xuébào (Shèhuìkēxué bǎn)* 16, no. 6 (2013); Barbara Schulte, *‚Zur Rettung des Landes': Bildung und Beruf im China der Republikzeit* (Frankfurt, New York: Campus, 2008), 157–92; Zuoyue Wang, "Saving China through Science: The Science Society of China, Scientific Nationalism, and Civil Society in Republican China," *Osiris, 2nd Series* 17 (2002).

83 Foreign Relations of the United States. FRUS, Office of the Historian; Chinese Eastern Railway. NARA/CER, RG 59/861.77, National Archives and Records Administration; Jìnshǐsuǒ dàng'ànguǎn [Modern History Archives]. JSDG, Academia Sinica.

84 Papers of John Ripley Freeman. PJRF, MC.0051, MIT Libraries Distinctive Collections; O. J. (Oliver Julian) Todd papers. OJTP, 73087, Hoover Institution Archives.

Unquestionably, biographical studies overemphasize those actors privileged enough to leave behind sources. As historian Will remarked in his study of two exceptional engineers, the "overall picture is blurred" when relying on selected, privileged few voices, whose discourse becomes "dominant, not to say invasive."[85] However, in the context of professionalization, these actors were also "imprinters" or "role modelers"[86]—those who influenced the profession more actively than the silent recipients. This also means that the actors had enough capital and institutional potency for their imaginaries to matter and, potentially, materialize. Moreover, their "dominant" sources include unheard voices: through the letter archives, previously forgotten junior engineers come to life and tell their stories. Their particular troubles represent the kind of troubles most members of the profession dealt with. Diversity also played a role when picking the case studies. The social backgrounds of the featured engineers range from the urban merchant to the son of an impoverished countryside teacher. Their fields of study encompass railroads, bridges, canals, and highways, and their areas of activities concern the Chinese eastern coast as much as the frontiers of the Southwest, Northwest, and Northeast.

4 Engineering and Its Discontents

An approach that looks at professionalization as a process must examine the activities and experiences of the emerging profession's members. Young professionals establish themselves through their deeds, standardized certification, and the exclusion and separation from other groups.[87] However, the development of a collective professional identity is not a narrow one-way street leading to a predefined destination. Rather, professionals adapt their habits to external political and social circumstances.[88] As sociologist Larson remarked, professionals are confronted with a dissonance "between theoretical knowledge and applications, between claims and reality."[89] Accordingly, the biographies of

85 Will, "The Emergence of the Modern Civil Engineer in China, 1900–1940," 3:107.
86 Johnson, "Professional Careers and Biographies," 258–59.
87 Siegrist, "Professionalization as a Process," 177.
88 Burrage, Jarausch and Siegrist, "An Actor-Based Framework for the Study of the Professions," 209; Jarausch, *The Unfree Professions*, 201–27; Le, "The Struggle for Autonomy from the State," 41; Torstendahl, "Introduction," 5; Torstendahl, "Essential Properties, Strategic Aims and Historical Development," 58–59.
89 Larson, "In the Matter of Experts and Professionals, or How Impossible It Is to Leave Nothing Unsaid," 43.

young civil engineers consisted of more than their successes. Without playing down the phenomenal feats of engineers like Mao Yisheng 茅以升 (1896–1989) or Zhan Tianyou,[90] this study sets out to recover the human elements of doubt and despair that members of the profession shared. Instead of showing how successful engineering in(ter)ventions reshaped societies at large, this book explores how civil engineers reflected on their role when they met resistance, or their projects failed, and analyzes how these experiences of discontent shaped the understanding of their profession.

In a broader perspective, "discontent" has been a productive category of analysis for over a hundred years. After Freud, in *Civilization and Its Discontents*, described the clash between individual desires and the repressions of society,[91] the concept has been adopted, for instance, in exploring the tensions of gender roles,[92] childhood,[93] genetically engineered food,[94] and postmodernity.[95] The first discussion of "professional discontent" is older than Freud's study. John P. Harrison, a professor of medicine, used the term in the title of an 1845 lecture. He problematized the "dissatisfaction with the pursuit of medicine as a calling, a dislike of its responsibilities, and an aversion to meet the trials incident to a just discharge of these responsibilities."[96] While Harrison's discussion primarily addressed professionals unwilling to fulfill their duties, professional dissatisfaction can also have other causes. Inversely, as the case studies will show, "professional pain" can arise when professionals strive to fulfill their duties but find themselves constrained from doing so.[97]

This anthropological approach requires taking the emerging engineers as seriously as they took their ideals. Consequently, the study must understand claims of professionalism as more than just tokens in a strategic quest to better one's own status in society. "Professional pride" is the result of genuine

90 Le, "The Struggle for Autonomy from the State," 13–102.

91 Sigmund Freud, *Das Unbehagen in der Kultur* (Vienna: Internationaler Psychoanalytischer Verein, 1930).

92 Judith Butler, *Gender Trouble* (New York, Oxon: Routledge, [1990] 2010).

93 Joseph Dunne and James Kelly, eds., *Childhood and Its Discontents: The First Seamus Heaney Lectures* (Dublin: The Liffey Press, 2003).

94 Rachel Schurman and Dennis Takahashi Kelso, eds., *Engineering Trouble: Biotechnology and Its Discontents* (Berkeley, Los Angeles, London: University of California Press, 2003).

95 Zygmunt Bauman, *Postmodernity and Its Discontents* (Cambridge: Polity, 1997).

96 John Harrison, *Sources, Evils and Correctives of Professional Discontent: An Introductory Lecture Delivered Nov. 4th, 1845* (Cincinnati: Ben Franklin Printing House, 1845), 4.

97 See Thijs Jansen, Jos Kole, and Gabriël van den Brink, "Introduction," in *Professional Pride: A Powerful Force*, ed. Thijs Jansen, Gabriël van den Brink and Jos Kole (Amsterdam: Boom, 2010), 15; Raymond Murphy, "Proletarianization or Bureaucratization: The Fall of the Professional?," in Torstendahl; Burrage, *The Formation of Professions*, 81.

dedication, of being content with one's practice, and of the public appreciating the value of one's work. When the imagined meaning of infrastructure is in dispute—for example, a civilian railway's abuse by militarist appropriation—it challenges the civil engineers' understanding of the meaning of their practice and creates discontent. Discontent emerges when actors fail to realize their ambitions, fail to meet their group's standards, or fail to receive the recognition they believe to deserve.[98] The idea of professional "pride" is older than its sociological formulation. In 1917, Gu Weijun 顧維鈞 (Vi Kyuin Wellington Koo, 1888–1985)—who changed his major from engineering to politics,[99] and later became prime minister—observed how the educated new elite imagined "a purpose in life, an ideal to attain" and suffered complaints and discontent when the "great occasion" to realize their "steadfast purpose" would not materialize.[100]

For the most part, professional discontent is seen as a side effect of the decline of an established profession in status, recognition, and autonomy.[101] Scrutinizing this premise, Le already demonstrated that, on the contrary, "conflicts and struggles reflected [...] the growth of a professional spirit."[102] The Republican era was a time when the professionalization of civil engineering in China (and elsewhere) was still an ongoing process. Consequently, this study explores how discontent may have been just as vital for the formation of a professional identity as the original ideals themselves.

Following the claim that sociotechnical imaginaries configure our understanding of space,[103] the civil engineers exhibited a large amount of "spatial literacy." Spatial literacy, the ability to recognize existing and imagine new spatial arrangements,[104] was an implicit skill the engineers gained through their education and practice of locating, surveying, and construction. In the case of the Chinese civil engineers, they were aware that their future infrastructures

98 Jansen, Kole and van den Brink, "Introduction," 16–17. See also the broader literature of the "sociology of failure," e.g., Adriana Mica, Mikołaj Pawlak, Anna Horolets, and Paweł Kubicki, eds., *Routledge International Handbook of Failure* (in preparation).

99 Hsu, *The Good Immigrants*, 59.

100 The Chinese Students Alliance, U. S. A., The Chinese Students' Monthly. CSM, 1906–1931, 13.1 (1917): Weijun Gu, Address Made at the Platform Meeting of the Chinese Student's Conference, Providence, September 6, 1917, 21–27.

101 Jansen, Kole and van den Brink, "Introduction," 15.

102 Le, "The Struggle for Autonomy from the State," 386.

103 Jasanoff, "Future Imperfect," 26.

104 Collaborative Research Center 1199, "Verräumlichungsprozesse unter Globalisierungsbedingungen: Finanzierungsantrag" (Leipzig University, Leipzig, 2019), 17. For a collection of definitions, especially in the field of education, see Helen King, "Understanding Spatial Literacy: Cognitive and Curriculum Perspectives," *Planet* 17, no. 1 (2006).

carried the potential to shape space to facilitate nationalist goals (chapters 1, 3, 7). The engineers' interpretation of their role in society depended on their understanding of "the nature of the universe, of man, and of society."[105] Engineers offered to use their discipline to solve the issues of the envisioned nation and state—by "saving the country through transport and communications" (*jiaotong jiuguo* 交通救國).[106] Like their US counterparts, they were "a new breed of man."[107] They fought for their deserved status in society by offering the public what they thought it needed most: the development of infrastructure. But while political scientists, educators, and sociologists have used the term "spatial entrepreneurs" to describe actors who utilize their high degree of spatial literacy to intervene in the structuring of spaces,[108] it remains to be seen if the Chinese civil engineers possessed the means to materialize their spatial imaginations or if the lack of doing so created further professional discontent. In this sense, the book's title *Engineering Trouble* has a two-fold meaning. On the one hand, "engineering trouble," as a compound noun, relates to the discontent the civil engineers developed through their professional frustrations. On the other hand, as a present continuous, the practitioners also actively engineered trouble.

In the case of civil engineers in China, I argue, it was the incapability of living up to professional values and serving the state that informed and shaped the profession's understanding of the self. Buck made a similar argument when discussing the "discontents" of US-educated scientists, not engineers, in China. According to his argument, US scientists understood the scientific method as a means not only to find solutions but also to define the problems, i.e., how to manage industrialization. Chinese scientists, however, imposed a "new meaning" to science by applying it to a pre-existing domestic problem, i.e., to solve the disorder of the Chinese state. By repurposing American science to an

---

105 Layton Jr., *The Revolt of the Engineers*, 53.

106 See, e.g., Gōngchuò Yè, *Jiāotōng jiùguó lùn* (Shanghai: Shāngwù yìnshūguǎn, [1924] 1926); Hóngxūn Líng, "Shù jiāotōng jiùguó lùn," *Guówén zhōubào* 2, no. 35 (1925).

107 Noble, *America by Design*, 33. Also Ruth Barton, "'Men of Science': Language, Identity and Professionalization in the Mid-Victorian Scientific Community," *History of Science* 41, no. 1 (2003); Robert Nye, "Medicine and Science as Masculine 'Fields of Honor'," *Osiris* 12 (1997); Erika Milam and Robert Nye, "An Introduction to Scientific Masculinities," *Osiris* 30 (2015).

108 See, e.g., Martin Lawn and Romuald Normand, "Introduction," in *Shaping of European Education*, ed. Martin Lawn and Romuald Normand (Oxon, New York: Routledge, 2015), 4; Ulf Engel, "Peace-building through Space-making: The Spatializing Effects of the African Union's Peace and Security Policies," *Journal of Intervention and Statebuilding* 14, no. 2 (2020): 224. Above definition derived from Collaborative Research Center 1199, "Verräumlichungsprozesse unter Globalisierungsbedingungen," 11.

unintended aim—the "saving of the country through science" (*kexue jiuguo* 科學救國)—the US-trained Chinese scientists added an unresolvable discontent to their craft.[109] This study expands the dimensions of discontent and applies them to the area of civil engineering.

5 Chapter Overview

The first chapter deals with the education of civil engineers in the US. It follows the actors of the later chapters through their student years. The engineering curriculum they received was not the only part of campus life that shaped their understanding of the profession. Extracurricular activities were breeding grounds for nationalist, republican, and cosmopolitan worldviews that likewise shaped the engineering practice by generating a sense of status and purpose within society. The students derived pride from their US biography and anticipated their role in reforming Chinese infrastructure. Their desire for government stability and "national" unification would remain recurring motives for the upcoming decades.

The second chapter follows a network of hydraulic engineers connected through the planned improvement of the Grand Canal (1918–1922). Contracted during the First World War, it was the first large infrastructural project on which the freshly graduated engineers worked hand-in-hand with their American colleagues. For the highly motivated group, the unfulfilled visions of the aborted project resulted in a continuous disillusionment about grand-scale technological visions vis-à-vis financial constraints. The project marked a point of origin and served as a foreboding of things to come, both for the personal makeup and the close collaboration of the Sino-American civil engineering segment, and for the professional discontent that was born out of frustration at this project, which continued for the remainder of the Republican era.

The third part follows up on another thread described in the first chapter. Wang Jingchun, who studied civil engineering in the United States, became an influential railway administrator of the early Republic. The chapter discusses his theoretical writing and analyzes how it translated into practice when he co-directed and later presided over the Chinese Eastern Railway (1919–1924). Wang realized that despite his ideal that railways should be managed efficiently and "purely commercial," everything about administering a large infrastructural project was interwoven with complex political entanglements. The

109 Peter Buck, *American Science and Modern China: 1876–1936* (Cambridge, London, New York: Cambridge University Press, [1980] 2010), 171–208.

disappointment that followed the revelation of these political constraints would further constrain the increasingly pessimistic attitude engineers developed about the limits of their profession.

Originating with the personnel behind the Grand Canal surveys of chapter 2 and with additional recruitment of railroaders such as Wang of chapter 3, the Association of Chinese & American Engineers (ACAE, 1919–1941) acted as the pivotal institution of US–Chinese civil engineering. As chapter 4 demonstrates, throughout the 1920s, the association and its journal enabled the Sino-American segment to promote their occupation as a recognized profession. Their transnational interactions enabled networking and led to genuine intimacy. They produced, diffused, and discussed knowledge but also shaped an understanding of the professional self by sharing their professional experiences of discontent.

Despite great ambitions, not every infrastructural project played out on a grand national stage. The fifth chapter provincializes the study by looking at highway projects of the China International Famine Relief Commission in Yunnan, Guizhou, and the Northwest (1926–1934). Engineers deployed in work relief projects dealt with local landowners and the local workforce to realize their visions. By analyzing the power structures at play, the chapter reveals that while engineers promoted their endeavors as aiding the country at large, they encountered strong opposition when expropriating lands from landowners and mobilizing the masses for compelled labor. There was less discontent in provinces where the monopoly of power was more concentrated. Consequently, the engineers were quick to reinterpret democratic ideals to suit the outcome of their professional work. The exclusion and exploitation of the common workforce demonstrate a process of distinction that strategically elevated the engineering profession from occupational groups that shared their emphasis on manual and physical work.

After the Nationalists consolidated their power in 1928, the new Nanjing government sped up infrastructural progress through ambitious development plans, supported by increased treasury income, potent state commissions, and the rise of a new bureaucratic technostructure. As chapter 6 demonstrates, Nanjing nationalization politics did not abruptly end the influence of the Sino-American engineering segment, but its dominant institutional position slowly faded. The ACAE, its heyday described in chapter 4, suffered from financial strains and structural hurdles. The association slowly unraveled under the impression of world affairs, expedited through a crisis of reputation originating in the failure of the Salaqi irrigation project (1929–1932). The segment no longer limited their discontent to other parts of society but extended their frustration to competing segments within the field of civil engineering.

At first glance, the Nanjing nationalization policies seemed to be of advantage to the engineering profession: the Guomindang established a relatively strong central state, moved to a planned economy, and heaved some engineers into leading positions. These developments appeared to create the work environment that engineers involved with the Grand Canal, the Chinese Eastern Railway, and the provincial highway and irrigation projects had yearned for and as they had expressed in the journal of the ACAE. However, as the seventh and final chapter demonstrates by analyzing Ling Hongxun's Nationalist-era highway and railroad projects (1932–1943), the government's reinvigorated potency did not empower civil engineers beyond the constraints of state authority. During the era of imminent and active warfare, civil engineers accomplished their projects out of fear and intimidation. Consequently, the new political environment resulted in a shift, but not a resolution, of professional discontent.

CHAPTER 1

Nationalism and the Cosmopolitical

Education Overseas, 1905–1918

This chapter traces the US campus lives of a handful of prospective civil engineers involved in the different case studies that follow. These recurring individuals include student leaders Wang Jingchun (chapters 3, 4) and Yang Baoling (chapters 2, 4, 6), excelling Boxer recipients Tan Zhen and Xue Zhuobin (both chapters 2, 4, 6), as well as Ling Hongxun (chapters 4, 6, 7), who championed an emphasis on practical training later held high by the Ford Motor Works interns in Detroit (chapter 5). Together, the students developed a unique Sino-American position that allowed them to, seemingly contradictory, derive pride from their foreign US educational background, and promote Chinese nationalism and national reconstruction. It was not only their academic syllabus but also their extracurricular experiences and impressions that molded their worldview, shaped their ideals and expectations, and set the ambitions that would eventually—after their return to China—evolve into frustrations.

1 Early Ways Into the United States

Following some of the introduction's basic assumptions—that professionalization usually requires shared knowledge, traditions, and concerns—the journey of civil engineering in China sets off with the first generation of Chinese who studied the subject at higher educational institutions. This development coincided with the abolishment of the civil examination system in 1905, as traditional means of certifying as a "scholar-gentleman" had evaporated. Given the vocational nature of early domestic schools,[1] the only recent founding of colleges in China, and the lack of depth of courses in Japan specifically marketed to the Chinese, many early professionals completed their tertiary qualification in Western countries. From 1905 to 1918, 14% of civil engineers attained

1 See, e.g., Georg Baur and Elisabeth Kaske, *China um 1900: Aufzeichungen eines Krupp-Direktors* (Weimar: Böhlau, 2005), 26–30.

 | DOI:10.1163/9789004549555_003

their degrees outside of China. Of these, 12% graduated in Japan, 20% in Europe, and 67% in the US.[2]

In 1905, the "Open Door Constituency" had laid the ground for making the United States the top destination for Chinese students to receive education abroad. The constituency, a lobby group of missionaries, businesspeople, and educationists, had petitioned the US government for the remittance of excess Boxer indemnity payments. President Roosevelt signed the bill in 1908, allowing students of the Qing Empire to study in the United States under a government stipend starting by the fall term of 1909.[3] However, while the Boxer stipends added consistency and quantity to the movement of studying in the States, the Boxer recipients were not the first overseas students dreaming of saving China through science and technology.[4] Rather, they added to those who came before them and who had created the cultural and institutional framework of campus activities. Several dozens of privileged students—funded privately or through provincial stipends—had already arrived in the States before 1909. Together with the Boxer recipients, they formed the first larger generation of Chinese civil engineering students in the US.

US civil engineering students largely centered on a few renowned colleges (table 1). In the years between 1905 and 1918, Cornell (38%), Michigan (9%), Purdue (9%), Illinois (8%), and Yale (8%) dominated the field.[5] Reflecting the distribution of these prestigious academic institutions, most Chinese studied on the East Coast, followed by the Midwest. The West Coast, where Chinese migrants from the gold rush and railroader era resided and where second-generation American Chinese slowly started enrolling in public colleges, did not play a larger role for the Chinese citizens who arrived to study at only the most prestigious institutions during these early years.

The reception and realization of foreign ideas was an activity in which a global elite, privileged through capital rather than nationality, participated.[6]

---

2 Pelzer et al., CERD, [{"t":"10"}, {"linked_to:940":[{"t":"58"}, {"f:9":"<=1918"}, {"f:9":">=1905"}, {"f:997":"9483"}, {"linked_to:964":[{"t":"57"}, {"linked_to:238":[{"t":"12"}, {"f:26":"-109"}]}]}]}].

3 Bieler, *"Patriots" or "Traitors"?*, 53–89; Madeline Hsu, "Transnationalism and the Emergence of the Modern Chinese State: National Rejuvenation and the Ascendance of Foreign-Educated Elites (*Liuxuesheng*)," in *A Century of Transnationalism: Immigrants and their Homeland Connections*, ed. Nancy Green and Roger Waldinger (Urbana, Chicago, Springfield: University of Illinois Press, 2016), 139–43.

4 See Iwo Amelung, "Science and National Salvation in Early Twentieth Century China," in *Discourses of Weakness in Modern China: Historical Diagnoses of the »Sick Man of East Asia«*, ed. Iwo Amelung (Frankfurt: Campus, 2020). Cf. Wang, "Saving China through Science," 292.

5 Pelzer et al., CERD, [{"t":"58"}, {"f:9":">=1905"}, {"f:9":"<=1918"}, {"f:997":"9483"}, {"linked_to:964":[{"t":"57"}, {"linked_to:238":[{"t":"12"}, {"f:26":"119"}]}]}].

6 Robert Culp, *Articulating Citizenship: Civic Education and Student Politics in Southeastern China, 1912–1940* (Cambridge, London: Harvard University Press, 2007), 15–16.

TABLE 1 US schools most popular with Chinese civil engineering students

	Share 1905–1918	Share 1905–1937
Cornell University, New York	38%	45%
University of Michigan, Michigan	9%	9%
Purdue University, Indiana	9%	9%
University of Illinois, Illinois	8%	8%
Yale University, Connecticut	8%	3%
Harvard University, Massachusetts	6%	3%
Rensselaer Polytechnic, New York	5%	4%
Columbia University, New York	3%	1%
I. C. S., Pennsylvania	3%	3%
M. I. T., Massachusetts	3%	7%

No matter whether studying domestically or abroad, with a stipend or without one, prospective college graduates came from privileged families. An analysis of domestic college files demonstrates students often came from families of merchant, qualified labor, and government official backgrounds.[7] Students who could afford passage to the States usually came from families with an even more ample supply of financial and educational capital.[8] As late as the early 1930s, a survey among ninety Chinese students in the US found 23% coming from merchants' and 18.8% from teachers' families. The next biggest groups were those of government officials and pastors.[9]

The applicants of the pre-1905 Chinese civil exams had mainly come from northern and central provinces, often building on family lineages of scholar-officials. In contrast, when modern students somewhat replaced this group, Christian families of the southern coastal areas spearheaded the trend of studying at the new technical colleges, especially studying abroad. A comparison between US-based and domestic civil engineering students of the time reveals that students departing for the US disproportionately came from the Cantonese coast (Figure 1).

7 Chén Liáng et al., "Mínguó Shànghǎi dìqū gāoxiào shēngyuán liànghuà chúyì," *Lìshǐ yánjiū* 3 (2017): 78–83; Yùnzhú Rèn, Chén Liáng, and James Lee, "Meritocracy and the Making of the Chinese Academe, 1912–1952," *The China Quarterly* 244 (2020): 957–63.

8 Wen-hsin Yeh, *The Alienated Academy: Culture and Politics in Republican China* (Cambridge, London: Harvard University Press, 1990), 59–66.

9 Tsung-kao Yieh, "The Adjustment Problems of Chinese Graduate Students in American Universities" (PhD Thesis, University of Chicago, 1934), 17.

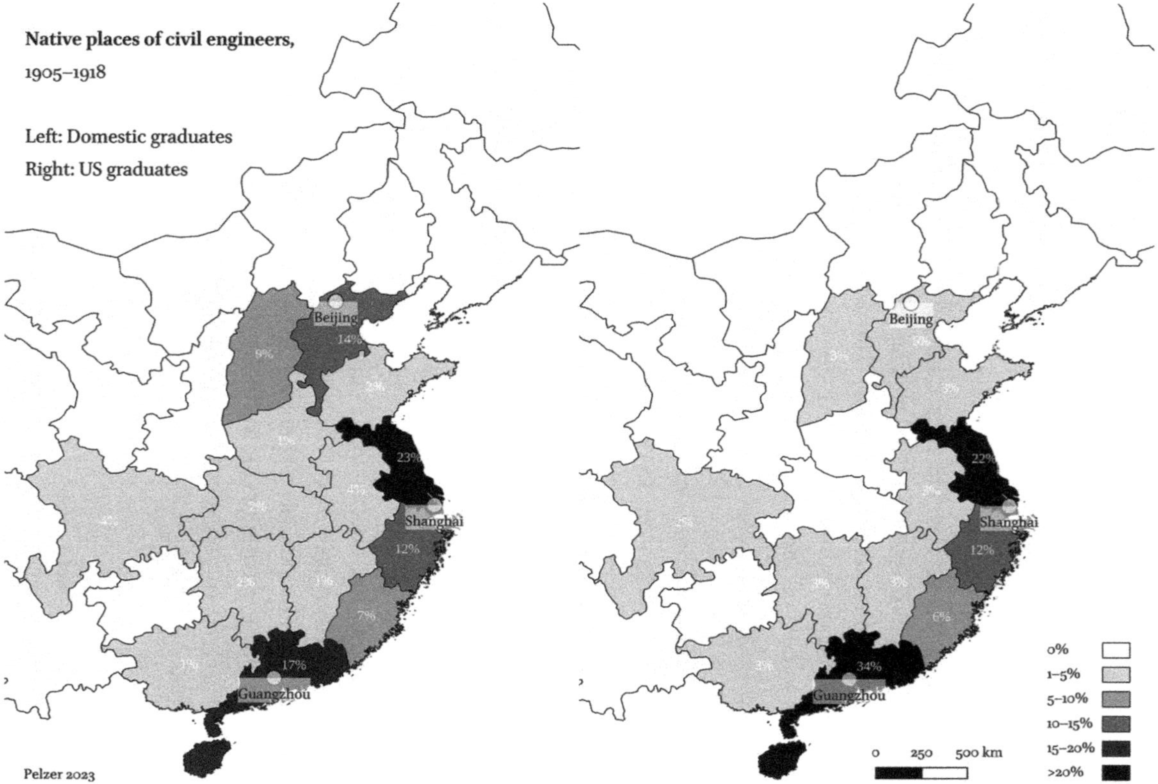

FIGURE 1 Native place of civil engineering graduates (data from CERD). Comparison of domestic graduates (left) and US graduates (right) between 1905 and 1918.

In the vicinity of the port cities, the 1842 Treaty of Nanjing had allowed protestant missions to generate an interest in Western, predominantly Christian, cultures. The offspring of privileged urban families, grown up in proximity to the international harbors and concessions, gradually came to share the homogenous features of an American-style class.[10] Temporary student migration to overseas universities created an environment where privileged students from different Chinese provinces met and lived on campus together, creating a transregional heterogeneity not comparable to domestic institutions of the 1900s and early 1910s.

Among the offspring of the Christian "American-style class" was Wang Jingchun (Figure 2). He was born in Luan County, in the administrative region Zhili, in 1882.[11] Located around Bohai Bay, Luan bordered Tangshan's Kaiping district, the coal-mining center of the time. The district was one of the country's first industrial powerhouses, hosting the Kaiping Mines since 1877, and the terminal station of the Kaiping Colliery Tramway, completed in 1881. According to a biographical dictionary co-edited by his junior colleague Ling Hongxun, Wang's father was a Christian pastor who worked at a Methodist school in Luan. This was where Wang received his early education and became proficient in English. In 1900, at 18 years of age, he fled the Boxers to Beijing, where he found work as an interpreter at the American delegation.[12]

In 1904, the city of St. Louis hosted the Louisiana Purchase Exposition, an international fair to commemorate the hundredth anniversary of the sale of Louisiana (1803). The expo, informally referred to as the St. Louis World Fair, served as the first opportunity for Wang to visit the United States. He did not encounter trouble with the Chinese Exclusion Act (1882–1943). The act excluded Chinese laborers and those suspected of being laborers from entering the country, but Wang had entered the US under the premise of the exhibition as part of the merchant class.

International exhibitions of technology were one of the strategies of modern nation-states to compete for technological superiority through public display.[13] However, far from displaying technological superiority, Wang's exhibition items had nothing to do with modern engineering. He represented a

10 Yeh, *The Alienated Academy*, 68–69.

11 Some sources refer to him by his courtesy name Wang Zhaoxi 王兆熙, while Western publications of his time used Ching-chun Wang or C. C. Wang.

12 Howard Boorman, ed., *Biographical Dictionary of Republican China: Volume 3: Mao–Wu* (New York, London: Columbia University Press, 1970), 366.

13 Ben Marsden and Crosbie Smith, *Engineering Empires: A Cultural History of Technology in Nineteenth-century Britain* (Houndmills, New York: Palgrave Macmillan, 2005), 226–34.

FIGURE 2
Wang Jingchun as a student (Yale *Class History*, 1908).

FIGURE 3
Yang Baoling as a student (*Purdue Debris* 23, 1911).

manual training school for artisan craftwork and sold cloisonné.[14] There were no innovations among the rest of the Chinese exhibits either. Exhibitions concentrated on traditional technologies of agriculture, weaving, and porcelain, as well as Orientalist displays of costumes and arts.[15] The Chinese displays may have done well to underscore the perceived Chinese technological inferiority during the era. If the St. Louis World Fair was a display of cultural imperialism,[16]

14 *The St. Louis Republic*, "Wong Ching Chun Visits Fair," May 3, 1904; *The St. Louis Republic*, "Keen Rivals in Making Cloisonne Art Ware," August 25, 1904. During these weeks, Wang also delivered his first documented public speech, at the Christian Endeavor Convention (*St. Louis Post-Dispatch*, "State C. E. Convention Begins Session Thursday," June 22, 1904).

15 Imperial Chinese Commission, *Catalogue of the Collection of Chinese Exhibits at the Louisiana Purchase Exhibition* (St. Louis: Shallcross Print, 1904).

16 The fair was "a celebration of Western history and progress," which "summed up much that was known […] and intentionally ignored almost as much." James Gilbert, *Whose*

then the Qing "Empire" made the impression of belonging to the side of the colonized.

Perhaps this showcase of backwardness provoked Wang to remain in the States after the exposition had ended. At 23 years of age, he enrolled in basic courses in science at Ohio Wesleyan University and then moved to New Haven, where he studied civil engineering at Yale. In 1908, Wang moved to the University of Illinois for a master's degree that specialized in railway administration. Self-supporting students opted for Illinois because of its lower tuition and living costs. This was how Urbana became a popular destination for privately funded students, while few Boxer recipients opted for it.[17] Although Wang, given his family background and former employment, was not poor by Chinese standards—and no Chinese college student entering the States during that time was—he nonetheless faced difficulties in financing his stay out of his savings. Therefore, one of his professors hired Wang as a lecturer for a course on "Orientalist culture"—as a form of scholarship during a time when the college did not provide fellowships to foreign students.[18] When the Qing diplomats Wu Tingfang 吳廷芳 (1842–1922) and Yan Huiqing 顏惠慶 (W. W. Yen, 1877–1950) visited the campus,[19] Wu met the Chinese students and awarded Wang a scholarship.[20]

Like Wang, his peer Yang Baoling 楊豹靈 (1887–1966, Figure 3) came from a Christian household. Yang's father had been a priest serving in Charles George Gordon's (1833–1885) Ever Victorious Army.[21] Yang studied at the Methodist Dongwu University in his hometown Suzhou. In 1907, he received one of Viceroy Tohoro Duanfang's 託濶羅端方 (1861–1911) scholarships that allowed him to study civil engineering at Cornell and Purdue. Engineering students did not disperse equally among the universities: while engineering was only one of many majors at Yale or Illinois, almost all Chinese students commencing their

Fair? Experience, Memory, and the History of the Great St. Louis Exposition (Chicago: University of Chicago Press, 2009), 49, 194.

17 Poshek Fu, "Trans-Pacific Cultural Connections: The University of Illinois and China," *Journal of American-East Asian Relations* 24, no. 1 (2017): 32–33.

18 Fu, "Trans-Pacific Cultural Connections," 35.

19 Summary of Visit, 1908, Box 17, 2/5/3, President E. James General Correspondence, University of Illinois Archives; Fu, "Trans-Pacific Cultural Connections," 31. See also Tíngfāng Wǔ, Why China and America Should be Friends, June 10, 1908, Box 17, 2/5/3, President E. James General Correspondence, University of Illinois Archives.

20 JSDG, 02-07-008-04-003: Wu Tingfang, February 17, 1909; Boorman, *Biographical Dictionary of Republican China*, 367.

21 Yuánxù Luó, *Dōng chéng* xī jiù: *Qī gè Huárén Jīdūjiào jiāzú yǔ Zhōng Xī jiāoliú bǎinián* (Hong Kong: Joint Publishing, 2012), 361.

study at Purdue picked engineering as their subject.[22] In this way, while there were fewer Chinese students in the Midwest than on the East Coast, Yang's contacts concentrated more on like-minded civil engineers. For instance, he made friends with his Purdue classmates Sa Fujun 薩福均 (Fuchuen Kenneth Sah, 1886–1955) and Wu Zhuopan 伍灼泮 (Pond Sheppon Wu).[23] After their return to China, all three would act in leading functions of the Association of American & Chinese Engineers (ACAE, chapter 4).

In the early years of Wang and Yang, Chinese reformers complained about the low number of overseas students studying engineering. Popular majors included law and political science. Others studied the liberal arts or the English language. This changed with the Boxer Indemnity Scholarship Program. The Qing government preferably issued stipends to students engaged in technical subjects, with 80% of Boxer Indemnity scholarships reserved for engineering, agriculture, and mining.[24] Of the first 47 Boxer students to arrive in the States in 1909, 43 pursued a major in a technological or scientific discipline. While there were just fifty Chinese students in the US in 1903, after the Boxer scholarship program started, the number grew to 239 in 1909 and to 830 in 1914.[25] By 1914, a law student complained that too many of them—three-quarters of new arrivals—commenced a study in engineering: while the quota followed the government's expectations, he was "doubtful" whether the numbers did not exceed the "present needs of China."[26] Within less than a decade, engineering had become the predominant subject for Chinese students in the US.

2 Engineering Education

US colleges offered state-of-the-art engineering curriculums that jointly taught the foundations of academic research and prepared for practical work. For instance, in their first year at Yale, civil engineering students took classes in differential and integral calculus, descriptive geometry and spherical trigonometry, theoretical mechanics, railway engineering, earth calculation, mineralogy, drawing and mapping, field and office work, English composition,

22 CSM, 10.2 (1914): "Club News," 184.
23 CSM, 6.3 (1911): "Club News," 313.
24 Weili Ye, *Seeking Modernity in China's Name: Chinese Students in the United States, 1900–1927* (Stanford: Stanford University Press, 2001), 53.
25 Hsu, "Transnationalism and the Emergence of the Modern Chinese State," 142–43.
26 CSM, 10.2 (1914): H. H. Wong, "The Chinese Student and the Study of Law," 144. See Köll, "The Making of the Civil Engineer in China," 158–59.

and German or French. In their second year, they further took courses in general civil engineering, geology, mechanics, mapping, adjusted observation, roads and pavements, astronomy, designing, and hydraulics.[27] In China, civil engineering subjects, which the Qing government first defined in their 1904 "Regulations for Polytechnic Schools" (*shiye xuetang tongze* 實業學堂通則), were very similar. However, it would take some time for domestic higher education institutions to adapt. For instance, at Nanyang in Shanghai, later renamed Jiaotong University, enrollment for the first regular civil engineering program started in 1907.[28] Before domestic programs gradually reached a higher level of expertise and their capacities reached the national demand, foreign universities had a head start.

At US colleges, faculties regularly highlighted the stellar performance of their Chinese students, foreshadowing the later "model minority" stereotype of Chinese migrants. For instance, Phillips Academy principal Alfred E. Stearns (1871–1949) attested to Chinese students' "splendid achievements" and a "pre-eminence in faithfulness and moral strength."[29] Likewise, MIT dean Alfred E. Burton (1857–1935) remarked on "the readiness and skill of the Chinese in participating in the social life of the students," as well as their good scholarship.[30] Student records of the time show a more differentiated picture.

Unquestionably, straight-A students existed. MIT Professor Harold Kilbrith Barrows (1873–1954) highlighted his "very bright" students Xue Zhuobin 薛卓斌 (Cho-pin Hsueh, 1895–1991, Figure 4) and Tan Zhen 譚真 (Chen Tan, 1899–1976, Figure 5).[31] Both were graduates of the government-run Tangshan College and had completed the Tsinghua preparatory course for overseas studies. Tan had completed his undergraduate studies at 18 years old and, at twenty years of age, became the youngest student to receive a master's degree from MIT. Xue and Tan would later become influential actors within the Sino-American engineering world (chapters 2, 4, 6).

---

27 Records of the Yale Alumni Records Office. YARO, Accession 2001-A-066/RU 709/RU 830, Yale University Archives, Engineering Student Cards 1908/1909.

28 Yǒuguāng Huò and Lìmín Gù, eds., *Nányáng gōngxué: Jiāotōng dàxué niánpǔ (1896–1949)* (Xi'an: Shǎnxī rénmín chūbǎnshè, 2002), 30.

29 CSM, 3.1 (1907): "News Column," 15.

30 Richard Maclaurin, "Report of the President," *Bulletin of the Massachusetts Institute of Technology Boston* 50, no. 2 (1915): 14; Alfred Burton, "Report of the Dean," *Bulletin of the Massachusetts Institute of Technology Boston* 50, no. 2 (1914): 32.

31 PJRF, 128: Harold K. Barrows to Freeman, August 5, 1918. See also David Pietz, *The Yellow River: The Problem of Water in Modern China* (Cambridge: Harvard University Press, 2015), 86.

FIGURE 4
Xue Zhuobin, c. 1917 (MITR, AC-0059, b. 4).

FIGURE 5
Tan Zhen, c. 1917 (MITR, AC-0059, b. 9).

Nonetheless, average performances were the rule and problematic cases also occurred. For instance, for sanitary engineering student Zhou Xiangxian 周象賢 (Ziang Yien Chow, b. 1890), MIT documented his education record as "unsatisfactory." His professor complained that he did not "get much out of him" due to language difficulties.[32] Scorecards from Yale likewise document mixed results, with Chinese students regularly concluding their classes with 2.x, and sometimes 3.x, grades on the 4.0 GPA scale.[33] In summary, while all students profited from a modern civil engineering curriculum, no generalized statement about their class performance can be made.

32 Office of the Dean for Student Affairs. Student Record Cards. MIT, AC-0058, M.I.T. Archives, 2: Chow, Ziang Yien.

33 YARO, 1908/1909 Score Cards.

3 Engineering Practice

Ling Hongxun came from a lineage of scholar-officials. The family once resided in Zhejiang but moved to Guangdong in the 18th century. He described himself as "born into the usual literary, impoverished scholar's family."[34] His father had failed the imperial examination and had become a local schoolteacher and gazette editor. The family's cultural capital remained high, but the economic capital was comparably low. As the fifth child and third son, Ling depended on a provincial Guangdong stipend to attend Nanyang College.

When in 1916, Ling Hongxun arrived in the States and took the Transcontinental Railroad from San Francisco to Chicago, he imagined what political leader Sun Yat-sen 孫逸仙 (1866–1925) and senior engineer Zhan Tianyou might have thought when taking the line. Ling argued that, as the Transcontinental Railroad had demonstrated that Chinese migrant workers could connect the people of a continent east and west, the Chinese should be able to repeat this feat in their own country.[35] The remark gave Chinese laborers more credit than they usually received in the United States.[36] It is also a hint that Ling's impoverished background (see below) may have resulted in more attention toward the actual labor involved in erecting infrastructure.

A strategic disassociation from the working class usually reinforced the students' skepticism against manual work, which was a vital part of the engineering education. When in the early 1930s, a Chinese doctorate interviewed US deans of engineering to find out about their opinion of students from China, one of his informants pointed out that although the students had "superior mental attributes," they lacked initiative and practical skills. He suggested having them "work with tools and machines, drive automobiles and to be thrown upon their own resources in dealing with life's problems." The interviewer explained that the regular Chinese student did not have the opportunity to ride a bicycle, drive a car, or "do much mechanical work" before arriving in the States.[37] However, the informant's remarks also hinted at the students' reluctance to get their hands dirty. The lifestyle of the arriving young Chinese men was at odds with the expectations of an engineer in the making. One student underscored that, in contrast to the traditional scholar-official ideal, the new generation had to

34 Hóngxūn Líng, *Qīshí zìshù* (Taipei: Sānmín shūjú, [1969] 1988), 3–4.

35 Líng, *Qīshí zìshù*, 23

36 See Gordon Chang and Shelley Fisher Fishkin, eds., *The Chinese and the Iron Road: Building the Transcontinental Railroad* (Stanford: Stanford University Press, 2019).

37 Yieh, "The Adjustment Problems of Chinese Graduate Students in American Universities," 90–91.

acquire practical knowledge that "cannot be acquired from college text-books."[38] They had to change their understanding of how a learned gentleman behaves and how he works to match the expectations of their field.

One of the first engineers to meet these expectations was Ling Hongxun. While authors would later concentrate on his ascription as a "Columbia man," Ling never attempted to conceal his limited extent of Columbia affiliation, which was confined to visiting evening classes. Whereas a decade earlier, Wang Jingchun had skipped his fieldwork classes at Yale,[39] Ling started his study in the States with an internship at the American Bridge Company in New York, Philadelphia, and Pennsylvania (Figure 6).[40] He worked a ten-hour day and received a monthly salary of $60 plus an additional $20 allowance, adding up to a sum similar to the Chinese government stipends.

Despite entering the States as a working student, Ling described his situation as similar to the experience of the college students. He attended Chinese student gatherings in Andover in 1916 and in Providence in 1917. He entered the same Chinese social circles and got accustomed to the American way of living. Nonetheless, he felt that interns and students had a different "situational consciousness."[41] More students gradually followed his model. For instance, the excelling MIT students Tan Zhen and Xue Zhuobin gained working experience in the Providence office of acclaimed civil engineer John R. Freeman (chapter 2). For the batch of students graduating before the end of 1918, US-trained civil engineers were more than twice as likely to commence

FIGURE 6
Ling Hongxun as an intern of the American Bridge Company, Pencoyd, PA, 1916 (Ling Hongxun papers at Columbia University).

38 CSM, 3.4 (1908): Huayan Yang, "Our Mission," 151–52.
39 YARO, C. C. Wang, 1908.
40 CSM, 11.6 (1916): "Club News," 448; Líng, *Qīshí zìshù*, 24–25.
41 Líng, *Qīshí zìshù*, 27.

an internship than their domestically trained counterparts.[42] Nonetheless, as a new trend, internships required some time to become the norm. The CERD sample contains no US internships before 1912 and 30 by 1918. For comparison, at least 109 engineering students would conduct a US internship in the 1920s.[43] By then, the Ford Trade School in Dearborn attracted Chinese students, which combined engineering coursework with internships at the Ford Motor Company. The practical experience eventually served as a competitive advantage that benefited their recruitment into US-led highway projects (chapter 5).

4 The Cosmopolitan Dimension

Apart from education in the classrooms and in the field, socializing took place on the campuses and their surroundings. In retrospect, Wang Jingchun's 1907 remark that "the age of seclusion has long passed by and that every nation, hence every creature, on the face of the globe has to deal with others" appears like an early discovery of globalization.[44] In any way, it is an example of the cosmopolitan mindset civil engineers adopted in the US and kept after their return to China.

Cosmopolitanism entailed an ambiguous hybridity: westerners continued to see the students as Chinese, while the Chinese back home found the returnees akin to Westerners.[45] On the outside, students cut off their queues, started wearing Western suits, and smoked cigars. They took part in student clubs and—in stark visual contrast to the image of traditional Chinese literati—joined the prestigious university sports teams.[46] Yale yearbooks emphasized their sociability and sketched them in favorable poses, such as trained swimmers, hard-working students, and fancy gentlemen. In the graduation polls, Chinese civil engineering students scored highly in categories such as "best student," "best natured," and "best dressed."[47] Western fashion was an act of

42 Pelzer et al., CERD, [{"t":"10"}, {"linked_to:940":[{"t":"58"}, {"f:9":"<=1918"}, {"f:997":"9483"}]}, {"linked_to:942": [{"t":"61"}, {"f:983":"實習"}]}].

43 Pelzer et al., CERD, [{"t":"10"}, {"linked_to:942":[{"t":"61"}, {"f:10":"<=1929"}, {"f:10":">=1920"}, {"f:983":"實習"}, {"linked_to:21":[{"t":"4"}, {"linked_to:134":[{"t":"12"}, {"f:26":"119"}]}]}]}].

44 CSM, 2.6 (1907): Jingchun Wang, "Prospects of the Coming Conference," 137.

45 Pao-chin Chu, *V. K. Wellington Koo: A Case Study of China's Diplomat and Diplomacy of Nationalism, 1912–1966* (Hong Kong: The Chinese University Press, 1981), 11.

46 CSM, 2.5 (1907): "News Column," 108; 3.1 (1907): Conference Committee on Athletics, "Conference Athletics," 39; 3.6 (1908): "Editorial," 209; Harvard Chinese Students' Club, Records of the Harvard Chinese Students' Club. HCSR, HUD 3277.3000, Harvard University Archives, March 25, 1911.

47 Lindell Bates, ed., *Class History 1910: Sheffield Scientific School* (New Haven: Yale University, 1910); Theophilus Hyde, ed., *Class History 1912: Sheffield Scientific School* (New Haven: Yale University, 1912).

republicanism and nationalism, a representation of liberty and equality—and an avant-garde display that only became commonplace in China in the 1930s.[48] Regardless of the outside display, US college socialization meant the reception and adaptation of progressive and democratic ideas.[49]

Engineering students in the US joined study societies (*xuehui* 學會), which had been a form of political participation from below since the late Qing era,[50] prominently advocated by reformer Liang Qichao in 1896.[51] They joined the American umbrella organizations of civil, mechanical, mining, and electrical engineers,[52] all of which would see Chinese counterparts in the next decades.[53] Apart from the professional study societies, prospective engineers joined a wide range of debating clubs, Greek letter organizations, the YMCA, and the fraternal Cosmopolitan Clubs.[54]

The Cosmopolitan Clubs were a central platform to connect with individuals from different world regions. A movement of the early 1900s, the clubs allowed local and foreign students a means of cultural exchange. The object of the Cosmopolitan Clubs, to quote the charter of the Illinois chapter, was to "cultivate social and intellectual intercourse among persons of different nationalities [and to] foster the spirit of universal brotherhood."[55] From 1908 to 1909, local clubs increased from 8 to 25. During the same time, Chinese members in these clubs increased from 33 to 89, making up the largest group besides Americans.[56] In this way, the Cosmopolitan Clubs became an important space of interaction between Chinese and American students.

48 Henrietta Harrison, *The Making of the Republican Citizen: Political Ceremonies and Symbols in China, 1911–1929* (Oxford, New York: Oxford University Press, 2000), 30–40, 49–60.

49 Stephen Craft, *V. K. Wellington Koo and the Emergence of Modern China* (Lexington: The University Press of Kentucky, 2004), 15–21.

50 Dikötter, *The Age of Openness*, 26–28.

51 Qǐchāo Liáng, "Lùn xuéxiào, 13: Xuéhuì," *Shíwù bào* 10 (1896).

52 Association of Chinese & American Engineers, *Journal of the Association of Chinese & American Engineers* [*Zhōng Měi gōngchéng xiéhuì yuèkān*], supertitled *The Oriental Engineer* between 1926 and 1927. JACAE, 1920–1941, Techn. 2055 s; TA4.A84; TA4.J865, Bavarian State Library; Cornell University Library; Smithsonian Libraries, 4 (10): Waddell, J. A. L., "The Functions of Both Pure and Applied Science in the Future Development of China," 17–21: 17. YARO, Chung, Daniel M.; OJTP, 31.03: Todd to Zhang Jichun, May 3, 1940.

53 Ye, *Seeking Modernity in China's Name*, 12–13.

54 E.g., *Purdue Debris: Volume XXIII* (Indianapolis: The Cheltenham-Aetna Press, 1911), 322, 334, 341. See Ye, *Seeking Modernity in China's Name*.

55 Association of Cosmopolitan Clubs Records. ACCR, 41/64/8, University of Illinois at Urbana-Champaign, 1: Constitution and By-Laws of the Cosmopolitan Club of the University of Illinois, Urbana, 1910: 3–4.

56 Wěi Lín, "Xué zuò shìjiè gōngmín: Liú Měi Zhōngguó xuésheng yǔ shìjiè huì yùndòng (1903–1914)," *Gāoděngjiàoyù yánjiū* 37, no. 3 (2016): 88–89.

Engineering students who became local Cosmopolitan Club leaders included Wang Jingchun, Yang Baoling,[57] and the Sichuanese hydraulic engineer Xu Zhaonan 許肇南 (Sawland J. Shu, 1886–1960). Philosopher Hu Shi 胡適 (1891–1962) translated the club's name as "Great Unity Society" (*datong hui* 大同會), alluding to Kang Youwei's 康有爲 (1858–1927) ideal society of the same name.[58] Local Chinese student clubs collaborated with the Cosmopolitan Clubs, where they hosted Chinese Nights, served Chinese food, showed lantern-slides of China, and played music.[59] On one such occasion, Wang gave an illustrated lecture about the city of Beijing,[60] and on another, one of his engineering colleagues at MIT, Xi Debing 席德柄 (Turpin Hsi, b. 1891), demonstrated the Chinese game of shuttlecock (*jianzi* 毽子) to the audience.[61] In 1910, Wang became the elected president of the nationwide Association of Cosmopolitan Clubs (ACC), whose motto was "Above all nations is humanity."[62]

Local Chinese student clubs also upheld the cosmopolitan spirit when meeting on their own. Not only was the vast majority of club business manners, interclub communication, protocols, and speeches held or written in English, as the college rules for officially recognized clubs stipulated. American students were surprised that contrary to the custom of Chinese diplomats, Chinese students wrote their surnames last, in accordance with the Western standard.[63] They also received English nicknames; for instance, Yang Baoling became "Pauline,"[64] and Wang Jingchun simply "C. C."[65] During internal Chinese club meetings, cultural boundaries between East and West stayed permeable. Chinese students played musical chairs and other games, sang "football songs," danced the Virginia reel, listened to the then-popular Honolulu Rag, ate ice cream and cakes, and hosted Christmas dinners.[66] In summary, the Chinese students not only socialized with the students of their host country but also actively performed the traditions they observed and developed a distinctively hybrid, cosmopolitan identity.

---

57 *Purdue Debris*, 174, 316.

58 Lín, "Xué zuò shìjiè gōngmín," 90.

59 CSM, 2.3 (1907): "News Column," 52; CSM, 3.5 (1908): "News Column," 183.

60 "News Column (no. 6)," *The Chinese Students' Monthly* 2, no. 6 (1907): 130.

61 MIT, 4: Hsi, Turpin.

62 *Evening Star Sunday Magazine*, "Rah! Rah! Round the World," April 3, 1910, 4.

63 *The Harvard Bulletin*, "The Harvard Chinese Club," March 31, 1909, 1.

64 *Purdue Debris*, 174.

65 YARO, C. C. Wang: News for the Weekly, October 22, 1925.

66 HCSR, October 24, 1908; November 4, 1908; December 24, 1908; June 17, 1911; October 14, 1911.

5 The National Dimension

Through their campus activities, the overseas engineering students not only received an internationally grounded understanding of their profession but also constructed the complex role in which they aimed to apply it: from early on, engineers charged their profession with social meaning, transgressing the technical world of blueprints and penetrating the space of politics. They positioned their discipline not just as a method to make industrial processes more efficient but explicitly and primarily as a means to "save" and rebuild the country. Although they found pride in belonging to an international class, they also pronounced their distinctive, reformed, nationalized version of Chineseness. This form of "national internationalism" was a "paradoxical" mode in which Chinese reformers sought to use Western ideas to defend their "nation" from intrusions of the countries from which these ideas originated.[67] There was no contradiction between internalizing Western discourse and using it for nationalist aims.[68] The students understood themselves as patriots without subscribing to the anti-foreign agenda that Chinese nationalists and communists popularized in the mid-1920s.[69]

In contrast to the Chinese students in Japan, where membership in associations concentrated on native-place societies, various revolutionary political networks, and a government-backed umbrella organization for overseas students,[70] students in the US directed their own nationwide umbrella association. They ran the organization independently, without direction from the government that sponsored some of their studies. They founded the association in 1905 and named it the Chinese Students' Alliance of the Eastern States (CSA).[71] Law student and eventual Prime Minister Gu Weijun and civil engineering student Wang Jingchun made up the early leading duo of the Alliance. They shaped the Alliance agenda on a long-term basis. Both were stern Chinese nationalists, vivid democratic republicans, Americophiles, and believers of the Christian faith.

The Alliance understood itself as a catchall association of Chinese students in the US.[72] It described its goals as laboring "for the general welfare of

---

67 Pan, "National Internationalism in Japan and China," 178–179.

68 Craft, *V. K. Wellington Koo and the Emergence of Modern China*, 15–21.

69 See Waldron, *From War to Nationalism*, 246–54.

70 Harrell, *Sowing the Seeds of Change*, 98–100. See also Peter Zarrow, *After Empire: The Conceptual Transformation of the Chinese State, 1885–1924* (Stanford: Stanford University Press, 2012), 215–16.

71 See Bieler, *"Patriots" or "Traitors"?*, 171–98.

72 CSM, 2.7–8: "Chinese Students' Conference," 157.

China," keeping "Chinese students in America in close touch with one another," and promoting "their common interests."[73] The organization lost its "Eastern States" affix in 1911 when it swallowed a few smaller clubs of the West and Midwest, but the East Coast remained the most relevant region through its advantageous infrastructure and greater member count.[74] According to the CSA directories, between two-thirds and three-quarters of Chinese students joined the Alliance over its 25 years of existence. With no distinct society for Chinese engineering students in the US founded before 1918, the CSA, its editorials, and its branches became the central platforms to influence the civil engineering students through the association's politics, rituals, and networks.

Wang remarked that "transacting [...] Alliance business" was "in itself an education."[75] While civil engineers, given their role in communications and infrastructure, could claim a distinctive role in national construction, they nonetheless perceived themselves as a common part of the political student body. As one engineering student remarked, engineers would have "just as direct an influence on the future of China as statesmen and diplomats."[76] Engineering students like Wang, without aiming at becoming politicians by profession, were actively taking part in the practice or "rehearsal" of modern politics through managing the Alliance affairs.

An example of such a "rehearsal" was the orderly club administration. The revised CSA constitution of 1907 made clear that all officers "shall be elected by popular and direct vote."[77] The benefits of a representative election were seen in its ability to summon a body of dependable, "able men,"[78] which by implication could be interpreted as mistrust in the Qing Empire's monarchic system to do so. Within their exile, the Chinese students were no longer passive subjects to a larger entity but both entitled and obliged to actively shape the environment they found themselves to be a part.[79]

A part of the Alliance experience was overcoming the regiolects that had marked socio-linguistic borders within the Qing Empire.[80] Discussing the

73 CSM, 2.7–8 (1907): Committee on the Revision of the Constitution, "Report of the Committee on the Revision of the Constitution," 167; CSM, 3.1 (1907): Jingchun Wang, "President's Message," 27.

74 CSM, 3.2 (1907): "Editorial," 55.

75 CSM, 2.6 (1907): Jingchun Wang, "Prospects of the Coming Conference," 138.

76 CSM, 12.5 (1917): K. Y. Mok, "The Technical Training of an Engineer," 264.

77 CSM, 2.7–8 (1907): Committee on the Revision of the Constitution, "Report of the Committee on the Revision of the Constitution," 170.

78 CSM, 2.7–8 (1907): "Chinese Students' Conference," 151.

79 CSM, 2.1 (1906): Jingchun Wang, "Individualism," 14.

80 See also HCSR, March 13, 1909; Ye, *Seeking Modernity in China's Name*, 19.

concept of patriotism, one member exclaimed "We are neither Pekinese nor Cantonese, but we are Chinese [...] we are members of the same body, the body is healthy only when all its organs act in harmony."[81] As a form of miniature nation-building,[82] annual Alliance conferences served the function of bringing students from the different colleges together. For a few days each year, the participants "forgot that [they] were in a foreign land,"[83] and served as a space of direct homosocial interaction between the future elite as members experienced "a feeling of brotherly love."[84] An announcement of the 1907 conference describes the annual event as "students from different provinces of our Empire" coming together "regardless of the difference in religion, school or age," aiming to "remove our narrow mindedness and harmful conservative sentiments."[85]

Students inside the US, coming from Cantonese, Wu, and Mandarin-speaking regions, got along so well that they were surprised themselves. As one local charter documented, the members were already "homogenous" in their "conceptions, ideas, likes + dislikes."[86] The trans-Chinese nature of the CSA meant practicing a national version of Chineseness that revolutionaries were imagining but which so far remained an unfulfilled project in China.

The students realized the discrepancy between the unity practiced abroad and the lack of domestic nation-building. Civil engineering students like Wang argued their profession could help this cause. To him, infrastructure not only physically connected places but also subjectively reduced distances in space and could therefore serve as a motor for nationalization.[87]

Overseas students performed civic engagement through debating competitions on political issues and speeches on questions of society.[88] US-based students, through their national structure, discussed all issues "Chinese." At the 1907 annual conference, Wang won a debating contest arguing, "it is advisable for China to use home capital alone to develop her natural resources."[89]

---

81 CSM, 2.5 (1907): C. T. C., "Patriotism," 111.

82 CSM, 3.6 (1908): Weijun Gu, "An Appeal to Chinese Students," 221.

83 CSM, 3.1 (1907): "Conference Notes," 30.

84 CSM, 3.1 (1907): "Conference Notes," 30.

85 CSM, 2.7–8 (1907): "Chinese Students' Conference," 149. Also CSM, 2.6 (1907): Jingchun Wang, "Prospects of the Coming Conference," 137.

86 HCSR, September 29, 1909.

87 See chapter 3 for a deeper analysis of Wang's thought. Similar in Hóngxūn Líng, "Tiělù gōngchéng de duōbiān rènshí [7]," *Xiāng Guì Qián xúnkān* 3, no. 30 (1948): 6. See also Wolfgang Schivelbusch, *The Railway Journey: The Industrialization of Time and Space in the Nineteenth Century* (Oakland: University of California Press, [1977] 2014), 33–44.

88 HCSR, March 13, 1909; April 24, 1909.

89 CSM, 3.1 (1907): "Conference Notes," 31–32.

In line with this, the Alliance supported independent Chinese funding of the Jiangsu–Zhejiang and Guangzhou–Kowloon railways, rejecting the foreign investments still prevalent at that time.[90] The Alliance openly supported the railway protection movement as early as 1907,[91] four years before the movement would trigger the Xinhai Revolution.[92] In 1909, Yang Baoling explored contemporary American socialism through a comparison with Wang Anshi's 王安石 (1021–1086) failed socioeconomic reforms of the 1070s.[93] Neither Yang's bourgeois background nor his narrow contact with other engineering students at Purdue hindered him from discussing the radical political currents of his time.

When Zhou Ziqi 周自齊 (Chow Tszchi, 1870–1923), a US alumnus, educator, and later influential cabinet member, addressed the students in 1907 and declared "unlimited demand for mechanical, mining or electrical engineers," he envisioned their task to cover the "twenty-two provinces with Mongolia and Tibet" with "a network of steel."[94] Echoing the territorial ambition of his senior, student leader Wang envisioned a Chinese market "from the extremities of Mongolia to the shores of Canton, from the mouths of the Yangtze and Yellow rivers to the heights of the Himalayas."[95] He described British presence in the "Tibet frontier" as an "invasion," complained about Japanese and Russian ambitions in Manchuria, and warned about English interests in Yunnan.[96] Therefore, the students not only practiced Chinese nationalism but also foreshadowed the continuation of the imperial borders, as laid claim to by all successive Chinese republics. They developed their vision of the nation while studying outside the borders of the imagined nation.

Wang Jingchun approached the Chinese working community in the US from a philanthropic angle. Gu Weijun, with whom he led the Alliance, described the Methodist interest in charity work as running contrary to the

90 CSM, 2.5 (1907): "News Column," 103; 3.2 (1907): "Business Column," 59; 3.3 (1908): Jingchun Wang, "The Kiangcheh Railway Case," 120–22.

91 CSM, 3.2 (1907): "Business Column," 59; CSM, 3.3 (1908): Jingchun Wang, "The Kiangcheh Railway Case," 120–22. Cf. CSM, 2.7–8 (1907): Heenan Shen, "Correspondence with Wellington Koo," 175.

92 See En-han Lee, *China's Quest for Railway Autonomy, 1904–1911: A Study of the Chinese Railway-Rights Recovery Movement* (Singapore: Singapore University Press, 1977).

93 CSM, 4.6 (1909): Baoling Yang, "Chinese Socialism," 395–399.

94 CSM, 3.1 (1907): "News Column," 11.

95 CSM, 3.6 (1908): Jingchun Wang, "China, Japan and the United States in the Far East Drama," 232.

96 Harvard Chinese Students' Club, Archives of the Harvard Chinese Students' Club, 1908–1928. 1911 Revolution. HCSC, 008955637, Folder 31, Harvard Yenching Library Special Collections, May 3, 1911.

"inborn nature or acquired disposition" of the ordinary Chinese. He argued that the principle of self-sacrifice was a personal virtue needed for China's revitalization, and that the required consciousness for social welfare could only be developed outside the cultural environment of China.[97] Together, the leading duo practiced welfare through crowd-funding campaigns for famine relief. For the engineers, the interest in flood prevention later translated into irrigation projects and famine relief work (chapters 2, 5). Furthermore, the students displayed an interest in the living conditions of overseas Chinese.[98]

The students carefully navigated around the contradiction that arose between compatriot sympathy and identification with the American bourgeoisie. They pitied the Chinese American lower classes and organized charitable activities, but pronounced their distinction from them. Wang's entry into the United States in 1905 had fallen right into a phase of tense US–Chinese hostilities. The tension between Americophilia and the Chinese Exclusion Act created its own form of discontent. Wang defended the Chinese Exclusion Act, claiming that the US had to defend itself against the Chinese "ignorant classes" to prevent what Theodore Roosevelt had called the Anglo-Saxon "race suicide."[99] Likewise, the *Monthly* described societies failing to upkeep class exclusivity as disorderly.[100]

Confronted with anti-Chinese sentiments, the Chinese students reinforced their strategy of keeping a distance from merchants and workers. As one American Harvard student remarked, the general perception of China was that of a "land of pagoda and junk [...] peopled by infinite duplicates of the undersized Cantonese laundry-man."[101] While there was sympathy for the non-academic migrants residing in the Chinatown ghettos, the students revised their club charters early to expel local merchants trying to participate in club activities.[102] Not all Chinese migrants were constrained to the laundry shops—but the families that succeeded in climbing up the social ladder usually lacked the financial capital to send their offspring to the exclusive institutions the Boxer students visited, opting for the public colleges on the West

97 CSM, 3.6 (1908): Weijun Gu, "An Appeal to Chinese Students," 221.

98 CSM, 3.1 (1907): Weijun Gu and Jingchun Wang, "Work and Progress of the Alliance," 20; CSM, 2.4 (1907): Jingchun Wang, "When the Chinese Exclusion Act Will be Repealed," 85.

99 Ching-chun Wang, "The Effect of the Revolution upon the Relations between China and the United States," *The Journal of Race Development* 3, no. 3 (1913): 284.

100 CSM, 2.4 (1907): "News Column," 83.

101 George Marvin, "Chinese Students in America: They Will Play Important Parts When They Return Home," *The Harvard Bulletin*, May 13, 1908, 3.

102 CSM, 2.3 (1907): "News Column," 51.

Coast instead.[103] Accordingly, when one civil engineering student remarked that he honors those Americans who "differentiate the lower and upper classes of our people,"[104] it meant that only by distancing themselves from their working-class compatriots, the Chinese students were able to gain recognition from the locals. After all, the students valued social interaction with Westerners highly. Publicly disassociating themselves from the lower Chinese classes was crucial to connect with fellow American students.

Before the revolution of 1911, students were financially dependent on stable Qing politics. Government-funded students, backed either by the provinces or later the Boxer program, required their monthly stipends, and privately funded students came from privileged families, which could lose their fortune in case of upheaval. An open display of republican enthusiasm would have inhibited career chances in China: the Qing bureaucracy, not the Han revolutionaries, had jobs to offer to the young professionals.[105] In contrast to the situation in Japan,[106] the Qing diplomats stationed abroad did not actively attempt to silence the students. However, through an overseas telegram, the Ministry of Foreign Affairs made clear that political actions of students met its displeasure.[107] While Wang and other frontbenchers claimed that the Alliance was "more or less political in character,"[108] other members urged that the society was "purely a social and educational institution."[109]

Pressure for political moderation ceased with the onset of the 1911 revolution. A group around the original leadership duo Gu and Wang, who had only recently ceded their positions, formed the Ai-Kwoh-Hwai (*aiguo hui* 愛國會, self-styled "Chinese National Union") to advocate aggressive action.[110] Wang petitioned local Chinese clubs "to give a blow to the government which has no longer the support of the people." Most local clubs immediately agreed to recognize the new Republic, by then four days old, and to lobby Chinese students in Europe via cablegram.[111]

---

103 Charlotte Brooks, *American Exodus: Second-Generation Chinese Americans in China 1901–1949* (Oakland: University of California Press, 2019), 43.

104 CSM, 4.7 (1909): Huayan Yang, "China and the United States," 446.

105 Brooks, *American Exodus*, 52–53.

106 Harrell, *Sowing the Seeds of Change*, 164; Ye, *Seeking Modernity in China's Name*, 12–13, 20; Zarrow, *After Empire*, 182.

107 CSM, 3.5 (1908): "News Column," 180.

108 CSM, 2.6 (1907): N. Yang et al., "Preliminary Report of Committee on W. C. S. F." 124.

109 CSM, 2.7–8 (1907): Heenan Shen, "Correspondence with Wellington Koo," 175.

110 HCSC, undated.

111 HCSR, undated; January 9, 1912.

The Xinhai Revolution had many conflicting aims and outcomes. It originated in a localist movement for railway autonomy and in radically Han-chauvinist sentiments. As the revolution ended, a united national government claimed the entire territory of the multi-ethnic Qing Empire while ruling through a fragile semi-presidential system. Perhaps echoing the administrative and technological focus of their study, the US-based students urged for "centralization and administrative supremacy" and highlighted the necessity of judicial checks to "guarantee a stable government."[112] This idea of an efficient, rule-based state complemented the engineering students' ideal of the professional self and continued playing a role after their return to China.

The "Shanghai–Canton–Hong-Kong axis" of coastal merchants, living in a cosmopolitan utopia of modern hygiene, infrastructure, education, and inspiration,[113] was the main profiteer of the 1911 revolution.[114] Accordingly, especially in the wake of the dynasty's demise, US-based students filed motions that were directly connected to political events. In the context of broader republicanism, not yet spearheaded by anti-foreign and anti-imperial activism, there was a genuine desire of the predominantly Christian civil engineering students to become politically involved.

6 Bringing the Profession Home

In 1902, Beijing Imperial University president W. A. P. Martin (1827–1916) translated the American civil engineer William Barclay Parsons' (1859–1932) speech "Engineering as a Profession" into Chinese.[115] It was the first Chinese language text to discuss the relationship between engineering and professionalism, at a time when neither the term "engineering" (*jiqi* 機器, later *gongcheng* 工程) nor the term "profession" (*daye* 大業, later *zhuanye* 專業) had a standardized correspondence in the language. Parsons' high standing in China originated in his early explorations in Hunan, including his surveying of the uncompleted Guangzhou–Hankou Railway (chapter 7).[116] Parsons stressed the

112 HCSR, May 10, 1913.

113 Dikötter, *The Age of Openness*, 40–42.

114 Marie-Claire Bergère, "The Chinese Bourgeoisie, 1911–37," in *Republican China 1912–1949: Part 1*, ed. John Fairbank (Cambridge, New York, Port Chester: Cambridge University Press, [1983] 2008), 721–27.

115 William Parsons, *Jīqì dàyè piān* (Shanghai: American Presbyterian Mission Press, 1902).

116 See William Parsons, *An American Engineer in China* (Cambridge: Cambridge University Press, [1900] 2013).

engineers' power to connect distant regions and thereby indirectly hinted at their nation-building potential. He claimed that telegraphy had "annihilated time," railroads and ships had reduced distances "to insignificant proportions," bridges and tunnels had removed natural barriers, and sanitation and mechanics had helped urbanization and prosperity.[117] Specialized, economical "thoroughness" should be the ideal engineers' guiding maxim.[118] According to Parsons, the engineer surpassed all other professions, including the theologian, in that they operate with "ascertainable truth" and work under "absolute and unchangeable" laws.[119] Engineers were to bring this economic rationality to all parts and world regions of society. They would transform "a sea of ignorance and superstition" into one "great modern machine shop" in which every engineer was to become their own heavily specialized machine.[120] The construction of a professional engineering identity was a source of debate not only for foreign practitioners in China but also for the Chinese engineering students. Many of their thoughts resembled those early ideas of Parsons.

These thoughts needed a platform. The main publication channel of the CSA was the *Chinese Students' Monthly*, whose contents would regularly be translated into Chinese and republished as the *Chinese Students' Quarterly*.[121] By definition, the *Monthly* was not an engineering journal. Neither the CSA nor the university-level Chinese clubs, most of which gradually recognized the Alliance as their umbrella organization, limited their membership to engineering students. However, given that from 1909 to 1914, one-third of Chinese students in the US studied some form of engineering,[122] as one *Monthly* author put it as late as 1917, "a very large percentage of the readers [of the *Monthly*] are engineers or would-be engineers."[123] In the early years, before the founding of the US-based China Institute of Engineering (*Zhongguo gongcheng xuehui* 中國工程學會) in 1918, the *Monthly* was the prime medium for Chinese students to share ideas about the engineer's status in society, discuss the discipline's role for reform in China, and construct their duties and ideals.

117 William Parsons, "Engineering as a Profession: Founder's Day Address" (Cornell University, 1901), 2.

118 Parsons, "Engineering as a Profession," 5–8.

119 Parsons, "Engineering as a Profession," 6.

120 Parsons, "Engineering as a Profession," 2, 5–6.

121 See Yuǎn Yáo, Ruì Wáng and Shùfēng Yáo, eds., *Zhōngguó jìndài kējì qīkān yuánliú 1792–1949* (Ji'nan: Shāndōng jiàoyù chūbǎnshè, 2008), 91–94.

122 CSM, 5.1 (1909): T. Chao, "English Secretary's Report," 30; Bieler, *"Patriots" or "Traitors"?*, 375.

123 CSM, 12.5 (1917): K. Y. Mok, "The Technical Training of an Engineer," 263.

Criticism of state bureaucrats was a widespread motif in the *Monthly*. In a 1907 address, US graduate Zhou Ziqi, perhaps speaking from his own outsider experience, had warned the US students that the "way things are done here may not do at all in China," lamenting the "prejudices and cherished ideas" of his compatriots.[124] In a 1908 *Monthly* guest commentary, Yan Huiqing, then serving as Minister of Education, pointed out some of the conflicts foreign-educated men were confronted with at home. Addressing the students in the "we" form and thereby temporarily claiming the shared identity of a US student rather than that of a loyal Qing minister, he lamented that returned students had no established position in society and were generally not trusted.[125]

The engineering students shared this pessimistic outlook of returning to China and facing established decision-makers. A *Monthly* editorial claimed that Qing authorities had "overlooked the importance of attending to matters near at hand and fundamental in nature."[126] In a contest-winning speech in 1911, Cornell civil engineering student Cai Guangji 蔡光勩 (Kwang-yi Char, b. 1889) asserted that Qing reforms had failed because the government was missing the "right kind of leaders."[127] He found Empress Dowager Cixi 慈禧 (1835–1908) lacking in knowledge and "tainted with selfish motives."[128] Painting the picture of a bureaucracy void of the engineering paradigms of rationalism and efficiency, he claimed that one "cannot find a single man holding a position of importance who puts the interest of the state ahead of his own."[129] A caricature the *Monthly* printed next to Cai's speech illustrates the frustration that the new foreign-education generation developed upon their return to China (Figure 7). Although schooled in modern techno-sciences and willing to apply them in the service of the nation, they depended upon the old elite and institutions to advance their careers and on the required capital to realize their projects. Cai's ideal servant to the state was a counter draft to the ruling official: prepared, unselfish, optimistic, moderate, and willing to make sacrifices.[130] His generation would have to commence "the difficult but glorious task of leading [the] nation into prosperity."[131] The foreign-educated students' abilities would result in responsibilities.

124 CSM, 3.1 (1907): "News Column," 13.

125 CSM, 3.5 (1908): Huiqing Yan, "Our Prospects and Responsibilities," 185.

126 CSM, 3.6 (1908): "Editorial," 211.

127 CSM, 7.1 (1911): Guangji Cai, "Leadership and Service," 61, 65.

128 CSM, 7.1 (1911): Guangji Cai, "Leadership and Service," 62.

129 CSM, 7.1 (1911): Guangji Cai, "Leadership and Service," 65.

130 CSM, 7.1 (1911): Guangji Cai, "Leadership and Service," 61, 65.

131 CSM, 7.1 (1911): Guangji Cai, "Leadership and Service," 62.

FIGURE 7
"Opportunity of the Returned Students. Servant: 'Gentlemen, can you wait a few hours longer? My master is not up yet.'" (CSM 7.1, 1911).

Wang's civil engineering classmate at Yale, Yang Huayan 楊華燕 (In Young, 1886–1948), was one of the *Monthly*'s more prolific writers. Like Cai, Yang found the ideal servant to the state uncorrupt, dutiful, unbiased, open-minded, and acting to the benefit of the public. These qualities were not limited to politicians but extended to all professions—including his own profession as an engineer.[132] Similar to Cai, Yang painted the ruling gentry as corrupt and anti-meritocratic. He criticized economic rent in the form of bribes and favoritism as hurting public wealth.[133] Yang's preferences in terms of an ideal government might have echoed his technocratic mindset trained in efficiency: contrary to the federalist vision popular at the time, Yang advocated a "strong central government."[134] The engineer's character traits were of greater importance than the curriculum. Yang claimed that "all the knowledge that the world holds" was insufficient to become "a leader among men" if the person did not hold a strong character.[135] He defined the leading qualities, such as standing above personal economic gain, as the "honor of manhood."[136]

The archetype that displayed this "honor of manhood" was, as the term suggests, normatively male.[137] Female student numbers slowly increased in the upcoming decades. The first Chinese woman to study (electric) engineering

132 CSM, 6.4 (1911): Huayan Yang, "True Citizenship," 359–60.
133 CSM, 6.4 (1911): Huayan Yang, "True Citizenship," 361.
134 CSM, 6.4 (1911): Huayan Yang, "True Citizenship," 359–60.
135 CSM, 3.4 (1908): Huayan Yang, "Our Mission," 153.
136 CSM, 6.4 (1911): Huayan Yang, "True Citizenship," 364.
137 CSM, 2.7–8 (1907): "Chinese Students' Conference," 153; 3.6 (1908): T. Chow et al., "News Abroad: New Haven Club," 215; 3.7–8 (1908): Jingchun Wang, "A Review of Our Conferences," 285; CSM, 3.7–8 (1908): Jingchun Wang, "A Review of Our Conferences," 285.

at MIT graduated in 1929,[138] and female students only started receiving *civil* engineering degrees in the 1930s.[139] For the earlier generation of overseas students, studying civil engineering was an exclusively male experience.

Engineers solely concentrating on their own field of study attracted the scorn of other students. One liberal arts student wrote that although he agreed with the government agenda that sought "trained engineers to improve and create means of transportation," an "over-anxiety to serve the country" resulted in a hastened graduation and lack of interest in the humanities.[140] Engineering students with little interest in the humanities must have been a widespread phenomenon, at least in the perception of those who studied the latter. An anonymous *Monthly* pamphlet complained that men were "graduating in engineering courses in the United States with an ignorance of literature, history and the other liberal arts so dense that no proficiency in science and technology can save them from the charge of being uncultured."[141] Civil engineering student Yang Huayan agreed; he advocated a combination of broad general knowledge about every field on the one hand and "deep, true knowledge of technicalities" on the other.[142] While students of the technical subjects emphasized that Chinese society depended on their fields of expertise to advance infrastructure and communications, not everybody remained ignorant about other subjects.

Students drew inspiration from political and corporate administration in the States. Echoing Viceroy Zhang Zhidong's dictum of taking "Chinese learning as substance, Western learning for application," Wang urged his fellow students to only "gather that which is best for our country and not defects." He analyzed the work processes of New York companies as being "orderly and systematic, hence rapid, progressive," and contrasted them to Chinese firms, which he thought of as steady but "in dire need of improvement."[143] A year after the revolution of 1911, he would go as far as to call for China to become "a New United States,"[144] showing his appreciation for both the US administrative and

138 Emma Teng, "1925: The First Chinese Woman Student at MIT," M.I.T., accessed May 4, 2021, https://earlychinesemit.mit.edu/1925-first-chinese-woman-student-mit.

139 Pelzer et al., CERD, [{"t":"10"}, {"f:20":"415"}, {"linked_to:940": [{"t":"58,59"}, {"f:9":"<=1935"}]]. Confirmed by CUSD-ROC: Rèn, Liáng and Lee, "Meritocracy and the Making of the Chinese Academe, 1912–1952," 955.

140 CSM, CSM, 3.3 (1908): Huai Tan, "Students' Problem of the Hour," 113–114.

141 CSM, 3.3 (1908): "Liberal and Technical Education," 120.

142 CSM, 3.4 (1908): Huayan Yang, "Our Mission," 152.

143 CSM, 3.4 (1908): Huayan Yang, "Our Mission," 151.

144 *Lawrence Democrat*, "Calls China a New United States," December 18, 1912, 15; Ching-chun Wang, "'The New China Will Be a New United States'," *The New York Times*, November 19,

political culture. Concerning the latter, Wang also argued that Chinese intellectuals had to adopt individualism. For him, the goal of individualism and independence was to cultivate a character better suited to look after national affairs.[145]

When taken as a whole, the character traits that the engineering students discussed all emphasized the civil engineer's abilities to serve the state. In consequence, when after their return to China, civil engineers failed to apply their practice to the benefit of the country, this challenged not only their identity as servants to the state but also their identity as apt engineering professionals. US civil engineering education, informed by overseas experience and practices, and including a set of values and norms, transgressed the profession and reached into the private sphere. The engineers' multifaceted identity, including a perceived Chinese cultural heritage and a cultivation influenced by US liberalism, would result in conflicts. For instance, when Yang Huayan returned to China in 1912, his family arranged for him to marry "a Chinese woman who neither reads, writes or speaks English." He felt unhappy and eventually divorced her. Out of consideration for his family, he would keep his second marriage to the sister of a former American classmate a secret throughout his life.[146]

7 Conclusion

Civil engineering students in the US closely associated the technical dimension with the political sphere. Most vocal US-based students were republicans and clearly understood their technical studies as a means to overcome old institutions. Engineering students not only received modern institutional knowledge in theory but also actively practiced new institutions through student organizations. Apart from the engineering curriculum, they brought home sociocultural experiences made within the American environment they found themselves in and within the Sino-American hybrid environment they created. While Chinese nationalism and the "invention" of the Chinese nation also took place domestically, the case of the foreign-trained students

---

1912.

145 CSM, 2.1 (1906): Jingchun Wang, "Individualism," 14.

146 YARO, 723, Young, In: Correspondence with Ethel A. Young, October 9, 1950. On the taboo of interracial marriage, see Emma Teng, *Eurasian: Mixed Identities in the United States, China, and Hong Kong, 1842–1943* (Berkeley, Los Angeles, London: University of California Press, 2013).

demonstrates that Chinese nationalism conversely emerged through civic nationhood constructed while—and under the impression of—being outside of China. The years of study in the US served as a training ground for later civil engineers in terms of nationalism, republicanism, and cosmopolitanism. Engineering students did not passively adopt foreign professional discourse. Rather, they actively shaped their own version through CSA interactions and publications. This included the construction of ideals and norms that would become the blueprint for professional codes of conduct. Their professional self-understanding was unique in that it centered on the engineer's responsibilities in a future Chinese state. They supported a republican government out of appreciation for US institutions. Nevertheless, they also supported the strong central administration of a continued Imperial territory and anticipated their role in constructing the required infrastructure. The engineering students' desire for government stability and perceived "national" unification are motives witnessed in all upcoming decades of engineering. How their understanding of the civil engineering profession was challenged upon their return to China will be the subject of the following chapters.

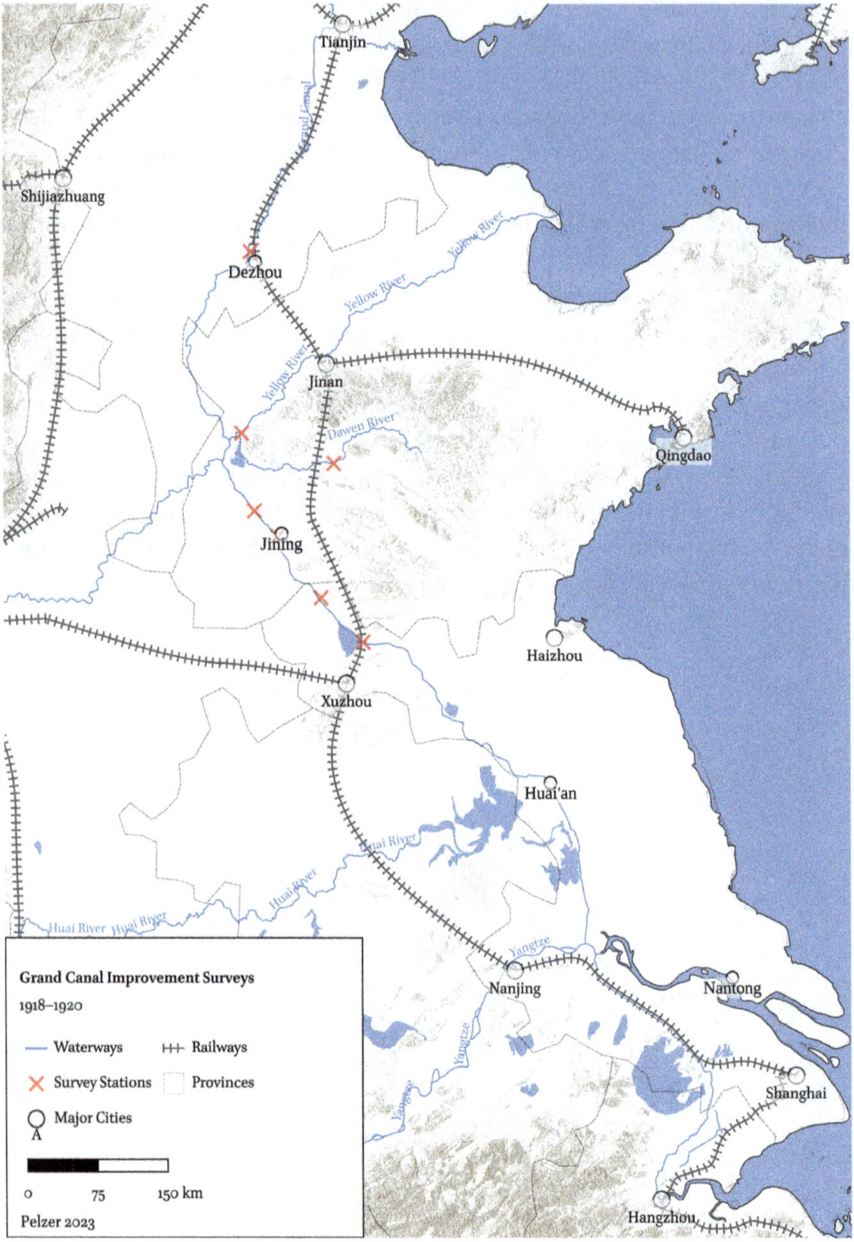

FIGURE 8 Map of the Grand Canal with markings showing the extent of Shandong surveys.

CHAPTER 2

Financial Constraint

The Grand Canal Board, 1918–1922

After their return to China, recently graduated civil engineers looked for opportunities to put their ambitions into practice. While many assumed positions in railway bureaus (chapters 3, 7), one of the first large-scale infrastructural projects in which young Chinese professionals had a chance to participate was the Shandong Grand Canal improvement scheme. The project, although it ultimately failed, turned out to be the central point of origin for two decades of Sino-American professional networking. Through private and work correspondence, the involved engineers documented and discussed their shared experiences of failure. The frustrations encountered during the project's course made the engineers realize the limits of their practice. Consequently, the discontent challenged and altered how engineers perceived their professional role.

1 A Transnational Venture

The Grand Canal, spanning over 1,700 km in length,[1] has its origins in three different periods. The mid-section from Zhenjiang to Qingjiangpu (Huai'an) is the oldest, built in the fifth century BCE. It makes use of the Gaoyou, Baoying, and Hongze Lakes. The stretch south of the Yangzi River from Hangzhou to Zhenjiang dates to the seventh century. Finally, the northern section from Qingjiangpu to Tianjin—the portion relevant to this chapter—measures almost 1,000 km and dates to the thirteenth century (Figure 8).

During the late Qing era, the Grand Canal became largely inoperable. After the Yellow River had changed its course in 1855, the flood flow and drainage direction of various tributaries changed, causing the canal to fill with silt. Major Yellow River floods in the 1880s and the abandonment of Grand Canal

1 The national government reported the Canal's length from Hangzhou to Beijing to measure 2,074 km, while Needham reports 1,770 km. H. Tseng and Jean Lyon, eds., *China Handbook 1937–1944: A Comprehensive Survey of Major Developments in China in Seven Years of War* (Chongqing: Chinese Ministry of Information, 1944), 7; Needham, Wang and Lu, *Physics and Physical Technology*, 307.

 | DOI:10.1163/9789004549555_004

maintenance work in the early 1890s left the system in hydraulic decay.[2] For example, the Canal no longer communicated with the Qiantang River, as no administrative body felt responsible for the upkeep of the sluice gate in-between the two streams. Unattended gates left the silk industry's mulberry groves volatile to intruding salt waters.[3] As the Canal depended entirely on rain for its water supply, any month-long drought halved the Canal's water level and seriously affected commerce and agriculture.[4] In the years 1906, 1909, and 1911, Hangzhou port regulators closed off the Canal because the water level was too high, while in 1903, 1904, 1907, and again also in 1911, they suspended traffic because the level was too low. The Canal's shallowness, as low as 15 centimeters in the extreme, forced junkmen to lighten their load by throwing cargo overboard. The resulting debris deteriorated the Canal's state further.[5]

Interest in restoring the Grand Canal developed out of the Huai River Conservancy (*dao Huai ju* 導淮局), an endeavor that the American National Red Cross (ANRC) and Nantong industrialist Zhang Jian 張謇 (1853–1926), then acting as Minister of Agriculture and Commerce, negotiated to reform in 1914.[6] That year, the former Shandong industrial commissioner and prospective Republican premier Pan Fu 潘復 (1883–1936) led some preliminary surveys on the Grand Canal.[7] In 1915, officials agreed to dissociate the Grand Canal improvement from the Huai River project to meet the demand of investors.[8]

The First World War started a "golden age" and "economic miracle" for Chinese industry. The value of silver rose, foreign competition declined, and

2 PJRF, 127: Paotung Tong, Preliminary Report, October 13, 1919, 6; 128: John Freeman, China Grand Canal Surveys Progress Report (Draft), February 19, 1919, 1; John Freeman, "Grand Canal Investigations 1918–1921: American International Corporation Review and Conclusions," University of Michigan Copy (1921), 5–6; Kenneth Pomeranz, *The Making of a Hinterland: State, Society and Economy in Inland North China, 1853–1937* (Berkeley, Los Angeles, Oxford: University of California Press, 1993), 213–21.

3 PJRF, 127: Paotung Tong, Preliminary Report, October 13, 1919, 3.

4 PJRF, 127: Paotung Tong, Preliminary Report, October 13, 1919, 7.

5 PJRF, 127: Paotung Tong, Preliminary Report, October 13, 1919, 6.

6 FRUS, 893.811/112: Paul Reinsch to Secretary of State, February 3, 1914. Lin-chun Wu, "Mínchū Měiguó Guǎngyì gōngsī yǔ Yùzhōng gōngsī duì Zhōngguó gōnggòng gōngchéng de tóuzī: Qǐyè, zhèngfǔ yǔ wàijiāo guānxì de kǎochá," *Dōnghuá rénwén xuébào* 14 (2009): 166–68. For Zhang Jian's early Huai River Conservancy, see David Pietz, *Engineering the State: The Huai River and Reconstruction in Nationalist China, 1927–37* (New York, London: Routledge, 2002), 23–38.

7 See Pomeranz, *The Making of a Hinterland*, 222–29.

8 Paul Reinsch, *An American Diplomat in China* (Garden City, Toronto: Doubleday, Page & Company, 1922), 207; Michele Fisher, "Service to China: The Career of the American Engineer, O. J. Todd" (PhD, Georgetown University, 1977), 28–29. For later Huai reconstruction efforts, see Pietz, *Engineering the State*.

the supply of machinery from the West improved. At a time when European powers started to withdraw from China, US interests expanded.[9] US engineers anticipated the end of the First World War as an opportunity for global reconstruction.[10] Particularly, the US had an interest in investing its surplus capital in China to counterbalance Japanese engagement.[11] With the Chinese economy and US finance doing well, the years between 1914 and 1920 were an excellent opportunity for Sino-American investment.

Accordingly, in 1915, the ANRC suggested that the American International Corporation (AIC) could get involved in funding.[12] The recently founded industrial banking group had high-risk capital invested in ventures in Brazil and Central America and was interested in entering the Chinese market.[13] Through the US Minister's mediation, the Siems-Carey Railway and Canal Company (*Yu Zhong gongsi* 裕中公司) started to act as the AIC's subsidiary in China. Pan Fu, together with the new Minister of Agriculture and Commerce Jin Bangping 金邦平 (b. 1881) and railway engineer Wang Jingchun (chapters 1, 3), negotiated on behalf of the Chinese side. In 1916, Siems-Carey signed a preliminary loan agreement granting approximately 2,414 km of railway concessions while agreeing to improve the Chinese waterway system.[14] A year later, the company completed its negotiations for the AIC in the form of two agreements. A 7% loan with the central government for the Huai River Conservancy and the Grand Canal's southern half,[15] and a 7% loan with the Shandong provincial government for the "South Grand Canal" ("south" meaning south of Tianjin: the lower part of the Canal's *northern* half).[16] Included in the agreement were the Lu Canal, running between the Yellow River in

9 Bergère, *The Golden Age of Chinese Bourgeoisie*, 67–70. See also George Mazuzan, "'Our New Gold Goes Adventuring': The American International Corporation in China," *Pacific Historical Review* 43, no. 2 (1974).

10 Layton Jr., *The Revolt of the Engineers*, 182–83.

11 Takeshi Matsuda, "Woodrow Wilson's Dollar Diplomacy in the Far East: The New Chinese Consortium, 1917–1921" (PhD, The University of Wisconsin-Madison, 1979), 100.

12 FRUS, 893.811/205: Robert Lansing to Paul Reinsch, January 10, 1916.

13 Matsuda, "Woodrow Wilson's Dollar Diplomacy in the Far East," 100–111.

14 The concessions consisted of the Hohhot–Lanzhou line, the Kaifeng–Suzhou (Anhui) line, and the Hangzhou–Wenzhou line. FRUS, 893.77/1542: Paul Reinsch to Secretary of State, May 17, 1916; Reinsch, *An American Diplomat in China*, 213–14. See also Wu, "Mínchū Měiguó Guǎngyì gōngsī yǔ Yùzhōng gōngsī duì Zhōngguó gōnggòng gōngchéng de tóuzī," 178–88.

15 Matsuda, "Woodrow Wilson's Dollar Diplomacy in the Far East," 139–51; Wu, "Mínchū Měiguó Guǎngyì gōngsī yǔ Yùzhōng gōngsī duì Zhōngguó gōnggòng gōngchéng de tóuzī," 168–72.

16 JACAE, 2 (1): Baoling Yang, "Digging a New Outlet for the Machang Canal," 5–11: 5.

the north and the Weishan Lake in the south, the Wen and Si River valleys, the Hai (Tianjin) River and marshes, and other tributaries of the Canal and reclaimed lands.[17]

The contract for the northern part met resistance from Japan. Under the framework of the 1912 sextuple agreement, US bankers could not organize loans without international consensus.[18] The adopted articles of the infamous "Twenty-One Demands" of 1915 likewise stipulated that the Japanese government had a say in Beijing accepting foreign loans. Complicating an agreement, the Japanese government had declared successive rights to Shandong after they had dispelled the German concessioners off Qingdao in 1914. The influential banker Ono Eijiro 小野英二郎 (1864–1927) proposed that, in order for the Japanese to agree to the US loan, the Imperial Bank of Japan should accede to the AIC–Shandong contract.[19] The public-sector bank would issue $2.5 million of a total six million US dollar loan.[20] The US side agreed, and the Chinese government reluctantly accepted, making way for the US-Japanese Grand Canal loan of 1917.[21] Despite the Japanese investment, surveying remained a Sino-American venture. According to the contract's stipulations, the chief engineer had to be American. An American auditor was to make sure that Chinese material had priority if the price was equal or lower to the material offered in other countries. With Japanese resistance resolved, the AIC began focusing on the renovation of the Shandong portion of the Canal, where provincial officials like Pan Fu were eager to go ahead.

2 Recruiting a Team

With the legal framework set, the project required structure and personnel. In 1918, the former Chinese premier Xiong Xiling 熊希齡 (1870–1937), then director-general of flood relief and conservancy, assumed directorship over

17 FRUS, 893.811/223: Pan Fu and Ernest Gregory, Agreement for Huai River Conservancy Grand Canal Improvement Loan, May 13, 1916.

18 Matsuda, "Woodrow Wilson's Dollar Diplomacy in the Far East," 111–29.

19 Matsuda, "Woodrow Wilson's Dollar Diplomacy in the Far East," 139–51; Pomeranz, *The Making of a Hinterland*, 229–32; Wu, "Mínchū Měiguó Guǎngyì gōngsī yǔ Yùzhōng gōngsī duì Zhōngguó gōnggòng gōngchéng de tóuzī," 172–77.

20 FRUS, 893.51/1896: Paul Reinsch to Robert Lansing, April 12, 1918.

21 FRUS, 893.811/281: Xiong Xiling and William Carey, The Chinese Government Grand Canal Improvement Seven Per Cent (7%) Gold Loan of 1917, November 20, 1917. See also Shirley Ye, "The Grand Canal in Republican China," *Journal of the Economic and Social History of the Orient* 62, no. 4 (2019): 737–39.

the newly formed Grand Canal Improvement Board (*duban Yunhe gongcheng shiyi chu* 督辦運河工程事宜處). In this function, he contracted the prospective engineers.

One of these engineers was Purdue graduate Yang Baoling (chapter 1). He played a pronounced role within the Board.[22] Upon his return to China in 1911, Yang had participated in the Revolution. Li Yuanhong 黎元洪 (1864–1928), then the Revolution's interim military governor, had appointed Yang as a member of the Board of Communications and chief secretary to the foreign office at Wuchang.[23] In 1912, Zhang Jian requested him to survey the southern portion of the Grand Canal and, a year later, made him the chief technical expert of his National Conservancy Bureau (*quanguo shuili ju* 全國水利侷). In 1918, Xiong established the Commission for the Improvement of the River System of Zhili (*Zhili shuili weiyuanhui* 直隸水利委員會) in response to the 1917 Zhili floods in the Tianjin area. With the establishment of the Grand Canal Board, Xiong arranged for Yang to spend half of his time working with the Zhili Commission and the other half consulting with the new Canal Board. In exchange, an American engineer of the Canal Board supported the Zhili Commission.[24]

At the Canal Board, Yang worked under the supervision of the project's consulting engineer. The AIC had opted for John Ripley Freeman (1855–1932) to fill this position. Freeman had previously made himself a name as the design engineer of the Panama Canal (1902), the Charles River Dam (1902), and the Hetch Hetchy Aqueduct (1905), and had influenced the design of the MIT Cambridge campus (1912). Freeman described the Canal improvement as "one of the most difficult hydraulic engineering problems of the world," and stressed the large number of people directly affected by the plans.[25] His idea was to direct the stream to have the water naturally dig a straightened channel on the intended lines.[26] He wanted to provide the Canal with a "modern system of locks" and, as a side effect, reclaim the land along its margin.[27] Swamp drainage and restored water irrigation would help navigation and improve

22 PJRF, 128: John Freeman, China Grand Canal Surveys Progress Report (Draft), February 19, 1919, 27.

23 CSM, 7.7 (1912): "The Student World," 579; M. Powell, ed., *Who's Who in China* (Shanghai: The China Weekly Review, 1925), 904.

24 PJRF, 128: John Freeman, China Grand Canal Surveys Progress Report (Draft), February 19, 1919, 28.

25 PJRF, 128: John Freeman, China Grand Canal Surveys Progress Report (Draft), February 19, 1919, 5.

26 PJRF, 125: Freeman to Oliver J. Todd, October 18, 1920; Freeman, "Grand Canal Investigations 1918–1921," 28–137.

27 PJRF, 123: Freeman to George C. Whipple, May 22, 1918.

agricultural resources.[28] Controlling the flow of water, as well as the related fields of irrigation farming and land reclamation, were not new to China. At the same time, the idea of proactively preventing famines and floods instead of reactively distributing relief also fit well with the conservationist movement then popular with US civil engineers.[29]

Freeman thought that the costs of the Shandong Canal improvement could pay for itself through land reclamation.[30] He imagined that a renovation would restore the Canal as the cheapest way to transport goods and reestablish prosperity in the local population. The financial benefits of a Canal renovation outweighed the profits of railroad construction.[31] Nevertheless, back in Freeman's time, the perceived technological backwardness of waterways questioned the project's future viability. Around the globe, the use of canals dwindled with the construction of railways.[32] With the partly parallel Tianjin–Pukou railway line completed in 1912 (Figure 8) and urban regions starting to explore the future of motorcars and airplanes, there were doubts about the timeliness of financing a costly hydraulic project.[33] The US students' *Weekly* (chapter 1) took the middle ground as it found canals a generally outdated, inefficient investment, but supported initial investigations of the Grand Canal, which it identified as the "only canal that possesses potentialities" for China's future.[34] In this regard, the 1880 Chinese Education Mission (CEM) veteran and railway engineer Kuang Jingyang 鄺景揚 (King-yang Kwong, 1863–1929) found that the costly Shandong Canal improvement would eventually save the government millions of dollars by stimulating trade and reenabling agriculture.[35] The Chinese government was also optimistic about the endeavor. For instance, according to the US minister, the Chinese Minister of Communications and Yale alumnus Liang

28 PJRF, 128: John Freeman, China Grand Canal Surveys Progress Report (Draft), February 19, 1919, 3–4.

29 See Samuel Hays, *Conservation and the Gospel of Efficiency: The Progressive Conservation Movement, 1890–1920* (Cambridge: Harvard University Press, 1959), 5–26, 91–121.

30 PJRF, 130: Freeman to Charles R. Crane, September 13, 1920; Freeman, "Grand Canal Investigations 1918–1921," 126–29.

31 Pomeranz, *The Making of a Hinterland*, 213.

32 van Laak, *Weiße Elefanten*, 70–72, 77, 225.

33 Elisabeth Köll, "Chinese Railroads, Local Society, and Foreign Presence: The Tianjin–Pukou Line in Pre-1949, Shandong," in *Manchurian Railways and the Opening of China: An International History*, ed. Bruce Elleman and Stephen Kotkin (New York: M. E. Sharpe, 2010), 124.

34 CSM, 11.7 (1916): C. H. Wang, "The Future of Canals as a Means of Transportation in China," 497–499.

35 PJRF, 125: Kuang Jingyang to Freeman, January 22, 1921.

Dunyan 梁敦彥 (1857–1924) enthusiastically remarked that he would soon be able to "go from Peking to Shanghai in a houseboat."[36]

When Freeman commenced his work on the Grand Canal, he had already established links to a remarkably wide range of high-ranking Chinese individuals, including political leaders, educators, and industrialists. His friend Charles Richard Crane (1858–1939), a businessperson, Arabist, and short-period US minister to Beijing, had awoken his late-career interest in the region.[37] Between 1916 and 1917, Freeman traveled to China and made various contacts. He completed his network through his second journey in 1918 and during his third stay from 1919 to 1920. Among his acquaintances were the Republic's President Li Yuanhong, whom he met at the YMCA,[38] and Li's secretary, the diplomat Guo Taiqi 郭泰祺 (Quo Taichi, 1888–1952). Further, he met senate president Wang Zhengting 王正廷 (Chengting Thomas Wang, 1882–1961), former acting premier Wu Tingfang, and YMCA leader Yu Rizhang 余日章 (David Z. T. Yui, 1882–1936).[39] He talked to former premier Tang Shaoyi 唐紹儀 (1862–1938),[40] Zhang Jian,[41] Xiong Xiling, and Pan Fu.[42] For the latter, Yang Baoling served as the interpreter.[43] In summary, Freeman was well acquainted with the Chinese who-is-who of the day, demonstrating both the receptivity of the Republic's upper stratum toward engineers, as well as Freeman's elevated position within the profession.

Freeman's selection of an engineering team for the surveys proved difficult. Although the contract did not stipulate the nationality of his engineers, given that the contracted company was American, and that Freeman had not yet gained experience working with Chinese experts, he initially attempted to recruit compatriot engineers. Thwarting his plan, many potential American candidates were serving in the World War. He ended up recruiting a mixture of former acquaintances and university professors on leave.[44] The limited pool

36 Reinsch, *An American Diplomat in China*, 170.

37 PJRF, 130: John Freeman, Memorandum for Address at Chinese Students Conference Brown University Providence, January 10, 1923, 1.

38 PJRF, 128: Photographic reference.

39 PJRF, 130: John Freeman, Memorandum for Address at Chinese Students Conference Brown University Providence, January 10, 1923, 1.

40 PJRF, 128: Banquet Tianjin, October 6, 1919.

41 PJRF, 130: Freeman to Zhang Jian, October 22, 1920.

42 PJRF, 128: John Freeman, China Grand Canal Surveys Progress Report (Draft), February 19, 1919, 6, 12.

43 PJRF, 128: John Freeman, China Grand Canal Surveys Progress Report (Draft), February 19, 1919, 26.

44 James Wallace Beardsley (1860–1944) acting as assistant chief engineer, Robert Hall Merrill (1881–1955) as principal engineering assistant, Ralph D. Goodrich (1878–1961) as

of available engineers may explain why Joseph Ripley (1854–1940), who had worked with Freeman on the Panama Canal,[45] was to act as chief engineer on site despite his advanced age and limited fitness.

After the first tedious appointments, the Beijing government asked Freeman to fill the remaining junior positions with recent Chinese graduates.[46] The request fell on sympathetic ears. Although Freeman was a racialist in that he thought that races had different mental capacities,[47] he believed the Chinese to have "natural intellectual endowments and memories" that made them industrious and receptive to education.[48] He thought that the Chinese radiated a "mental inheritance" of old and genetically brought with them a "genius of brain."[49] Freeman's racialist but Sinophile worldview explains why he was open-minded toward hiring and befriending young prospective Chinese engineers.

In his search for Chinese civil engineers, Freeman outlined his Grand Canal plans to colleagues at his alma mater, MIT, and asked for their recommendations. MIT Dean Burton and Professor Barrows referred to their students.[50] These included Lin Quancheng 林全誠 (Homer Ling, b. 1891), Tan Zhen, and Xue Zhuobin (both chapter 1). The latter two approached Freeman humbly. They composed a letter asking Freeman for book recommendations and mentioned that they had heard of the upcoming Grand Canal survey.[51] Freeman invited Xue to his home in Providence and proposed that he could spend time with his son, Evert Wendell Freeman (1898–1955), who also studied engineering at MIT at that time.[52] In this early phase, Freeman was unsure of the extent of the Chinese students' qualifications. He designed library research tasks as

---

the chief hydrographer, and Howard Benjamin Merrick (1871–1926) as the principal assistant. Michigan University professors Hugh Brodie (1871–1932) and Clifton O'Neal Carey (1875–1963) joined the delegation while on leave. PJRF, 128: John Freeman, China Grand Canal Surveys Progress Report (Draft), February 19, 1919, 7.

45 J. David Rogers, "The American Engineers That Built the Panama Canal," in *Engineering the Panama Canal*, ed. Bernard Dennis Jr. (2014), 196–97.

46 PJRF, 123: Freeman to L. B. Hastings, January 10, 1919.

47 PJRF, 130: John Freeman, Memorandum for Address at Chinese Students Conference Brown University Providence, January 10, 1923, 2.

48 PJRF, 130: Freeman to H. C. Hwang, July 27, 1920. See also Ye, "The Grand Canal in Republican China," 748.

49 PJRF, 130: John Freeman, Memorandum for Address at Chinese Students Conference Brown University Providence, January 10, 1923, 2, 6.

50 PJRF, 128: Alfred E. Burton to Freeman, July 16, 1918; 128: Harold K. Barrows to Freeman, August 5, 1918; 128: Freeman to Frank E. Winsor, July 2, 1919. See also Pietz, *The Yellow River*, 86.

51 PJRF, 128: Xue Zhuobin and Tan Zhen to Freeman, April 5, 1919.

52 PJRF, 128: Freeman to Xue Zhuobin, May 8, 1919.

a "preliminary tryout" to recruit students for the Canal expedition.[53] At the Library of Congress, Lin had to research the history of floods, the Grand Canal, and the Yellow River. Xue and Tan had to prepare information about Xiong Xiling.[54] By the end of 1918, Freeman and his Chinese assistants had produced 1,200 typewritten pages of library research.[55] Satisfied with the outcome, Freeman confirmed that Lin could accompany the US engineers on their way to China and work as a Canal Board secretary.[56] Tan and Xue did not depart for China immediately, as the fresh graduates wanted to look for US internships before returning to their home country. But they remained on Freeman's radar.

Tan and Xue had originally intended to receive at least one year of professional experience in the US before returning to China. However, they lamented the difficulty of finding internships in the US.[57] The post-World War I recession of 1918–1919 had left many experienced American engineers jobless, leaving little room for Chinese interns. Although Freeman used his contacts to recommend Xue and Tan to various enterprises,[58] he was unable to find employment for the two.[59] While Stone & Webster in Boston eventually employed Xue in August 1919, the contract only ran for a few months. Xue still depended on money from his family.[60] As Freeman could not help them find any better internships, he suggested the two could accompany him to China.[61] He was unable to make promises as to what positions they would take inside the Grand Canal organization or say anything definite about the length of employment.[62] Despite the lack of security, Tan finally agreed to accompany Freeman to China, but Xue decided to continue his search in the States.[63] The unstable US job market forced Xue and Tan to make difficult career decisions, forcing them to temporarily part ways.

When it came to salary, the customs of the US engineering profession clashed with the expectations of Chinese graduates. Freeman claimed that

53 PJRF, 128: Freeman to George Won, July 19, 1918.

54 PJRF, 128: Xue Zhuobin and Tan Zhen to Freeman, July 1, 1919.

55 PJRF, 128: John Freeman, China Grand Canal Surveys Progress Report (Draft), February 19, 1919, 6.

56 PJRF, 128: Freeman to Lin Quancheng, August 5, 1918.

57 PJRF, 128: Xue Zhuobin and Tan Zhen to Freeman, July 5, 1919.

58 PJRF, 128: Freeman to Frank W. Winsor, June 28, 1919; 128: Freeman to Xue Zhuobin and Tan Zhen, July 18, 1919.

59 PJRF, 128: Freeman to Xue Zhuobin, October 6, 1921.

60 PJRF, 128: Xue Zhuobin to Freeman, August 2, 1919.

61 PJRF, 128: Freeman to the Secretary of the Chinese Legation, July 23, 1919.

62 PJRF, 128: Freeman to Tan Zhen, July 29, 1919.

63 PJRF, 128: Tan Zhen and Xue Zhuobin to Freeman, July 20, 1919; 128: Xue Zhuobin to Freeman, August 11, 1919.

he offered $100 a month to any fresh graduate, regardless of their nationality, and that this salary already exceeded his own entry wage. While the sum was low, he argued that networking and experience compensated for it.[64] Against Freeman's expectations, most Chinese engineers did not see the opportunity of gaining experience in a grand infrastructural project and working under a renowned engineer as a selling point that could substitute a befitting compensation. The gained experience could not outweigh the "loss of face" associated with accepting a low salary and position.[65] In contrast to Freeman's customs, Shandong industrial commissioner Pan Fu paid young Chinese engineers better than what was the standard for young American engineers commencing work in the US, displaying an appreciation for the returned experts. Pan intended to make the engineering profession as attractive as possible and did not see gaining first practical experience as a unit of value, hence the higher salaries.[66]

Apart from their desire to have the salary reflect an appreciation for their work, the young Chinese graduates also had material reasons to be unimpressed by Freeman's initial offers. While Tan Zhen agreed with Freeman's reasoning that the canal work helped him gain experience and knowledge "which cannot be obtained in school"[67] and declined the salary altogether as he was affluent enough through his family, Lin Quancheng could not afford to accept the position even if he wanted to. Lin explained that he understood that gaining experience under the leadership of senior engineers would make up for a "low compensation,"[68] but pointed out that the salary had become unrealistically low: the post–World War I recession had seen the US dollar, fixed on gold, halve its value against the Beiyang silver dollar.[69] Without affluent family support, the amount became insufficient to cover the standard of living Lin Quancheng found adequate for a professional.[70] While some engineers eventually found pride in a frugal lifestyle (chapter 7), the objections demonstrate that a poorly paid entry into professional networks was unacceptable

64 PJRF, 128: Freeman to Lin Quancheng, August 5, 1918; 128: Freeman to B. E. Chiu, October 26, 1918.

65 PJRF, 128: John Freeman, China Grand Canal Surveys Progress Report (Draft), February 19, 1919, 34–35.

66 PJRF, 128: John Freeman, China Grand Canal Surveys Progress Report (Draft), February 19, 1919, 37–38.

67 PJRF, 123: Tan Zhen to Freeman, August 10, 1919.

68 PJRF, 128: Lin Quancheng to Freeman, July 16, 1918.

69 Arthur Salter, "The Silver Problem," *Political Science Quarterly* 46, no. 3 (1931): 322; Lawrence Officer and Samuel Williamson, "The Price of Gold, 1257–Present," MeasuringWorth, accessed July 2, 2021, http://www.measuringworth.com/gold/, Gold/silver price ratio 1915–1925.

70 PJRF, 128: Lin Quancheng to Freeman, August 7, 1918.

to the young engineers if they felt it displayed a lack of appreciation or if they were not privileged enough to afford it.

The Grand Canal Board eventually paid its salaries in silver and based them on the rates the Zhili Commission had already established. There were large gaps between the ranks. As assistants at the Grand Canal Board, Lin received $210 silver, and Yang, working part-time but in a senior position, $190.[71] The US senior designing engineer and the assistant chief engineer earned $513.30 each, while chief engineer Ripley received a salary of $2,000. The nominal salary of director general Xiong Xiling was $1,000 silver and his department chief earned $480, although they received half of their wages since they did not commit full-time.[72] For reference, messenger-boys at the Board received just $10 silver per month,[73] roughly the standard for unskilled laborers during that time.[74] For comparison with railway engineers, according to a study the Siems-Carey Company conducted at the time, the various railroad contractors each paid about $700–800 silver to foreign division engineers, $400–600 to senior assistant engineers, and $250–450 to junior engineers.[75] The latter slightly exceeded the Canal's entry salary. Therefore, gaining practical experience under a renowned lead remained part of what made the deal agreeable.

3 Surveying an Unstable Environment

The Canal Board surveyed the Grand Canal along its way through Shandong province and past its crossing with the Yellow River. Pan's 1914 survey notes served as a starting point for the team. However, Freeman deemed them incomplete and not up to professional standards.[76] In total, seven different teams were to survey, from south to north, the areas around Hanzhuang (near Weishan Lake), Tengzhou, Jining, Nanwang (Wenshang), the Yellow River

71 PJRF, 128: Freeman to Hardy Cross, August 3, 1918; 123: Pay Roll, Engineering Department, April 1919.

72 PJRF, 123: Pay Roll of Directions Department. Although the source does not clarify the currency, a comparison with average salaries of US engineers shows that it has to be silver ("The Compensation of Civil Engineers," *Engineering News* 73, no. 1 (1915); "Statistics of the Earnings of Graduate Engineers," *Engineering News* 74, no. 11 (1915)).

73 PJRF, 123: Pay Roll of Headworks Bureau, April 1919.

74 "Costs of Living in Peking: Wages," *Chinese Social and Political Science Review* 10, no. 3 (1926): 100; cited and discussed in Robert Allen et al., "Wages, Prices, and Living Standards in China, 1738–1925: In Comparison with Europe, Japan, and India," *The Economic History Review* 64, S1 (2011).

75 PJRF, 128: F. C. Hitchcock to W. F. Carey, November 1, 1917.

76 PJRF, 128: John Freeman, China Grand Canal Surveys Progress Report (Draft), February 19, 1919, 26.

crossing at Dongping Lake, Dawenkou (Tai'an), and Dezhou (Figure 8).[77] Each survey team consisted of an American lead engineer and Chinese engineering assistants.[78] The surveying was to take eight months, involving 6–7 American and 20–30 Chinese engineers.[79] The work included surveying reservoirs, drainages, and flood channels, measuring rainfall and evaporation, planning locks, drains, and gates, and studying the problems of the Yellow River that traversed the Grand Canal in the north (Figures 9, 10).[80]

FIGURE 9 Surveyors of the Grand Canal Improvement Board, between 1919 and 1921 (OJTP 65.5).

During the surveying, the engineers experienced the political instability of the era firsthand. In the first four years of the Republic, leading democrats had been expelled or assassinated, while President Yuan Shikai 袁世凱 (1859–1916) had attempted to declare himself emperor—only to pass away soon after. A new civil government had since formed in Beijing—but in the provinces, numerous military strongpersons, later called "warlords," engaged in civil warfare. The Canal team experienced the side effects early on. After the core group departed from San Francisco, their ship arrived two weeks late because active warfare was blocking civil maritime transport. The presence of anti-government troops and highwaymen around Shandong forced the Board to arrange for stronger

77 PJRF, 125: Table of Hydraulic Districts.
78 PJRF, 128: John Freeman, China Grand Canal Surveys Progress Report (Draft), February 19, 1919, 23.
79 PJRF, 128: Freeman to E. J. Tudsbury, July 19, 1919.
80 OJTP, 33.06: Tan Zhen to Todd, January 15, 1946.

FIGURE 10 A gauge, marked with pencil, installed along the Yellow River banks near the Tianjin–Pukou Railroad Bridge, November 1918 (PJRF 129).

protection. At all times, the Canal surveyors received protection from Chinese military convoys.[81] Regardless, some engineers refused to enter unsafe areas. The team shared many anecdotes about their experiences in these regions. On one occasion, a group of government soldiers mistook a Canal team for "bandits" and opened fire. On another occasion, the engineers saw 16 heads of killed oppositional soldiers hanging from the trees in front of the Yangzhou railway station.[82] A further episode had chief engineer Ripley intending to return from Jining to Tianjin when fights interrupted the Tianjin–Pukou train service. To reach Tianjin, he had to embark on a 2,000 km detour, heading back south and then returning north via the Beijing–Hankou Railway.[83] In total, civil warfare had the Board lose two crucial months of good weather.[84] The connection between political stability and infrastructure must have been striking for the Board members, as the lack of both hindered their surveying work.

81 PJRF, 128: John Freeman, China Grand Canal Surveys Progress Report (Draft), February 19, 1919, 25.

82 PJRF, 128: John Freeman, China Grand Canal Surveys Progress Report (Draft), February 19, 1919, 31; Freeman, "Grand Canal Investigations 1918–1921," 20–21.

83 PJRF, 128: John Freeman, China Grand Canal Surveys Progress Report (Draft), February 19, 1919, 32–33.

84 PJRF, 128: John Freeman, China Grand Canal Surveys Progress Report (Draft), February 19, 1919, 29.

In 1919, Freeman visited the Grand Canal Board to check if everything was in order. In a personal letter to Charles A. Stone (1867–1941), president of the AIC, Freeman expressed his devastation about the Board's lack of progress.[85] He found that the progress had stalled not only because of the political situation but also because of the personnel. For Freeman, the American engineering assistants, especially the professors on leave, had turned out to be a disappointment, as they lacked practical experience in the fields.[86] He sent them home by early 1920.[87] Freeman further alleged that Xiong Xiling's assistant Wei Yi 魏易 (T. S. Wei, 1880–1932), former secretary of the national oil and salt administrations, had recruited an unnecessarily large number of assistants to "reward friends and to pay old scores."[88] However, Freeman was most frustrated with the work performance of chief engineer Ripley, whom he had still described as "the most experienced man in the world" one year earlier.[89] Freeman criticized Ripley's refusal to venture into the countryside and take part in practical work, instead conducting "swivel-chair engineering" from his office.[90] He portrayed Ripley as physically unfit, as a "Smiling Buddha" who "sat all day long and smiled and did nothing else."[91] Through the criticism, Freeman propagated the professional ideal of the practical-oriented, physically able engineer (chapter 4). He felt offended by Ripley's weight gain and claimed his overweight "slowed down the energy of his once active brain." Accordingly, he advised the AIC to recall Ripley to save "everybody's face."[92] According to his own account, Freeman worked between 12 and 14 hours a day to save the American engineering world from further embarrassment.[93] To straighten up the Board's operations, Thomas Hollis Wiggin (1873–1964), with whom Freeman had served on the New York City Board of Water Supply, replaced Ripley, while Oliver Julian Todd (1880–1974), one of Freeman's former Hetch Hetchy assistants, became principal assistant engineer.[94] Pan Fu succeeded Xiong Xiling as director of the Board.

85 PJRF, 128: Freeman to Charles A. Stone, July 23, 1919.
86 PJRF, 130: Freeman to Charles R. Crane, September 13, 1920.
87 PJRF, 123: Freeman to Bogert, February 5, 1920.
88 PJRF, 130: Freeman to Charles R. Crane, September 13, 1920. The Canal's junior hydraulic engineer Tang Baotong (b. 1887) was a relative of Tang Shaoyi.
89 PJRF, 128: Freeman to Wong Kwong, November 14, 1918.
90 PJRF, 130: Freeman to Charles R. Crane, September 13, 1920.
91 PJRF, 123: Freeman to Bogert, February 5, 1920.
92 PJRF, 128: Freeman to Charles A. Stone, July 23, 1919.
93 PJRF, 130: Freeman to Charles R. Crane, September 13, 1920.
94 Freeman, "Grand Canal Investigations 1918–1921," 14.

4 The Money Goes Astray

While the late replacements in management proved historically significant in that they kick-started Oliver J. Todd's influential, two decades spanning career in China (chapters 4, 5, 6),[95] they could not turn around the Canal Board's imminent fate. The Board quickly ran out of money. Overhead expenses, including salaries of commissioners, secretaries, accountants, translators, and soldiers for protection, amounted to twice the sum of the $151,800 originally designated for the surveying work.[96] The unfavorable exchange rate between US gold and Beiyang silver dollars worsened the situation further.[97] It became clear that the Board had to concentrate on a smaller portion of the Canal than envisioned.[98]

By January 1920, $60,000 of the surveying budget remained, sufficing for less than three more months.[99] Freeman recommended reducing the workforce to an absolute minimum so that the Board could keep its office running until the construction phase would commence.[100] At that time, the Board's average payment for an American engineer was $974 silver monthly, while a Chinese engineer earned $217 on average. As US engineers assumed senior positions while recent Chinese graduates started on the assistant level, the pay gap did not necessarily reflect racial discrimination. On the contrary, the unavoidable wage cutting had American engineers' salaries decrease by $340, while the average Chinese salary slightly increased by $58,[101] adjusting the wages despite gaps in prior work experience. The wage increase did not help many, as the reductions also included job cuts: the AIC reduced the Board from the then remaining five American and eleven Chinese engineers to three of each nationality. Shortly after these reforms, the post-World War I recession had worn off, and Freeman was once again hopeful that the Canal Board could receive the AIC's continued financial support.

95 Pietz, *The Yellow River*, 83–84.

96 PJRF, 128: John Freeman, China Grand Canal Surveys Progress Report (Draft), February 19, 1919, 24, 44.

97 PJRF, 128: John Freeman, China Grand Canal Surveys Progress Report (Draft), February 19, 1919, 20; Ye, "The Grand Canal in Republican China," 758.

98 PJRF, 128: John Freeman, China Grand Canal Surveys Progress Report (Draft), February 19, 1919, 45.

99 PJRF, 128: John Freeman, China Grand Canal Surveys Progress Report (Draft), February 19, 1919, 25.

100 PJRF, 128: Freeman to Pan Fu, January 3, 1920.

101 PJRF, 128: Freeman to Pan Fu, January 3, 1920.

To the team's frustration, a month after the situation had recovered, the US deflationary Depression of 1920–1921 started casting its long shadows over the enterprise. In November 1920, Freeman anticipated that the Board was again "on the eve of a general drastic wage reduction."[102] The AIC suffered severe financial losses. Its stock fell from $116 to $22. While the stock recovered to $40, the depression still resulted in a loss of $30 million for the AIC. A transition from the completed Canal surveys to actual construction became increasingly unlikely. By early 1921, the AIC noted that the world financial situation did not permit the immediate commencement of construction.[103] Still, Freeman believed the AIC was "earnestly desirous" to find the money.[104] In addition, he believed Siems-Carey was still interested as the Canal served as a "stepping stone" for future railway concessions.[105] Yang Baoling was less optimistic. He became increasingly frustrated with the AIC, which did not provide the Board with a definite statement about its future. He found the situation hopeless, describing the Canal Board affairs as having become "more like a monastery than an engineering office."[106]

Yang's frustrations about the dependency on world finance become understandable when comparing the fate of the Canal's northern portion with that of the southern part. While Siems-Carey and the AIC had prepared for a future extension of the improvement work to Zhang Jian's southern base in Jiangsu, the Nantong industrialist did not wait for US investment to commence. Instead, in 1920, Zhang organized the Jiangsu Grand Canal Improvement Board in Yangzhou through local land taxes.[107] Yang Baoling acted as an adviser for this sister project, which, although also depending on stable politics and a growing domestic economy, did not rely on foreign investment.[108]

In contrast, the Shandong Canal Board formally halted operations on March 1, 1921. For Lin Quancheng, the closing felt rushed.[109] Freeman submitted his final review of the Grand Canal around this time, concluding his business with the AIC. He ended his involvement on a bad note. Siems-Carey's vice-president Philip Walter Henry (1864–1947) had proposed to complete the

102 PJRF, 125: Freeman to L. J. Buckner, November 22, 1920.

103 PJRF, 125: Freeman to Yang Baoling, February 25, 1921. See also Mazuzan, "'Our New Gold Goes Adventuring'," 230–32.

104 PJRF, 125: Freeman to Lin Quancheng, June 4, 1921.

105 PJRF, 128: Freeman to Yang Baoling, January 16, 1922.

106 PJRF, 130: Yang Baoling to Freeman, January 21, 1921.

107 Pomeranz, *The Making of a Hinterland*, 234; Ye, "The Grand Canal in Republican China," 755.

108 JACAE, 1 (1): 32.

109 PJRF, 125: Lin Quancheng to Freeman, April 24, 1921.

Canal contract in the cheapest way possible. In Freeman's eyes, Henry merely wanted to reopen the old Canal without any larger improvements, ignoring its potential for flood relief, irrigation, and land reclamation.[110] Freeman urged the Chinese government not to accept this lower budget plan.[111] Likewise, Pan Fu still desired construction to commence as originally planned. Emulating Zhang Jian's path, he proposed starting the work using $250,000 of local Shandong relief funds.[112]

Soon afterward, the financial crisis that had brought the AIC's default led to the depreciation of silver and the bankruptcy of the Beijing government.[113] As the Beijing government ran out of money to pay its civil servants in 1922, domestic subsidies were out of the question. The Canal Board, already put on halt, permanently closed its doors in September.[114] There was no money left. Yang Baoling pressed Xiong Xiling and Pan Fu to secure a budget of $200 a month to maintain the recording gauges and rainfall stations,[115] but apart from this minimal service, the Board never recovered. If the fate was not sealed already, a domestic economic crisis hit the Chinese market in 1923,[116] dealing a deathblow to any fantasies of financing the project locally.

For the journalist Dong Xianguang 董顯光 (Hollington K. Tong, 1887–1971), then working for the Zhili Commission, the disappointment resulted in bitterness toward the US and aroused pro-Japanese tendencies.[117] His colleague Yang Baoling likewise explained that the AIC's decision to cancel the Canal improvement scheme left the Chinese engineers "greatly disappointed." His disappointment derived from being "one who received his education in America."[118] Yang, upon commencement of the surveys, had written about his confidence that the Canal project would not end up like previous US ventures, such as the surveys of the Huai River, the Zhuzhou–Jinzhou Railway, and the Shaanxi oil fields, which had "resulted in nothing more constructive [...] than copies of well-printed reports."[119] Ironically, in the case of the Canal

110 OJTP, 52.01: John R. Freeman to Todd, June 1, 1921; PJRF, 130: Freeman to Yang Baoling, November 25, 1921; 128: Freeman to Yang Baoling, January 16, 1922.

111 PJRF, 130: Freeman to Oliver J. Todd, February 16, 1922.

112 OJTP, 25.12: Freeman to T. H. Wiggin, October 2, 1920; OJTP, 13A.2: File Copy of Letter to the ACAE, presumably by Pan Fu, undated, c. 1921.

113 Bergère, *The Golden Age of Chinese Bourgeoisie*, 84–91.

114 PJRF, 129: Yang Baoling to Freeman, August 30, 1922.

115 PJRF, 129: Yang Baoling to Freeman, August 30, 1922.

116 Bergère, *The Golden Age of Chinese Bourgeoisie*, 84–91.

117 PJRF, 130: O. J. Todd to Freeman, September 16, 1921.

118 PJRF, 125: Yang Baoling to Freeman, May 18. 1921.

119 Pao-ling Yang, "Improvement Begins on China's Famous Grand Canal," *Millard's Review of the Far East*, September 28, 1918, 164. Similar doubts from Xiong Xiling and Pan Fu, see Ye,

Board, the budget did not even suffice to print a useless report.[120] The extent of underperformance must have been devastating. With some bitterness, he sent Freeman the Chinese five-colored flag as a reminder that he had left the Canal work unfinished.[121]

In contrast to Yang, Freeman believed the AIC had been "sincerely and earnestly" desirous to the end and blamed the difficult conditions in Beijing.[122] The AIC, he conceded, was temporarily unable to provide funding due to the economic depression. Nevertheless, Freeman primarily blamed the country's lacking integrity, not the AIC, for the suspension of the Grand Canal funding.[123] He remarked that China seemed "to be its own worst enemy," as Chinese civil wars made government bonds drop to disastrous prices.[124] He identified "the selfish military rascals" as the prime factor for China's desolate situation.[125]

Given the Japanese co-investment in the project, anti-Japanese sentiments may also have played a role in the reluctance of national politicians to support the project more wholeheartedly.[126] However, the involved Chinese engineers came up with different reasons for the project's failure than later economic historians.[127] While contemporary historians primarily blame domestic politics,[128] the involved Canal personnel emphasized insecure world finance, unstable currency exchange rates, and unwilling investors.

The assessment of whom to blame for the end of the Canal Board differed between the Chinese and American engineers. But in both cases, the premature closure of the Board served as a first disillusionment about the feasibility of realizing infrastructural visions when operating under difficult political-economic conditions.

5 Solidarity—The Yellow River Bridge Controversy

The halt and eventual closure of the Canal Board, under the impression of global economic crisis and domestic bankruptcy, led to disappointment on

"The Grand Canal in Republican China," 754.

120 Oliver Todd, *The China That I Knew* (Palo Alto: s. n., 1973), 19; Fisher, "Service to China," 36.

121 PJRF, 129: Yang Baoling to Freeman, August 30, 1922.

122 PJRF, 130: Freeman to Yang Baoling, November 25, 1921.

123 PJRF, 130: Freeman to Wong Kwong, December 14, 1921.

124 PJRF, 125: Freeman to Yang Baoling, June 13, 1921.

125 PJRF, 129: Freeman to Yang Baoling, November 7, 1922.

126 Pomeranz, *The Making of a Hinterland*, 234.

127 Ye, "The Grand Canal in Republican China," 734, 758–759.

128 Mazuzan, "'Our New Gold Goes Adventuring'," 230–32; Pomeranz, *The Making of a Hinterland*, 234.

all sides. However, the professional network of the involved Chinese and American engineers lived on. Freeman explained to Tan that neither his memories nor his interest in China suffered from the way the Canal improvement project immaturely concluded.[129] For Freeman, "the lure of the Far East" was still upon him.[130]

In 1921, when the Canal activities had already been frozen, the project's core group got involved in another controversy. Notwithstanding the troubled national treasury, the Ministry of Communications tendered the construction of a Yellow River Bridge for the Beijing–Hankou Railway. The 3 km long original bridge, built in 1906, could not bear the heavy railway carriages in the long term and was prone to erosion.[131] The Investigation Committee for the Planning of a New Yellow River Bridge (*Huanghe xin qiao sheji shencha hui* 黃河新橋設計審查會), consisting of Chinese government consultants from Belgium, France, Japan, the UK, and the US, was to judge international submissions and award prizes to the three best entries.[132] One of the jury members was the US-Canadian civil engineer John Alexander Low Waddell (1854–1938, chapters 4, 6, 7), whom the young Ling Hongxun (chapters 1, 4, 6, 7) assisted in auditing the calculations of the competitors.[133] Freeman took part in the contest. He claimed that discoveries during the Canal Board surveys had led him to a new vision for the bridge.[134] His idea was to "tame" the Yellow River into a straight, narrow course, allowing for a very short bridge one-fifth the length of the 2.8 km long Belgian design,[135] thereby being approximately five million Beiyang dollars cheaper than the most popular competitor.[136]

The controversy started when the jury disqualified Freeman's entry because they had received the application a day late.[137] Instead, the jury, by a majority of the European judges, voted to award the first prize to the Belgian company's design, with estimated costs of $10.8 million silver. French companies received the second and third prizes. The Ministry of Communications argued that

129 PJRF, 125: Freeman to Tan Zhen, January 4, 1922.

130 PJRF, 130: Freeman to Charles R. Crane, September 13, 1920.

131 JACAE, 2, no. 8 (1921): "The Yellow River Bridge Tenders," 12.

132 See Lin-chun Wu, "Mínchū Měiguó qǐyè duì huánghé tiěqiáo hé nánfāng dàgǎng de tóuzī: Qǐyè, zhèngfǔ yǔ wàijiāo guānxì de kǎochá," *Táidà lìshǐ xuébào* 47 (2011): 64–81.

133 Líng, *Qīshí zìshù*, 35–36.

134 PJRF, 130: Freeman to Thomas Meloy, October 6, 1921; Canal Board engineers had surveyed the site, as reported in Hugh Brodie, "Reconnaissance Trip through Honan and Shantung Provinces, China," *The Michigan Technic* 33, no. 4 (1921): 200.

135 PJRF, 130: Freeman to Le Directeur-Administrateur, Ligne de Kin-han, January 23, 1924.

136 PJRF, 130: Freeman to O. J. Todd, September 29, 1921; 130: Freeman to O. J. Todd, October 13, 1921.

137 PJRF, 130: Roger M. Freeman to Freeman, December 22, 1921.

the Commission acted in accordance with the rules of bidding, disqualifying Freeman for late entry,[138] although, according to one informant, the jury acknowledged that his design was technically better than the awardees.[139] Freeman, however, believed his disqualification to be a "gross injustice."[140] He argued that the Ministry had decided to plunder the national treasury for a wasteful bridge design, attending to neither justice nor Chinese interests.[141] For him, the verdict was uneconomical and excluded considerations of river control and famine safeguarding.[142] The rejection of his design led to his further disillusionment with Chinese government institutions.

Canal Board engineer Oliver Julian Todd reminded Freeman that his bridge scheme was "too daring" for the "very conservative" Chinese decision-makers. The remark points to the discontent with non-engineering officials in charge. Todd reported rumors that there was a conspiracy between the three European judges and the Chinese vice-chairman of the commission to award the contract to the Belgian company.[143] Lin Quancheng likewise described the procedure as a "political intrigue" to benefit Belgian concerns. From the affair, Lin concluded that it was difficult to "sift the politics from the technical truth." This realization was representative of the disillusionment that arose from experiences in which politics hindered engineering feats. Lin called for the engineers who studied in the US to organize a joint petition.[144] The plan shows that there was a feeling of group belonging among the returned US-educated Chinese engineers, as well as a felt responsibility to support an American concern.

Freeman joined Todd's and Lin's belief that he had become the victim of an international conspiracy and the scheming of rivals.[145] He complained to Tang Shaoyi, referring to a plot by "someone who wanted to build the longest possible bridge."[146] For him, awarding the costlier bridge undermined the Chinese testimony of requiring the remainder of the US Boxer Indemnity to fight poverty.[147] Unsuccessfully, he tried to intervene in the Indemnity payments

138 PJRF, 130: Freeman to Senator Le Baron B. Colt, October 15, 1921.
139 PJRF, 130: Thomas Meloy to Freeman, February 25, 1922.
140 PJRF, 128: Freeman to Xue Zhuobin, October 6, 1921.
141 PJRF, 130: Freeman to Thomas Meloy, December 14, 1921.
142 PJRF, 129: Freeman to Shi Zhaoji, September 15, 1921.
143 PJRF, 130: O. J. Todd to Freeman, September 16, 1921.
144 PJRF, 125: Lin Quancheng to Freeman, September 16, 1921.
145 PJRF, 130: Freeman to O. J. Todd, September 29, 1921.
146 PJRF, 129: Freeman to Tang Shaoyi, August 29, 1922.
147 PJRF, 129: Freeman to Fred M. Dearing, September 23, 1921.

through acquainted journalists, US senators, and the US Department of State.[148] Further, he thought he could use his "great friendship" with the elder political leaders of China—Xiong Xiling, Zhang Jian, and Pan Fu—to lobby his cause[149] and asked Todd to speak to the latter.[150]

Yang Baoling tried to comfort Freeman, stating that his "friends in the Ministry" would do their best to resolve the case.[151] He called the procedures a "rotten arrangement" but defended the Ministry. For him, the European powers used the Chinese internal disarray to influence procedures to their liking. He saw the outcome as proof that there was "too much international jealousy and scheming" interfering with domestic Beijing politics.[152] Like in his judgment of the Canal Board closure, Yang primarily blamed foreign institutions and portrayed domestic corruption as an afterthought.

Apart from what the Grand Canal group perceived as a conspiracy of the European majority within the commission, the engineers also fiercely criticized Waddell for sitting on the jury. Waddell had previously presented his own version of an ideal bridge at an engineering conference that the Ministry of Communications had hosted in Hankou.[153] As the Japanese submission bore a close resemblance to Waddell's design, his sitting on the jury was, in Freeman's opinion, "unpardonable."[154] The British Institution of Civil Engineers reprimanded Waddell for his role in the procedures. Waddell claimed he did not realize the Japanese scheme was a reproduction of his own design[155] and found his "professional integrity" attacked.[156]

When Waddell advised delaying the construction of the Yellow River Bridge and reissuing its tender,[157] the group of Sino-American civil engineers made

148 PJRF, 130: Freeman to O. J. Todd, October 13, 1921; 130: Freeman to Senator Le Baron B. Colt, October 15, 1921; 130: Petition of John R. Freeman in re: China Yellow River Bridge, to the United States Department of State, August 24, 1921; 130: Freeman to Thomas Meloy, December 14, 1921.

149 PJRF, 130: Freeman to O. J. Todd, September 29, 1921.

150 OJTP, 52.01: John R. Freeman to Todd, May 7, 1921; PJRF, 130: Freeman to O. J. Todd, September 26, 1921.

151 PJRF, 130: Yang Baoling to Freeman, April 23, 1924.

152 PJRF, 130: Yang Baoling to Freeman, October 30, 1921.

153 PJRF, 130: Yang Baoling to Freeman, October 30, 1921.

154 PJRF, 130: Freeman to O. J. Todd, October 13, 1921.

155 John Waddell, "Dr. Waddell Flatly Repudiates Charges Made Against Him by Colleagues on Yellow River Bridge Board," *The Peking Leader*, September 1921, 5.

156 Waddell, "Dr. Waddell Flatly Repudiates Charges Made Against Him by Colleagues on Yellow River Bridge Board," 1.

157 PJRF, 130: Freeman to O. J. Todd, February 16, 1922; JACAE, 2, no. 10 (1921), 44. Ling, Waddell's assistant at the time, mentions no conflict of interest and explains the delay with insufficient funds for construction (Líng, *Qīshí zìshù*, 36).

peace. The Association of Chinese and American Engineers (ACAE), of which Yang was vice-president at that time (chapter 4), backed Waddell and portrayed him as a victim of the European jury.[158] Freeman supported the ACAE's defense of Waddell and, despite his discontent, described him as "absolutely honest at heart."[159] A few years later, Freeman received an invitation from the Beijing–Hankou Railway administration to join a retry of the Yellow River bridge competition. However, Freeman was too frustrated from his previous experience and chose not to participate.[160] He did not believe authorities would commence construction when warfare had severely strained the Chinese treasury.[161] He would continue to advertise his bridge scheme through the last years of his life, but to no avail.[162]

Through the Yellow River Bridge affair, after the AIC had already failed to deliver the Canal improvement, Yang's disappointment in foreign powers grew further, while Freeman grew ever more disappointed in the Chinese institutions. Regardless of whether there was a conspiracy and who was to blame for it, the episode demonstrates that the Canal engineering group had formed a professional network of mutual support that extended beyond the Grand Canal endeavor. Taken as a whole, the Yellow River Bridge contest and its aftermath was the first instance in which the Sino-American engineers that made up the core of the Association of Chinese and American Engineers (chapters 4, 6) cooperated as an allied group. It was a precursor to the professional solidarity observed during the Salaqi controversy ten years later (chapter 6).

6 Frustration—Huai River Improvement Schemes

Despite the setbacks with the Canal Board and the Yellow River Bridge, Freeman assured Yang Baoling that he continued to have a "lively interest in China" and would like to return despite health concerns.[163] He claimed his interest in China was no longer financial but entirely private, and that he would return as

158 PJRF, 130: O. J. Todd to Freeman, September 16, 1921; Waddell, "Dr. Waddell Flatly Repudiates Charges Made Against Him by Colleagues on Yellow River Bridge Board," 5.

159 PJRF, 130: Freeman to Yang Baoling, November 25, 1921.

160 PJRF, 130: Freeman to Yang Baoling, January 22, 1924.

161 PJRF, 130: Freeman to Tan Zhen, February 5, 1924.

162 PJRF, 129: Freeman to Yang Baoling, December 19, 1930.

163 PJRF, 125: Freeman to Yang Baoling, February 25, 1921. Also PJRF, 129: Freeman to Wong Kwong, August 29, 1922.

a tourist.[164] In the ensuing 1920s, a number of opportunities emerged for the Canal Board team to regroup and continue their hydraulic work. The fact that the engineers actively turned these invitations down demonstrates that their earlier frustrations affected their pessimistic outlook in the long term.

The Great Huabei Famine of 1920–1921 (chapter 5) was the starting point of Freeman's private, philanthropist interest in the Chinese river systems, particularly the Yellow and Huai Rivers, which he identified as potential grounds for a "radical" and "novel" river taming theory.[165] His idea was to have the stream of water dig itself gradually deeper along a newly defined straight line. On each side, river sediments would create banks at least 1.6 meters thick to reinforce the dikes against the maximum height of flooding.[166] Through its new, narrower course, the river would gain enough power to transport its silt to the sea instead of clogging the Grand Canal.[167]

At first, Freeman believed his scheme to force the river to excavate itself to be "a new discovery in river hydraulics."[168] However, Yang reminded Freeman that his idea to utilize natural forces to assist in the excavation of a river was not novel. Chinese architects had described the method in the past but did not develop the scheme in a scientific way.[169] Freeman asked Tang Shaoyi to send him information about ancient hydraulic methods, including information on the legendary ruler Yu the Great 大禹 and on Li Bing 李冰 (third century BCE), the man behind the Dujiangyan irrigation system in Sichuan.[170] While the modern civil engineers found that ancient methods lacked scientism, they were not ignorant about them.[171] Xu Shiying 許世英 (1873–1964), then civil governor of Anhui, also remarked that Freeman's plans had precursors in

164 PJRF, 130: Freeman to O. J. Todd, February 16, 1922; PJRF, 129: Freeman to E. W. Lane, September 1, 1922.

165 PJRF, 130: Freeman to Xiong Xiling, October 19, 1920; Pietz, *Engineering the State*, 35, 47; Pietz, *The Yellow River*, 83.

166 PJRF, 130: Freeman to Yang Baoling, October 9, 1920.

167 PJRF, 130: John Freeman, Memorandum for Address at Chinese Students Conference Brown University Providence, January 10, 1923, 8.

168 PJRF, 130: Freeman to Wong Kwong, October 18, 1920.

169 PJRF, 130: Yang Baoling to Freeman, January 21, 1921.

170 PJRF, 129: Freeman to Tang Shaoyi, August 29, 1922. Interest in hydraulic proto-engineers spread in foreign engineering circles, see Joseph Ehlers, *Far Horizons: The Travel Diary of an Engineer* (New York: Carlton Press, 1966), 29–30, 47.

171 For attempts to integrate Chinese traditions into scientific history, see, e.g., Amelung, "Die ‚vier großen Erfindungen'"; Iwo Amelung, "Lokalität und Lokalisierung: Zur Entwicklung der Wissenschaften im China des späten 19. und frühen 20. Jahrhundert," *Jahrbuch zur Europäischen Überseegeschichte* 14 (2014).

Chinese antiquity. As historical figures such as Pan Jixun 潘季馴 (1521–1595) had applied similar ideas,[172] Xu found it easy to believe in Freeman's scheme.[173]

Freeman believed Zhang Jian, progressive and interested in engineering, to be his best chance of financing a test of the scheme,[174] but in 1922, Zhang ran into financial trouble.[175] That year, the US Consul and the Red Cross in Nanjing asked Freeman to study the flood problems in Jiangsu.[176] Freeman was interested in leading the Huai River Conservancy and already envisioned reemploying his former assistants Tan Zhen, Oliver J. Todd, and Xue Zhuobin.[177] The latter's interest quickened from the fact that he considered the Huai his home.[178]

However, Yang Baoling strongly advised Freeman to reject the invitation to join the Commission. To Yang's frustration, the river had been "surveyed, and resurveyed, and resurveyed" without ever reaching the stage of construction. He thought that new surveying would damage the reputation of the engineering profession further. Yang reiterated his claim that the American Red Cross' fruitless previous surveying of the Huai, Freeman's abortive improvement of the Grand Canal, and an inconclusive surveying of the Zhuzhou–Jinzhou Railway had led officials to conclude that US-educated, "modern" engineers were unable to perform actual work.[179] Todd criticized Yang for his opposition, claiming that Yang exhibited "poor logic in the oriental mind" for insisting on greater funding before proper surveying work had been performed.[180] Freeman found more preliminary surveys of the Huai necessary[181] but shared Yang's suspicion that the offer in question would be a waste of time and unlikely to bear fruit. He did not want to spend his time making "paper plans, with no immediate prospect of making the dirt fly."[182] For Freeman, two things hindered surveys from evolving into actual construction work. The first was the officials' "extreme conservatism" and "lack of appreciation of the scientific

172 Eduard Vermeer, "P'an Chi-hsün's Solutions for the Yellow River Problems of the Late 16th Century," *T'oung Pao* 73, 1/3 (1987).

173 PJRF, 130: Yu Mingde to Freeman, June 21, 1923. Also Pietz, *The Yellow River*, 83.

174 PJRF, 130: Freeman to Yang Baoling, October 9, 1920; 130: Freeman to Xiong Xiling, October 19, 1920.

175 PJRF, 130: Xue Zhuobin to Freeman, April 10, 1923; Qin Shao, *Culturing Modernity: The Nantong Model, 1890–1930* (Stanford: Stanford University Press, 2004), 198–214.

176 PJRF, 129: Freeman to Shi Zhaoji, June 1, 1922.

177 PJRF, 130: Freeman to Xue Zhuobin, November 8, 1922.

178 PJRF, 130: Xue Zhuobin to Freeman, October 4, 1922.

179 PJRF, 129: Yang Baoling to Freeman, August 30, 1922.

180 PJRF, 130: Freeman to O. J. Todd, November 8, 1922.

181 PJRF, 129: Freeman to Yang Baoling, November 7, 1922.

182 PJRF, 130: Freeman to H. S. Sung, December 19, 1923.

method." The second was the lack of a stable government and, consequently, the lack of investors.[183]

Rather than instigating against the Huai assignment,[184] Yang probably acted under the impression of his own disillusionment with the government. He was frustrated that political instability hindered his employer, the Zhili River Commission, in its construction efforts[185] and criticized the Beijing government for neglecting the field of river conservancy.[186] Yang claimed the dependence of his profession on the state affected his nerves.[187] Similarly, his former colleague Lin Quancheng developed frustration at his new assignment with the Fuzhou waterworks. He wrote to Todd that "political uncertainty" and the lack of cooperation from the local military strongperson Zang Zhiping 臧致平 (1869–1944) meant that no progress could be achieved with his project.[188] For Yang, frustration with the state institutions led to his interest in private business as he founded the Chinese Engineering & Development Company (*Dachang* 大昌) with his Chinese and American engineering colleagues (chapter 4).[189]

After Freeman had attended the Washington Naval Conference in 1921—a conference in which the Chinese government only reluctantly accepted, and never implemented, international policing of its railways (chapter 3)—his outlook on China had reached an all-time low. In 1924, Jiangsu's military governor Qi Xieyuan 齊燮元 (1885–1946) again requested Freeman to take charge of surveying the Huai River.[190] Two Chinese junior engineers who had recently interned at his office attempted to soothe the Freeman's pessimistic outlook. They assured him that engineers would not run a risk surveying, as Qi's armies would protect them, and emphasized the project's profitability through land reclamation.[191] Unimpressed, Freeman declined and recommended German hydraulic engineer Hubert Engels (1854–1945) to take the position instead.[192] He received a third invitation to work on the Huai River in 1926, this time

183 PJRF, 129: Freeman to W. W. Peter, June 29, 1922.

184 See PJRF, 130: William E. Souter to Freeman, July 10, 1923.

185 PJRF, 129: Yang Baoling to Freeman, August 30, 1922.

186 PJRF, 125: Yang Baoling to Freeman, May 18, 1921.

187 PJRF, 129: Yang Baoling to Freeman, August 30, 1922.

188 OJTP, 29.06: Lin Quancheng to Todd, December 17, 1923.

189 PJRF, 130: Yang Baoling to Freeman, April 23, 1924.

190 PJRF, 130: Telegram from June 15, 1924.

191 PJRF, 130: Yu Mingde to Freeman, March 5, 1924; PJRF, 130: H. S. Sung to Freeman, May 18, 1924.

192 PJRF, 130: Freeman to Wm. E. Souter, September 3, 1924. For Engels in China, see also Iwo Amelung, "Der Gelbe Fluß in Deutschland: Chinesisch-deutsche Beziehungen auf dem Gebiet des Wasserbaus in den 20er und 30er Jahren des 20. Jahrhunderts," *Oriens Extremus* 38 (1995): 158–59.

through Sun Chuanfang 孫傳芳 (1885–1935), the "Nanjing warlord" of Jiangsu. This time, his former assistant Tan Zhen advised Freeman to turn down Sun's invitation. Tan Zhen, Oliver J. Todd, and Xue Zhuobin were all critical of the Jiangsu authorities.[193] A last time in 1929, Chinese ambassador Li Jinlun 李錦綸 (Frank William Lee, 1886–1956) asked Freeman to become technical advisor for the newly appointed Huai River Conservancy Commission.[194] Freeman responded that he could come for a month and attached his salary expectation of $150–200 per day.[195] Freeman explained to Yang that he did not intend "to be personally avaricious."[196] Instead, similar to how the young Chinese graduates had turned down Freeman's salary offers ten years earlier, he felt that if the government could not agree to the high salary, it lacked appreciation for his services.[197] The government was unable to meet the demands.

7 Conclusion

For the Canal Board's engineers—on the Chinese side Lin Quancheng, Tan Zhen, Xue Zhuobin, and Yang Baoling—the experiences with the project and its termination resulted in a continuous disillusionment about grand scale technological visions vis-à-vis (inter-)national *realpolitik*. The Canal Board had run off course: the engineers realized that global finance and national unrest hindered them in realizing their engineering projects. To follow the waterway metaphor, the engineers depended on the political and economic currents of the time. For the young graduates, who returned to China to use their profession in service to the nation, this realization was frustrating. The lack of financial capital served as a reminder that civil engineers were not autonomous "spatial entrepreneurs" who could freely shape the national infrastructure to their liking, regardless of the excellent training they had received to do just that. The young professionals had little hope that the state could help them in their craft in the near future. The way in which Freeman, through consultation with his closest former engineering assistants, dismissed the various government offers to renovate the Huai demonstrates that, after the initial Grand Canal failure, the disappointment with the Chinese political climate advanced to a stage where when presented with the chance to realize grand

193 PJRF, 129: Tan Zhen to Freeman, March 14, 1926.

194 PJRF, 129: Li Jinlun to Freeman, January 11, 1929. See Pietz, *Engineering the State*, 41–60.

195 PJRF, 129: Freeman to Li Jinlun, January 17, 1929.

196 PJRF, 129: Freeman to Yang Baoling, March 27, 1929.

197 PJRF, 130: Freeman to A. J. Brown, February 2, 1923.

infrastructural projects, the engineers rejected the opportunities in disbelief. The naiveté of the earlier years was gone. For the Canal Board members who blamed international dependencies, domestic unrest, and a general lack of political will for the Canal Board's failure, the experiences would form a basis for reappraising the liberal ideals of their student days and gradually voicing support for governments with a greater concentration of power (chapters 5, 6). Although the Canal Board's aspirations never materialized, the project was a starting ground for a professional association and long-lasting friendship (chapter 4). In this way, despite all hardships, the project would—at least partly and temporarily—evolve into a successful starting ground for transnational cooperation.

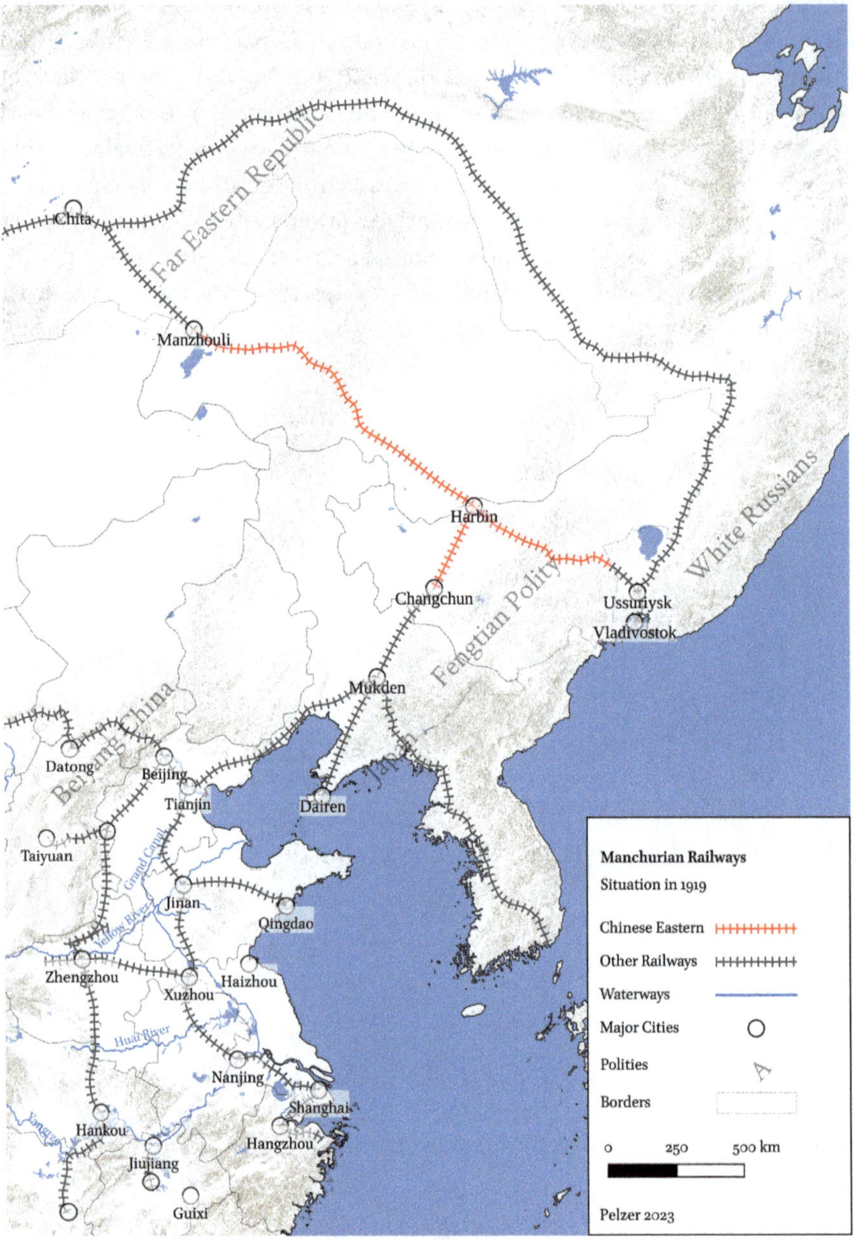

FIGURE 11 The Chinese Eastern Railway and its environs, 1919.

CHAPTER 3

Political Dependency

Wang Jingchun at the Chinese Eastern Railway, 1919–1924

After his return to China in late 1911, Yale and Illinois-educated civil engineer Wang Jingchun (chapter 1) began his career as a railway administrator in the new Beijing Republic. After a swift rise to influence inside the Ministry of Communications, he began his most cumbersome undertaking in 1919. The Chinese Eastern Railway (CER) was a Russo-Chinese enterprise. It became a test for key assumptions that Wang had put forward in a large range of theoretical writings.[1] Set prior to the proclamation of the Nanjing Republic in 1927, the story provides an insight into the limits of early technocracy and scientific management. Technocracy, here, is used in its 1920s understanding: not as a state rule of "technocrats,"[2] but as expert management of government institutions based on the technological principles of rationality and efficiency.[3] Wang attempted to apply these paradigms to the politically charged administration of the CER. Ultimately, his ideals conflicted with his practical experience at the railway. An inescapable net of political entanglements seriously dampened the engineer's republican optimism and his idea of efficient, apolitical management.

1 The Nationalist Paradigm

During his student years, Wang had discovered that his fields of work—civil engineering and railroad administration—held their own nationalist potentials. For him, railways fulfilled three public objectives: they could develop the country, augment wealth, and unite the people.[4] Railways were not just

1 Wang published many of his treatises in both Chinese and English. To make references accessible to a larger readership, the English versions of his texts are cited. As to how publishing in English may have been a performative action to mark one's belonging to the Sino-American engineering segment, see chapter 4.

2 See Duncan McDonnell and Marco Valbruzzi, "Defining and Classifying Technocrat-led and Technocratic Governments," *European Journal of Political Research* 53, no. 4 (2014): 656–58.

3 See ideas of Thorstein Veblen (1857–1929) and Frederick Taylor (1856–1915) discussed in Akin, *Technocracy and the American Dream*, 1–45.

4 Ching-chun Wang, "Some Dangers of Railway Development in China and How to Prepare against Him [*sic!*], Part 1," *The Chinese Social and Political Science Review* 1, no. 4 (1916): 48;

 | DOI:10.1163/9789004549555_005

commercial but social, economic, and political enterprises.[5] In this respect, he referred to former US vice-president Henry Wilson (1812–1875). Wilson had remarked that costs were secondary for the Transcontinental Railroad (const. 1869). Primarily, the line aimed to connect and "bind" the American people of the West and East Coasts.[6] Apart from enhancing the flow of money, dispersing production, diversifying labor, and creating new commerce, railways enabled the diffusion of ideas, the movement of people, and the standardization of languages.[7] Accordingly, infrastructure was a tool for the new national state to bring its people closer together.

His graduation thesis on the "Economic Significance of Railway Location," submitted and printed in 1909, was where Wang first outlined his idea of using infrastructure to restructure and nationalize space. In it, he argued that location was the central factor to assess when planning a railroad.[8] Rhetorically, Wang tried to connect this theory with Chinese antiquity: he remarked that the *Book of Mengzi* (c. 4th century BCE) already stated, "good times are not as important as good location."[9] Wang pointed out that railways were the first artificial means of transportation for which the location was determined more by man than by nature.[10] This meant that the natural advantages of a city—for instance, the proximity to a coastline—would become secondary to the city's connection to the railway system.[11] In contrast to natural rivers, governments could strategically locate railway lines to shape centers and link them with peripheries.[12] Railway locators had the power to shape and change space, as their decisions would move city centers, increase the price of land, and ultimately turn "the fate of nations."[13] To make his point, Wang referred to recent

---

Ching-chun Wang, "Economic Significance of Railway Location" (A. M., University of Illinois, 1909), 5. Wang would later use this argument to successfully argue for using the British Boxer Indemnity Fund to construct the Guangzhou–Hankou Railway.

5 Wang, "Economic Significance of Railway Location," 26. Also Ralph Huenemann, "The Dragon and the Iron Horse: The Economics of Railroads in China, 1876–1937" (PhD, Harvard University, 1982), 300.

6 Wang, "Economic Significance of Railway Location," 6. Original in John Rives, ed., *The Congressional Globe: Senate, 37th Congress, 2nd Session* (Washington, D. C.: Blair & Rives, 1862), 2754.

7 Wang, "Economic Significance of Railway Location," 6–9, 16. Also Wang, "Some Dangers of Railway Development in China and How to Prepare against Him [sic!], Part 1," 49.

8 Wang, "Economic Significance of Railway Location," 24–26.

9 Wang, "Economic Significance of Railway Location," 73–74. The original is in the *Book of Mengzi*, "Gongsun chou xia," verse 10 (*tianshi bu ru dili* 天時不如地利).

10 Wang, "Economic Significance of Railway Location," 47.

11 Wang, "Economic Significance of Railway Location," 65–71.

12 Wang, "Economic Significance of Railway Location," 59.

13 Wang, "Economic Significance of Railway Location," 10, 30, 104.

US and Chinese history. Between Missouri and Illinois, St. Louis had lost its relevance against the railway hub of Chicago. Likewise, after the Imperial Chinese Railway had reached Beijing in 1897, the Tongzhou district had become a "deserted town" despite its strategic proximity to the Bai River. In relation to nation-building, an optimal railway layout helped consolidate the political center, with branches of tracks spreading out to the peripheries.[14] In their essence, railways had a staggering nationalist potential, as they were something "every individual person as well as every national" felt.[15]

Wang argued that railways could not only unite the country but also bring different nations closer together[16] and help make his Christian belief dominant in China.[17] His cosmopolitan socialization never saw a contradiction between nationalism and internationalism (chapter 1), as he claimed the overthrow of the Manchu dynasty (1911–1912) to have been an "American revolution."[18] In accordance with this belief of a positive impact of US culture on Chinese republicanism, Wang had no colonial qualms when he marketed a railroaded China as an investment utopia in which US traders could enrich themselves through selling material, machinery, and tools in a fair and open market.[19] In a US journal, almost a decade before Sun Yat-sen published his famous development plan,[20] Wang advertised that China would soon become an international "workshop" of railways, mines, and industries, keeping "all factories of the world busy to cope with her demands."[21]

In summary, Wang understood the potential of railways to incorporate and structure space in a nation's interest: they could link culturally and linguistically diverse regions to have its people grow a national consciousness, while also strengthening the national economy through integration into an international market. This understanding was not a new revelation. As outlined above, the US Transcontinental had demonstrated the importance of railways for national development. Likewise, his close friend at the Ministry, the bureaucrat

---

14 Wang, "Economic Significance of Railway Location," 59.

15 Wang, "Economic Significance of Railway Location," 7.

16 Wang, "Economic Significance of Railway Location," 17.

17 Wang, "The Effect of the Revolution upon the Relations between China and the United States," 269–70.

18 Ching-chun Wang, "China's Revolution and Its Effect," *The North American Review* 197, no. 687 (1913): 192, 196, 200–201; cf. Wang, "The Effect of the Revolution upon the Relations between China and the United States," 282; Wang, "'The New China Will Be a New United States'."

19 Wang, "China's Revolution and Its Effect," 199; Wang, "The Effect of the Revolution upon the Relations between China and the United States," 268–69.

20 Yat-sen Sun, *The International Development of China* (Shanghai: Commercial Press, 1920).

21 Wang, "China's Revolution and Its Effect," 198.

Ye Gongchuo 葉恭綽 (1881–1968), described the potential of railways to "adjust boundaries, change terrain, and create links" in his 1924 essay on how transport and communication could save the country.[22] However, Wang's case was special in that, different from Ye, he was a civil engineer, and earlier than his professional peers, he reached powerful positions in which he could attempt to apply his theory to practice. His reflection of how his work could actively change the perception of space, an awareness also called "spatial literacy," together with his capital of profession and office, had him be cut out for the category of the "spatial entrepreneur." Wang exhibited a strong spatial literacy in that he knew what he wanted railways to be used for: he was a spatial entrepreneur with nationalist intentions.

2 The Efficiency Paradigm

Apart from his nationalist aim, Wang followed economic ideals. The two seemingly contradicting dimensions existed next to each other. They created a unique strand of nationalist technocracy in which political convictions were fought for using technical considerations of efficiency.

To discuss how engineers conceived society, profession, and the discontents that emerged from them, it makes sense to differentiate between the terms "technocracy" and "solutionism" without denying potential interrelations between them.[23] The promise of "problem-solving," real or symbolic, surrounds many professions.[24] While the popularity of the term "technological solutionism" is a recent development,[25] the idea is not. Historians of engineering have described the engineers' role conception as following universal laws to manipulate the environment to solve social problems.[26] Social issues become mathematical equations that have an optimal solution. Though the

---

22 Yè, *Jiāotōng jiùguó lùn*, 8.

23 Cf. Victor Seow, "Carbon Technocracy: East Asian Energy Regimes and the Industrial Modern, 1900–1957" (PhD, History and East Asian Languages, Harvard University, 2014), 8.

24 Torstendahl, "Introduction," 2–4.

25 Evgeny Morozov, *To Save Everything, Click Here: The Folly of Technological Solutionism* (New York: PublicAffairs, 2013).

26 Buck, *American Science and Modern China*, 205; Layton Jr., *The Revolt of the Engineers*, 58–60, 74; van Laak, *Weiße Elefanten*, 40; Robert Zieger, "Herbert Hoover, the Wage-Earner, and the 'New Economic System,' 1919–29," in Hawley, *Herbert Hoover as Secretary of Commerce*, 88.

solution may still be unknown, reaching the solution becomes a depoliticized,[27] technically reasonable process with a permanent conclusion.[28] Engineering became an ideology with "selfish and technocratic overtones,"[29] as professionals strategically attempt to establish themselves above the rest of society due to their unique problem-solving abilities. In the early 20th century, US engineers reasoned that they, not politicians or bureaucrats, were the experts truly fit for managing the economy.[30] Wang displayed these tendencies as well.

Efficiency, although rational in nature, takes its form in relation to the pursued goal.[31] Civil engineers believed professional efficiency could serve a higher cause. For instance, secretary of commerce and later US President Herbert Clark Hoover (1874–1964), whose essays the Association of Chinese and American Engineers (chapter 4) often reprinted, had championed efficiency as a method benefiting charitable causes when he headed famine relief efforts in Belgium and Russia.[32] In the Sino-American context as well, scientific management was a means to achieve external political goals. In Wang's case, the cause for which to apply efficiency was Chinese nationalism.

In his final year in the US, Wang composed an essay on the abolition of the queue. His argument differed greatly from the well-known writings of intellectuals such as Lu Xun, who popularized the conception that cutting the legally mandatory queue was an expression of liberation and political revolt. Instead, Wang looked at the queue from a national economic standpoint. He argued that shorter hairstyles were simply more hygienic and economically efficient, calculating that the unproductive time spent combing and braiding China's queues translated into an annual national loss of $365 million.[33]

---

27 Larson, "In the Matter of Experts and Professionals, or How Impossible It Is to Leave Nothing Unsaid," 32–40; Uekötter, "Techniker an der Macht," 396.

28 Layton Jr., *The Revolt of the Engineers*, 66; Zieger, "Herbert Hoover, the Wage-Earner, and the 'New Economic System,' 1919–29," 88. Also Akin, *Technocracy and the American Dream*, 3.

29 Layton Jr., *The Revolt of the Engineers*, viii.

30 Larson, "In the Matter of Experts and Professionals, or How Impossible It Is to Leave Nothing Unsaid," 40–41; Layton Jr., *The Revolt of the Engineers*, 5–6. See also Wilson, "Democracy and Efficiency," 291–292.

31 Freidson, *Professionalism*, 171. Also Camprubí, *Engineers and the Making of the Francoist Regime*, 160; Jarausch, *The Unfree Professions*, 99–200; Renneberg and Walker, "Scientists, Engineers and National Socialism," 6, 11.

32 Robert Murray, "Herbert Hoover and the Harding Cabinet," in Hawley, *Herbert Hoover as Secretary of Commerce*, 24. Also Akin, *Technocracy and the American Dream*, 12–13; Layton Jr., *The Revolt of the Engineers*, 189–93.

33 Ching-chun Wang, "The Abolition of the Queue," *Atlantic Monthly* 107 (1911): 815.

This economic logic expanded to the realm of infrastructure. For Wang, the ideal railway administration was bound to public goals, but had to operate under commercial logic. He did not want to privatize critical infrastructures but, following liberal economics, argued that running them *like* commercial enterprises would ensure efficient management in the best public interest.[34] The railway economic theory of his professor, Yale president Arthur Twining Hadley (1856–1930), clearly acted as an influence. Hadley had conducted a comparative analysis between railway administrations of different countries and concluded that the industrial magnates of the *laissez-faire* gilded age were ultimately the most efficient means to ensure the upkeep of railways in the interest of the public.[35]

While Wang agreed with Hadley's ideas about an efficient railroad administration offside political bureaucracy, he added an important modification by adding his findings about railway location. He had studied the English railways' financial collapse of 1845 and the 1907 Knickerbocker Crisis[36] and had become critical of the location of railways constructed under *laissez-faire* policies, as competing, "socially unnecessary" lines proved inefficient.[37] To him, the national mesh of railway lines was a sensible, unstable system that needed careful, centralized balancing and control.[38] Therefore, he called for a centralized, national administration to regulate railway locations in the public interest. The central government would be best suited for this task due to its impartiality.[39] This deliberately contradicting quote of Wang perhaps best sums up his ideal of commercially operating railroads bound to nationalized supervision: "The railway is a common carrier, but it still remains a private business. [...] It is a public utility; but it is not and cannot be a public utility."[40]

Under the influence of social Darwinism, Wang explained that biological and social isolation in the Qing Empire, paired with the ritualized imperial

34 Wang, "Economic Significance of Railway Location," 6, 19–21, 59.

35 Arthur Hadley, *Railroad Transportation: Its History and Its Laws* (New York, London: The Knickerbocker Press, [1885] 1886), 236–58.

36 Wang, "Some Dangers of Railway Development in China and How to Prepare against Him [sic!], Part 1," 51–52; Wang, "Economic Significance of Railway Location," 97.

37 Wang, "Economic Significance of Railway Location," 39, 60, 73–74. See also Ching-chun Wang, "The Administration of Chinese Government Railways," *The Chinese Social and Political Science Review* 1, no. 1 (1916): 70–71.

38 To this regard, Wang also advocated fiscal and through traffic reforms. For instance, he urged the abolition of *lijin* 釐金 tax, an inland trade barrier obstructing railway traffic (Ching-chun Wang, "Some Dangers of Railway Development in China and How to Prepare against Them, Part 4," *The Chinese Social and Political Science Review* 2, no. 3 (1917): 59–60; Huenemann, "The Dragon and the Iron Horse," 301).

39 Wang, "Economic Significance of Railway Location," 110–11.

40 Wang, "Economic Significance of Railway Location," 19.

examination system, had led to technological stagnation, as there was neither internal nor external incitement for innovation.[41] He was an outspoken critic of the incompetence and corruption he saw prevalent among leading officeholders. In his eyes, the establishment was "slow, stupid, conservative," guided by ignorance, prejudice, fear, and superstitions.[42] To him, his predecessors inside the Ministry were "expectant taotais [*circuit intendants*] who might never have had a ride on a railway," and whose "knowledge was confined to writing poetry about them."[43] The large tail of gentry administration was like a "parasity" [*sic!*] costing money and lacking efficiency, fostering corruption, disorganization, and degeneration.[44]

Wang lamented that frequent shifts and rotations inside railway bureaus had resulted in a lack of continuity in developing expert knowledge.[45] In this respect, he defended the role of foreign engineers in China. He explained that Chinese personnel were not well enough trained. Consequently, replacing all foreign hands with Chinese engineers would do "more harm than good."[46] More important than an engineer's nationality was their expertise.[47] The paradigm of efficiency was the superordinate standard—the "only distinctive value"—of engineering.[48]

Despite his goal to use the railway infrastructure for nationalist aims, when it came to the management, Wang called for railway administration to be "properly safeguarded from the influence of politics," explaining that "railways will increase the corruption in politics and politics, in turn, will corrupt the railways."[49] However, although Wang criticized that unqualified bureaucrats interfered too much with the work of engineers, he did not yet go as far as

---

41 Wang, "The Effect of the Revolution upon the Relations between China and the United States," 272. See also Wang, "China's Revolution and Its Effect," 189–190, 195.

42 Wang, "The Effect of the Revolution upon the Relations between China and the United States," 280; Wang, "The Administration of Chinese Government Railways," 69.

43 Wang, "The Administration of Chinese Government Railways," 84–85.

44 Wang, "Some Dangers of Railway Development in China and How to Prepare against Them, Part 4," 58–59, 63.

45 Reinsch, *An American Diplomat in China*, 112. Wang claimed that under his direction, the number of foreign-returned employees in the department rose from three to 14 out of 50, including eight with practical experience (Wang, "The Administration of Chinese Government Railways," 85).

46 Ching-chun Wang, "Some Dangers of Railway Development in China and How to Prepare against Them, Part 2," *The Chinese Social and Political Science Review* 2, no. 1 (1917): 104.

47 Wang, "Economic Significance of Railway Location," 18, 79, 108–9; Wang, "Some Dangers of Railway Development in China and How to Prepare against Them, Part 4," 58.

48 Freidson, *Professionalism*, 171. Also Noble, *America by Design*, 34.

49 Wang, "Some Dangers of Railway Development in China and How to Prepare against Them, Part 4," 58.

to call for engineers to become the sole administrators. Rather, he called for the creation of a national board composed of political economists, political scientists, and civil engineers.[50] He understood himself as one of those capable men,[51] a "reliable and properly qualified" railway administrator.[52]

In the 1910s, Wang's plans to reform railway administration under the strict paradigm of efficiency were still avant-garde. An undergraduate program in railroad administration at Jiaotong College only opened in 1918,[53] while Wang had already received a PhD in the field in 1911. In the US, the idea that engineers could solve social problems by unraveling the scientific laws supposedly hiding behind them had found its formulation in Taylorism, i.e., the ideology of scientific management.[54] Chinese businesses only slowly started to adapt its principles, and their appliance remained limited to foreign-educated merchants until the late 1920s.[55] The translation of efficiency paradigms from business to politics through elements of technocratic governance would not become a wider phenomenon until the 1930s (chapters 6, 7) and again after the Mao era.[56] In other words, Wang's assumptions had not been tested out before him. The question remained if efficiency and nationalism could be brought together in the field of railway administration in China, or if a failure to do so would result in disillusionment and discontent.

3 Wang Jingchun's Ascent to Power

In a pictorial retelling for the Yale alumni office, Wang claimed that as soon as he had heard about the Xinhai Revolution (1911–1912), he had become "an enthusiastic revolutionist" and at once made his way to Nanjing, where he

50 Wang, "Economic Significance of Railway Location," 80, 108–9.

51 Wang, "Some Dangers of Railway Development in China and How to Prepare against Them, Part 4," 68. See also Wang, "Economic Significance of Railway Location," 79.

52 Wang, "Some Dangers of Railway Development in China and How to Prepare against Them, Part 4," 68.

53 Brent Jesiek and Yi Shen, "Educating Chinese Engineers: The Case of Shanghai Jiao Tong University During 1896–1949," in *Engineering, Development and Philosophy: American, Chinese and European Perspectives*, ed. Steen Christensen et al. (Dordrecht: Springer, 2012), 131.

54 Layton Jr., *The Revolt of the Engineers*, 134–36.

55 Stephen Morgan, "Transfer of Taylorist Ideas to China: 1910s–1930s," *Journal of Management History* 12, no. 4 (2006): 409–12.

56 See Hong Lee, *From Revolutionary Cadres to Party Technocrats in Socialist China* (Berkeley: University of California Press, 1991); Joel Andreas, *Rise of the Red Engineers: The Cultural Revolution and the Origins of China's New Class* (Stanford: Stanford University Press, 2009).

had "at least a dozen narrow escapes from the sound of the bullet."[57] While Wang may have witnessed the final weeks of upheaval, he was probably not the staunch "veteran of the battles" he made himself out to be.[58] In contrast to his fellow students who left the States to join the revolution,[59] Wang had not been in a hurry. After he had graduated in 1911, he sailed for Europe and spent some time in England to study at the British Board of Trade. In fall and winter, when the revolution was in its final stages, Wang had gone on vacation to Switzerland.[60]

Regardless of the extent of his revolutionary action, in 1912, Wang returned to China and started working at the Ministry of Posts and Communications (*youchuan bu* 郵傳部). Although the Ministry formally changed its name to Ministry of Communications (*jiaotong bu* 交通部) later that year, its personnel, administrative system, and responsibilities—which included overseeing 5,702 km of government railways—remained the same as in the Imperial era. The Ministry first appointed Wang as vice-director of the Beijing–Mukden Railway and, a year later, as vice-director of the Beijing–Hankou Railway. These early high-ranking appointments were unusual as, during these early years, the old Imperial bureaucracy still controlled much of the Ministry. The new generation of returned students usually climbed the career ladder in smaller steps. Apparently, officials at the railway bureaus believed hiring Wang would be beneficial since he was the first Chinese citizen to hold a PhD in railroad administration.[61] The expertise and prestige associated with the title resulted in a credit of trust.

Wang evidently proved himself, as the Ministry appointed him to a more autonomous position later that year. He became acting chairperson of the Commission on the Unification of Railway Accounts and Statistics (*tongyi tielu kuaiji hui* 統一鐵路會計會), which he co-founded with one of his closest colleagues within the Ministry, the progressive bureaucrat Ye Gongchuo.[62] From studying US railway administration, Wang understood that the lack of

57 YARO, Box 723, Wong Ching-chun.

58 Wang, "'The New China Will Be a New United States'."

59 Y. S. Tsao, "The Relation of the Returned Students to the Chinese Revolution," *Journal of Race Development* 4, no. 1 (1913): 103–4.

60 YARO, Box 723, Wong Ching-chun.

61 During his student days, he received a stipend from Wu Tingfang (chapter 2) and after his return to China, very briefly worked as a foreign advisor in Sun Yant-sen's provisional government. Wang had influential connections, but it is not evident that they helped him into his positions.

62 Gēnglín Guān, ed., *Jiāotōng shǐ lùzhèng biān vol. 5*, Jìndài jiāotōng shǐ quánbiān 22 (Beijing: Guójiā túshūguǎn, [1935] 2009), 4–13. For Ye's thoughts on efficient accounting reform, see Yè, *Jiāotōng jiùguó lùn*, 12–16, 83–87. See also Línghé Mǎ and Fāng Wú, "Běiyáng shíqí jìshù guānliáo zài chǎnyè xíngzhèng zhōng de dìngwèi: Yǐ tiělù guǎnlǐ zhìdù biàngé wèi

a unified system of accounting and statistics made figures unreliable and cost the government money.[63] To help with the problem, he invited the renowned statistician Henry Carter Adams (1851–1921), who had previously directed unification efforts in the US Interstate Commerce Commission. To make the railways more efficient and comparable to each other, the commission standardized accounting, regulated statistics, and streamlined budgets.

In 1914, the Ministry of Communications saw a major restructuring, giving accounting and statistics a more prominent role. This included the establishment of the Standing Committee on the Unification of Railway Terms (*tielu jishu biaozhun weiyuanhui* 鐵路技術標準委員會) and the institutionalization of the beforementioned Commission on Accounts, which became a standing committee.[64] Wang received a promotion and became director of the Department of Railway Accounts and Finance (*tielu kuaiji si* 鐵路會計司), to which the latter committee belonged. In 1917, after another Ministry reform had dissolved his department, Wang returned to the Beijing–Mukden and Beijing–Hankou lines, this time as director general. A few years later, Wang's emphasis on efficient, streamlined accounting, as witnessed in many of these early assignments, would play a role in his efforts and frustrations administering the Chinese Eastern Railway.

Parallel to his ascend inside the Ministry, Wang represented the new Republic on the international stage. Wang often moved between the blurry lines of technocrat and diplomat. He returned to the US in the fall of 1912 to act as a delegate to the Fifth International Congress of Chambers of Commerce in Boston,[65] and again for the 1915 Panama-Pacific International Exposition.[66] When the Ministry of Communications prepared to offer railway concessions to the US Siems-Carey Railway and Canal Company to attract loans (chapter 2), Wang's Yale senior Liang Dunyan appointed him to carry out the

---

bèijǐng," *Shèhuì kēxué yánjiū* 6 (2019): 140–41; Wang, "The Administration of Chinese Government Railways," 84.

63 Wang, "Some Dangers of Railway Development in China and How to Prepare against Them, Part 2," 83.

64 See Guān, *Jiāotōng shǐ lùzhèng biān vol. 5*, 13–25.

65 Boston Chamber of Commerce, ed., *Fifth International Congress of Chambers of Commerce and Commercial and Industrial Associations* (Boston: The University Press Boston, 1912), 66.

66 Ching-chun Wang, "A Plea for Recognition of the Chinese Republic," *The Atlantic Monthly* 61 (1913): 45; *The San Francisco Call*, "China Formally Takes Place on 1915 Exposition Map: New Republic's Flag Floats Over the Site for Building," October 25, 1912, 147. See Carol Huang, "The Soft Power of U.S. Education and the Formation of a Chinese American Intellectual Community in Urbana-Champaign 1905–1954" (PhD, University of Illinois at Urbana-Champaign, 2001), 186–95.

negotiations. Together with Minister Cao Rulin 曹汝霖 (1877–1966), Wang offered what the US Minister thought to be the most beneficial of all foreign concessions,[67] again pointing at Wang's pro-American sympathies that existed in parallel to his nationalist conviction.[68] Wang also participated in the 1919 Paris Peace Conference. Although he formally acted as technical advisor for the Ministry of Agriculture and Commerce, he participated in the Commission on the International Régime of Ports, Waterways, and Railways,[69] a commission whose concerns aligned with those of his actual employer, the Ministry of Communications.[70] His techno-diplomatic assignments—the Panama expo, the Siems-Carey negotiations, and the Paris conference—demonstrate that, despite attempts to distance oneself from the old elite through technical expertise, the boundaries between engineering and politics were fuzzy.

Not every engineer was as eager as Wang to climb the Ministry's hierarchy: When in 1917, Liang Dunyan offered Zhan Tianyou the position of Minister of Communications, Zhan declined and explained that he wanted to remain a practicing engineer.[71] In contrast, Wang's ascend within the bureaucracy continued rapidly. In 1920, upon his return from Paris to China, he became a ministry chief counselor (*canshi* 參事). In 1921, he received his final promotion to chief of the Ministry railway department (*tielu sizhang* 鐵路司長),

67 The concessions consisted of the Hohhot–Lanzhou line, the Kaifeng–Suzhou (Anhui) line, and the Hangzhou–Wenzhou line. See FRUS, 893.77/1542: Paul Reinsch to Secretary of State, May 17, 1916.

68 See also Wang, "China's Revolution and Its Effect," 202; Wang, "A Plea for Recognition of the Chinese Republic," 45.

69 FRUS, 180.03401/122, IC–176G: Arthur Sifton, Supplementary Report Presented by the Commission on the International Régime of Ports, Waterways, and Railways to the Peace Conference, April 25, 1919, Document 23, Vol. V, Paris Peace Conf., 266.

70 The Chinese delegation consisted of many of Wang's former and later acquaintances. His student day contacts Gu Weijun (chapter 1) and Wang Zhengting acted as plenipotentiary delegates, just as London ambassador Shi Zhaoji 施肇基 (Alfred Sze, 1877–1958), who would become his superior 15 years later. Other technical advisors included the geologist Ding Wenjiang, with whom he would work closely together during the British Boxer Indemnity Commission in 1926, and Guo Taiqi, with whom he would work in London after 1934 (FRUS, 182/70: Composition of the Conference, April 1, 1919, Document 1, Volume III, Paris Peace Conf., 33–4). Wang also used his stay in Europe to study the railway administrations of England, in connection to his work at the standardization commission (Ching-chun Wang, "China's Railway Ambitions: Expansive Views of Mr. C. C. Wang," *The North-China Herald and Supreme Court & Consular Gazette*, August 16, 1919, 412 JACAE, 1 (3): 1–2). Wang would infamously lose a briefcase with classified documents on this trip, as the press reminded years later (NARA/CER, 23, 3532: Japan Adventure, "Chinese Railroad Head to See U. S." June 22, 1924).

71 Le, "The Struggle for Autonomy from the State," 44, 61.

which included directorship over the through traffic administration.[72] When the Ministry requested Wang to preside over the Chinese Eastern Railway, he had the best prospects inside the Ministry in Beijing and was at the height of his career.

4 The Chinese Eastern—"A Railroad Born in Sin"

After political upheavals like the Second Revolution (1913), the Bai Lang Rebellion (1913–1914), and the failed monarchic restoration (1915–1916), in 1917, Wang imagined the central government to have finally arrived at a state that allowed efficient railway administration and development.[73] However, the railway he got involved with in 1919 and ended up presiding over in 1922 was the far opposite of this imagination. The Chinese Eastern Railway (CER), contracted in 1896, was a Russian extraterritorial railroad concession that extended the Trans-Siberian Railway through Northern Manchuria to its terminal Vladivostok.[74] While in 1916, the Czarist government had completed a longer Northern route that circumvented Chinese territory (Figure 11),[75] the straight Manchurian route remained the preferred passage and a bone of colonial contention.

The terms of the CER's original contract were unequal, and the railway continued to be at the center of political stir.[76] It was an obscure imperial remnant in that, in contrast to the other European railway concessions in China—which secured a semi-colonial zone of influence but also catered toward domestic Chinese transportation interests—the CER was primarily a shortcut to transport goods between Siberia and the Russian Far East. At times, the CER was also a tool to secure Russian influence over Northern Manchuria, but it was never a logical line for the national Chinese infrastructure. As the US consul in Moscow stated frankly, the CER was "an integral part of the Russian railway

72 JACAE, 2 (10): 41.

73 Wang, "Some Dangers of Railway Development in China and How to Prepare against Them, Part 4," 67. Ironically, and just contrary to Wang's observation and expectation, today's historiography usually portrays the Republic's first years as relatively stable and 1916 as the start of the chaotic "Warlord Era."

74 See Sören Urbansky, *Beyond the Steppe Frontier: A History of the Sino-Russian Border* (Princeton, Oxford: Princeton University Press, 2020), 39–44.

75 Sören Urbansky, *Kolonialer Wettstreit: Russland, China, Japan und die Ostchinesische Eisenbahn* (Frankfurt, New York: Campus, 2008), 107.

76 This includes the Russo-Japanese War (1904–1905), indirectly leading to the Russian Revolution (1917), as well as the Sino-Soviet Conflict (1929) and, eventually, the Mukden "Incident" (1931) and the Japanese invasion of Manchuria.

system" that "would never have been built as a part of a Chinese railway system."[77] The CER had nothing in common with the type of railways Wang had imagined beneficial for the young republic. When in his theoretical writing, he had warned, "an ill-located railway may prove not much better than no railway at all,"[78] he might have had the CER in mind. Wang later referred to the track as "a railroad born in sin," and as a railway that had "bred trouble ever since it was conceived."[79]

The Russian Civil War (1917–1922), fought between Bolshevik revolutionaries and "White" resistance troops, had brought more trouble to the CER. With the advancement of the Bolshevik Army, the Czechoslovak Legion had been stranded in the Russian Far East behind the Amur, where remnants of the White movement roamed as well. In 1918, US President Wilson agreed to join the anti-Bolshevik Japanese Siberian Intervention to escort the Czechoslovaks out of Siberia. In January 1919, the involved states signed the Inter-Allied Railway Agreement. The agreement called for an Inter-Allied Technical Board (IATB, *jishu bu* 技術部) to control and manage the CER to help the Intervention.[80] Officially, the mission of the IATB was, as one of the chairpersons put it, "technical and absolutely non-political,"[81] but for the US, it was also a means to counter Japanese influence in the region.[82] The civil engineer John Frank Stevens (1853–1943), with whom John R. Freeman (chapters 2, 4) had worked on the Panama Canal, acted as the IATB's chairperson. For a short time, Zhan Tianyou represented the Chinese side.[83] When he passed away a few months later, the Chinese government appointed Wang Jingchun as the new representative.[84] The IATB encroached on Chinese and Russian sovereignty, but as its period of operation had a clearly defined end—the IATB would withdraw once political stability in the Far East was restored—at first, both the Chinese and

77 NARA/CER, 12, 2445: D. C. Poole to Department of State, December 19, 1921.

78 Wang, "Economic Significance of Railway Location," 33.

79 Ching-chun Wang, "A Railroad Born in Sin," *World's Work* 58, no. 12 (1929): 62; Ching-chun Wang, "The Sale of the Chinese Eastern Railway," *Foreign Affairs* 12, no. 1 (1933): 57.

80 See Carolyn Grubbs, "American Railroaders in Siberia, 1917–1920," *Railroad History* 150 (1984); Leo Bacino, *Reconstructing Russia: U. S. Policy in Revolutionary Russia, 1917–1922* (Kent, Ohio, London: The Kent State University Press, 1999), 129–72.

81 Benjamin O. Johnson Papers. BOJP, 73070, Hoover Institution Archives, Johnson to Charles H. Smith and F. M. Clarke, January 10, 1921.

82 Bacino, *Reconstructing Russia*, 132–46; Urbansky, *Kolonialer Wettstreit*, 92–93.

83 South Manchuria Railway Co., ed., *Tō Shi tetsudō wo chūshin to suru Ro-Shi seiryoku no shōchō* (Dalian: Minami Manshū tetsudō, 1928), 356–57.

84 JSDG, 03-32-227-03-025: Ministry of Communications to Foreign Ministry, December 15, 1919.

the Russian sides generally found it to be a useful buffer to stop bilateral contentions until the Russian Civil War was settled.[85]

The international administration of the CER was not the only grievance during the Russian Civil War. In March 1920, the regional militarist Bao Guiqing 鮑貴卿 (1867–1934) discharged Dmitry Leonidovich Horvat (1858–1937), White Russian associate director of the CER, and gained control over the administration and police of the railway zone.[86] Without a permanent White Russian government, the CER lacked half of its administrative regime. To save the railway from political unrest, Ye Gongchuo's Ministry of Communications and the Russo-Asiatic Bank signed an agreement in October 1920, providing for the joint management of the railway as a "commercial enterprise."[87] The bank was the next best thing to a contractual partner, as it had some history with the railway. However, the bank could not produce evidence that it still had a stake in it. The problematic outcome was, as government adviser John Earl Baker (1880–1957) described it, that the railway was now "in the hands of one who is not the owner," meaning there was "no incentive [...] to promote the welfare of the railway."[88]

Neither the bank nor the ungoverned White Russians had a reason to secure the railway budget, streamline its operation, or boost its efficiency. This situation left the railway vulnerable to corruption. The annual costs of the CER amounted to about as much as all other government railways combined. To put the mismanagement into perspective, while the Beijing–Hankou Railway contented itself with a staff of 300, the office in Harbin counted 3,500 people.[89] As all profits from the railway were applied against a potential Chinese redemption of the railway,[90] only the Chinese side had a good reason to optimize its revenue. With the 1920 contract, Wang (Figure 12) *ex officio*

85 NARA/CER, 12, 2494: Jacob Gould Schurmann to Secretary of State, March 28, 1922.

86 Chu, *V. K. Wellington Koo*, 101.

87 FRUS, 893.00/4112: Quarterly Report of the Legation in China, October 1, 1920; Economical Bureau, C. E. R., *The Chinese Eastern Railway and Its Zone* (Harbin: Economical Bureau, C. E. R., 1923), 25; Chu, *V. K. Wellington Koo*, 102; Sow-theng Leong, *Sino-Soviet Diplomatic Relations: 1917–1926* (Canberra: Australian National University Press, 1976), 103–7; Hóngxūn Líng, *Qīshí nián lái Dōng Qīng, Zhōng dōng, Zhōng Cháng tiělù zhī jīngguò: Zhōng piān* (Taipei: Jiāotōng bù jiāotōng yánjiū suǒ, 1965), 9–13.

88 NARA/CER, 20, 3160: John Earl Baker to William Cullen Dennis, June 8, 1923. See also Bacino, *Reconstructing Russia*, 166–72. Cf. Economical Bureau, C. E. R., *The Chinese Eastern Railway and Its Zone*, 25.

89 JACAE, 1 (3): Wang, Jingchun, Address at First General Meeting of the ACAE, Tianjin, October 6, 1920, 8–11: 8.

90 According the stipulations, the Chinese government could buy the CER in 1932, or, alternatively, wait until 1976 for the railway to transfer automatically.

joined Bao Guiqing's Board of Directors (BoD) as a representative of the IATB,[91] but he continued to fulfill other functions within the Ministry. Evaluating the state of the railway, he found it had apparently been the goal to "make the railway as unprofitable as possible in order that China might never be able to pay it back."[92] The new enterprise inherited a $100,000,000 debt, encumbered through a mode of operation that had followed politics rather than the laws of economic management.

Meanwhile, the political situation surrounding the CER deteriorated. In November 1920, the Army of the Bolshevik-supported Far Eastern Republic defeated the White Russian troops of Vladimir Kappel (1883–1920) and Grigory Mikhaylovich Semyonov (1890–1946).[93] The Kappel Army was stranded on the CER concession and, together with the Japanese Army, which had supported Kappel, occupied parts of the area.[94] The Allied administration, regardless of their original mission, would not leave before the Japanese troops had also left the area. The allies further weakened trust in the temporal limitation of the IATB with the signing of the Second Consortium agreement. It aimed at institutionalizing international control over the CER, a plan put on the agenda of the 1921 Washington Naval Conference.[95] The Chinese side vehemently refused to sign the agreement. After the attempted internationalization, the tone between Stevens and Wang during IATB meetings became rougher, as Wang frequently questioned the IATB's extent of authority.[96] On top of the political trouble, the second pneumonic plague epidemic (1920–1921) was sweeping the Manchurian and Siberian regions.[97]

Wang's path to presidency opened when corruption allegations forced the incumbent CER president (*duban* 督辦), the retired military governor Song

---

91 Leong, *Sino-Soviet Diplomatic Relations*, 103.

92 JACAE, 1 (3): Wang, Jingchun, Address at First General Meeting of the ACAE, Tianjin, October 6, 1920, 8–11: 8.

93 BOJP, Paraphrase of telegram from Benjamin Oliver Johnson to the Secretary of State, November 23, 1920.

94 BOJP, Johnson to Secretary of State, December 9, 1920; Johnson to Secretary of State, December 30, 1920.

95 Bacino, *Reconstructing Russia*, 172–84; Roberta Dayer, *Finance and Empire: Sir Charles Addis, 1861–1945* (Houndmills, London: MacMillan, 1988), 126–7, 133, 142; NARA/CER, 12, 2446: Confidential information from D. C. Poole, January 20, 1922.

96 NARA/CER, 9: Minutes of Meeting of the Technical Board, April 19–21, 1921. See also NARA/CER, 13, 2581: John F. Stevens to Secretary of State, May 10, 1922.

97 BOJP, Telegram paraphrase from Benjamin Oliver Johnson to the Secretary of State, March 9, 1921; Lien Wu, "The Second Pneumonic Plague Epidemic in Manchuria, 1920–21: A General Survey of the Outbreak and Its Course," *The Journal of Hygiene* 21, no. 3 (1923); Mark Gamsa, *Harbin: A Cross-Cultural Biography* (Toronto, Buffalo, London: University of Toronto Press, 2020), 156–57.

Xiaolian 宋小濂 (1863–1926), to step down.[98] At first, the Chinese President, the civilian politician Xu Shichang, had appointed his Minister of Agriculture, Wang Naibin 王迺斌 (b. 1870), as the successor.[99] In the vein of the previous CER presidents, he would have been a northern politician, on friendly terms with the regional Fengtian government. There were various rumors as to why a few weeks later, before ever having reached his new office, Wang Naibin resigned from the position. These rumors included a sudden fallout with Fengtian powerholder Zhang Zuolin 張作霖 (1875–1928) or the fear that past corruption allegations would emerge.[100] According to a report from the IATB, both Wang Naibin and a variety of other "old school Chinese" ended up declining the call because the limited executive power would do their career more harm than good.[101] With no willing elderly political leaders in sight, the Ministry of Communications, which had recently fallen back into the hands of Ye Gongchuo, nominated Wang Jingchun for the position.[102]

Apart from being a candidate of last resort, Wang may have been a suitable compromise within the political power structure in early 1922. In Beijing, civilian president Xu Shichang, who had come into his position by nomination of the Anhui Clique, balanced a grand coalition of the Zhili and Fengtian factions. At that time, Zhili politicians were generally skeptical of bureaucrats in the orbit of the Communications Clique (*jiaotong xi* 交通係), an opaque network that navigated somewhere between loosely defined coterie and political party. The Communications Clique leaders had a history of affiliation with the Fengtian government in the north and the Guomindang (Nationalist Party, GMD) government in the south—ideologically contradicting liaisons that had in common their opposition to the Zhili military faction.[103] Wang was not a big name in the clique,[104] but for a group that defined itself through personal connections rather than shared political values or membership, his association

---

98 NARA/CER, 12: Translation of Tsin pao, February 11, 1922; 22, 3104: Ostrooemov, B. V., Strictly Confidential Memorandum about the Russo-Chinese Relations, undated, 1923.

99 *Dàgōng bào,* "Dàzǒngtǒng lìng," January 5, 1922, 2.

100 *Dàgōng bào,* "Wáng Nǎibīn cí Dōnglù dūbàn," March 12, 1922, 6.

101 NARA/CER, 12, 2451: Telegram from B. O. Johnson to the Secretary of State, March 9, 1922.

102 *Dàgōng bào,* "Wáng Jǐngchūn dūbàn Dōng lù zhī yuányīn," April 6, 1922, 7.

103 See Shúcūn Jiǎ, "Sūn Zhōngshān jítuán yú jiāotōng xì de ēnyuàn," *Yúnmèng xuékān* 31, no. 1 (2010).

104 It is impossible to assume clique "membership" on basis of association with the Communications Ministry, Communications Bank, or Railroad Union despite personal overlaps between their leaderships and the clique leadership; otherwise, Sun Ya-tsen and thousands of others would have been part of the clique. Wang's "membership" was rather the result of his affiliations than a driver for his actions. See Shúcūn Jiǎ, *Běiyáng jūnfá shíqī de jiāotōng xì* (Zhengzhou: Hénán rénmín chūbǎnshè, 1993), 32–34; Andrew Nathan, *Peking Politics, 1918–1923: Factionalism and the Failure of Constitutionalism* (Berkeley, Los Angeles, London: University of California Press, 1976), 245–51.

with Communications Clique powerbroker Ye Gongchuo may have rendered him an acceptable compromise to Zhang Zuolin, who had until recently sought to "eliminate" Wang as IATB delegate.[105] Similarly, Zhili military leader Wu Peifu 吳佩孚 (1874–1939) had recently accepted Zhou Ziqi as prime minister because his US education rendered him unsuspicious to Japanese infiltration and his Shandong native place counterbalanced his Communication Clique association.[106] In the same way, Wang Jingchun's US background and his native place located firmly within the Zhili heartland made him an agreeable candidate to the Zhili party as well. He inherited a mélange of troubles but also met the opportunity to implement the reforms he had outlined in his writings.

FIGURE 12 Portrait of Wang Jingchun (JACAE 2.7, 1921).

105 FRUS, 861.77/1907: B. O. Johnson to Acting Secretary of State, December 30, 1920; 861.77/1930: B. O. Johnson to Secretary of State, January 29, 1921.

106 Jiǎ, "Sūn Zhōngshān jítuán yú jiāotōng xì de ēnyuàn," 52.

When Wang accepted the call early in April, he was 39 years of age—young compared to the past gentry presidents. He may have made up for his age through his technical expertise, which past presidents had lacked. To this effect, a newspaper in the orbit of the Anhui Clique portrayed Wang as the ideal candidate, as he would bring his talent and a "new style of wisdom" to the CER's administration.[107] Wang was at his height of international recognition, as well. The China Society of America included him in a volume of biographies of 24 contemporary Chinese leaders—the only engineer in the collection.[108] US minister Charles Crane also evaluated Wang most positively. He argued Wang's increase in influence would benefit the States due to Wang's "pro-American sympathies and upright character."[109]

The first issue Wang had to deal with was the Allied presence. Tolerance for the IATB had evaporated when, despite the Japanese withdrawal from Siberia, Stevens showed little will to dissolve the Board as promised. Stevens found an exclusively Chinese takeover of the CER "a childish suggestion" as it would violate Russian and Allied rights, pointing out that according to Inter-Allied stipulations, the IATB acted in the interest of Russia, not China. However, the IATB had no Russian side for a handover, as the last provisional White governments had fallen, and the Soviet government remained unrecognized. Stevens was aware of Wang's desire to dissolve the IATB, but he did not take these plans seriously.[110] Instead, he painted a dark picture of China, under the impression of continued anti-Christian and anti-foreign sentiments and excessive corruption.[111]

The perceived betrayal of Chinese sovereign rights and the Wilsonian ideal of national self-determination led to street demonstrations in Harbin.[112] At the height of the campaign, on April 27, 1922, two thousand Chinese paraded the city. US intelligence pointed to Wang as the instigator of the protests.[113]

107 *Dàgōng bào*, "Wáng Jǐngchūn dūbàn Dōng lù zhī yuányīn," 7.

108 Merle Walker, *Contemporary Chinese Leaders: Brief Sketches*, China Society Pamphlets 3 (New York: The China Society of America, 1922), 24–25.

109 BOJP, Charles R. Crane to Secretary of State, December 15, 1920: 2.

110 NARA/CER, 13, 2581: John F. Stevens to Secretary of State, May 10, 1922.

111 PJRF, 129: Freeman to Charles K. Edmunds, June 28, 1922.

112 NARA/CER, 12, 2452: Consul General to Secretary of State, February 8, 1922; 12: Translation of Tsin pao, February 11, 1922. See also NARA/CER, 12, Translation of The International, January 1, 1922; Dayer, *Finance and Empire*, 126–27; Adam Tooze, *The Deluge: The Great War and the Remaking of Global Order, 1916–1931* (London, New York: Allen Lane, 2014), 106–7, 406.

113 NARA/CER, 12, 2469: G. C. Hanson to Secretary of State, February 20, 1922; 13: G. C. Hanson to Secretary of State, April 27, 1922; FRUS, 861.77/3633: Memorandum by Douglas Jenkins of the Division of Russian Affairs, Department of State, May 1, 1922. See also James Carter,

Stevens claimed that Wang addressed the crowd in person, assuring "he would do all he could to have the Technical Board eliminated."[114] Soon after, the IATB announced its intention to retreat from the CER at the end of the year.[115] Regardless of the exact extent of Wang's role in the anti-IATB campaign, the experiences with the IATB's interference with the CER put a damper on his positive view of US international policies.

Wang had to juggle between Chinese interests of stopping internationalization, Japanese territorial encroachment, and an expansion of Soviet influence.[116] The first attempt at playing this game proved unsuccessful: right after the IATB had withdrawn from Manchuria, the Soviet-sponsored Far Eastern Republic took over the port of Vladivostok, thereby taking control over the Far Eastern part of the Trans-Siberian Railroad and seizing property of the CER.[117] Ultimately, the seizure helped the Japanese more than the Soviets, as the CER now ended its traffic liability at the Manchurian border and all traffic previously destined for Vladivostok had to be diverted to the Japanese harbor of Dairen (Figure 11). While the South Manchurian route had always been more profitable for the CER,[118] the loss of the Vladivostok competition meant the CER now had a weaker bargaining power in freight negotiations with Japan.

At the bottom line, while Wang ultimately succeeded in getting rid of the IATB, it enabled the Soviet seizure of the CER's eastern terminal, meaning Wang exchanged one trouble for another. The seizure, enabled by the withdrawal of the IATB and therefore a consequence of his failure in political meddling, further aggravated his negative views against the Soviets and the Russians in general.[119] The US consul was astonished about Wang's lack of diplomatic aptness, recounting that he was "unable to constrain himself" after the Soviets had conquered Vladivostok.[120] After the fiasco, Wang called sick

"Struggle for the Soul of a City: Nationalism, Imperialism, and Racial Tension in 1920s Harbin," *Modern China* 27, no. 1 (2001): 100–101.

114 NARA/CER, 13, 2581: John F. Stevens to Secretary of State, May 10, 1922.

115 Bacino, *Reconstructing Russia*, 172, 188–189.

116 See FRUS, 861.77/1930: B. O. Johnson to Secretary of State, January 29, 1921. Wang had warned about Japanese railway imperialism as early as 1912 (Wang, "'The New China Will Be a New United States'") and continued to do so throughout the Second Sino-Japanese War (Ching-chun Wang, *Japan's Continental Adventure* (New York: Routledge, [1940] 2011)).

117 NARA/CER, 17, 2944: G. C. Hanson to Secretary of State, December 9, 1922.

118 NARA/CER, 17, 2951: G. C. Hanson to Secretary of State, December 12, 1922.

119 NARA/CER, 17, 2962: Translation from The International, "An Important Interview with Dr. C. C. Wang," December 10, 1922.

120 NARA/CER, 17, 2962: G. C. Hanson to Secretary of State, December 21, 1922.

because of dysentery and had the board meetings held in his house.[121] His political commentary paused as well. In an interview, he proclaimed that he no longer wanted to comment on Soviet politics, stating that as "a railway man," he wanted "to concentrate [his] attention on railway matters."[122]

5 The Wang–Ostroumov Dyad

In his position as CER president, Wang worked with Boris Vasil'evich Ostroumov (1879–1944), a civil engineer who had previously administered the Bukhara, Yuzhsib, and Kolchuginsky railways.[123] The Board of Directors (BoD), following the Russo-Asiatic Bank's nomination, had appointed him as the railway's general manager (*guanli juzhang* 管理局長) in February 1921.[124] For Wang, who had frequently lamented the lack of expert knowledge inside government institutions,[125] the outlook of a presidency and management formed by a technocratic dyad must have been inviting. For the first time in its history, the CER's administration was in the hands of two professional civil engineers.

Wang pursued his life-long motif of efficiency. The previous troubles of the railway and its "inefficiency," he explained, were due to a "mix-up of the railway with politics."[126] He announced a five-point plan to reorganize the railroad: traveling had to become more convenient, transportation improved, the railway police reorganized, freight fees lessened, and railway accounts be disclosed to the public.[127] He stressed a business-like handling of all affairs. Anti-corruption reforms included the introduction of a public tender system operating under an efficiency and waste reduction logic.[128] Wang appointed a Chinese West Point graduate to help guard the railway from ongoing bandit

121 NARA/CER, 17, 2962: Translation from The International, "An Important Interview with Dr. C. C. Wang," December 10, 1922.

122 NARA/CER, 17, 3000: The Weekly Review, "The Legal Status of the Chinese Eastern Railway," January 20, 1923.

123 South Manchuria Railway Co., *Tō Shi tetsudō wo chūshin to suru Ro-Shi seiryoku no shōchō*, 520–21.

124 NARA/CER, 7, 1940: B. O. Johnson to Secretary of State, February 7, 1921; Bacino, *Reconstructing Russia*, 174.

125 Reinsch, *An American Diplomat in China*, 112.

126 NARA/CER, 13: Extract from Peking and Tientsin Times, April 11, 1922.

127 *Dàgōng bào*, "Zhěngdùn Dōng tiě zhī jìhuà," May 24, 1922, 7; JSDG, 03-32-249-03-024: Wang Jingchun, March 1922.

128 NARA/CER, 13: Extract from Peking and Tientsin Times, April 11, 1922.

attacks.[129] He introduced audits for the railway police[130] and dismissed ticket sellers that squeezed money.[131] Through the CER educational department, he also opened Chinese schools along the railway zone.[132] Furthermore, he started investigating budget expenses to diminish rents and shrink the oversized Russian police force operating under the railway's payroll.[133]

In accordance with his theory, Wang used the efficiency paradigm to support the nationalization of the line. For instance, Wang had earlier claimed that a line's passenger traffic was not bound to predetermined, economic necessity but could be increased beyond the current demand by making travel more accessible.[134] To ease accessibility, Wang introduced the Chinese language to the Eastern. Instead of resorting to nationalist arguments, he explained that the current situation was "unbusiness-like," as the Russian language was "inconvenient" for Chinese travelers. He claimed that the reforms were not about national "sympathies" but about "greater efficiency."[135]

Wang was also able to realize his plans for monetary reform, a topic he had explored since 1911.[136] Arguing with convenience, he replaced the Czar gold ruble—out of circulation everywhere but the CER—with the Chinese silver standard for passenger tickets and freight, against protest from Japanese and Western observers.[137] Wang called this "a step conducive to the national sovereignty,"[138] revealing that his alleged reforms of accessibility also contained a political agenda. Accordingly, general manager Ostroumov warned that what Wang introduced as methods of efficient, standardized accounting might prepare for an eventual Chinese redemption or illegal takeover of the railway. He portrayed Wang as a conspiring character, displaying a "shovinist [*sic!*] and

---

129 The West Point graduate being Wen Yingxing 溫應星 (1887–1968). NARA/CER, 12, 2401: Consul General to Secretary of State, February 1, 1922.

130 NARA/CER, 13: Extract from Peking and Tientsin Times, April 11, 1922.

131 NARA/CER, 17, 2962: Translation from The International, "An Important Interview with Dr. C. C. Wang," December 10, 1922.

132 Blaine Chiasson, *Administering the Colonizer: Manchuria's Russians under Chinese Rule, 1918–29* (Vancouver, Toronto: UBC Press, 2010), 188.

133 BOJP, Report by Charles Richard Crane to the Secretary of State, December 15, 1920: 3.

134 Wang, "Economic Significance of Railway Location," 98.

135 NARA/CER, 13: Extract from Peking and Tientsin Times, April 11, 1922.

136 Ching-chun Wang, "The New Chinese Currency," *The North American Review* 193, no. 667 (1911).

137 BOJP, Report by Charles Richard Crane to the Secretary of State, December 15, 1920: 4. See Bacino, *Reconstructing Russia*, 176; NARA/CER, 22, 3453: The Manchurian Daily News, "Adoption of Large Dollars," April 25, 1924; NARA/CER, 23, 3507: Extract from Zarya, May 20, 1924.

138 NARA/CER, 17, 2962: Translation from The International, "An Important Interview with Dr. C. C. Wang," December 10, 1922.

haughty" attitude, and as being "ill-disposed and half-disdainful" toward the Russian nation.[139]

Wang found himself in a situation where, though he was nominally the president, he had only limited power in administering the railway's business. Decisions were bound to a majority vote of the BoD, and half of the board consisted of Russian delegates. According to press reports, the Russian directors met "in secret" before the board meetings and continued to administer the CER in the same way as before the 1920 agreement.[140] The accusation points to a fundamental problem of Wang's presidency: politically, not economically, grounded bilateral stipulations limited his actions. For instance, according to the 1920 agreement, a Chinese position should be created for every Russian staff member, and so, while Wang could decrease railway employee numbers to some extent, he was legally unable to solve the situation of redundant and unnecessary positions.[141]

Russian skepticism was not the only obstacle in Wang's administration. Shortly after he assumed presidency, the short-lived Zhili–Fengtian coalition in Beijing collapsed. The Zhili faction defeated the Anhui forces in April and repressed the Fengtian Army in June. Zhang Zuolin's Fengtian domain, officially the "autonomous government of the three Eastern provinces," fell just short of declaring outright independence. It controlled Northern Manchuria and the city of Harbin, where the railway administration resided. Wang, as the CER's compromise president, had no affiliation with the Fengtian government. Although stationed in Fengtian-controlled Harbin, he remained under the payroll and order of the Ministry of Communications in Beijing. Wang found himself in the unpleasant situation of having to console between sometimes-conflicting orders from the regional Fengtian government and the Ministry. In the meantime, many of his student-era associates returned to the capital to serve in the new "able persons government" (*hao ren zhengfu* 好人政府).[142] The reformist Beijing cabinet added to Fengtian powerholder Zhang Zuolin's suspicions against Wang.

---

139 NARA/CER, 22, 3104: Ostroumov, B. V., Strictly Confidential Memorandum about the Russo–Chinese Relations, undated, 1923. An eventual redemption was certainly always Wang's goal, but unaffordable.

140 NARA/CER, 13: Translation of The International, March 22, 1922.

141 NARA/CER, 17, 2988: M. de Bunting, "Chinese Exaction on C. E. R." The North China Herald, December 30, 1922; Chiasson, *Administering the Colonizer*, 62; Urbansky, *Kolonialer Wettstreit*, 103–4.

142 They included fellow student-era activist Gu Weijun and Yale alumni Wang Chonghui 王寵惠 (1881–1958). See Craft, *V. K. Wellington Koo and the Emergence of Modern China*, 71–76.

For the time being, the Fengtian government and Ostroumov had a common enemy. Their circles frequently spread rumors of an imminent replacement of Wang or of Wang having resigned.[143] Furthermore, Ostroumov spent money to instigate a newspaper campaign in favor of his management of the CER. The railroad town's population had recently surpassed 100,000 residents as Russian refugees fled the civil war. The controversy between Ostroumov and Wang unfolded in the five Russian and Chinese language newspapers fighting over the city's emerging public, circulating several thousand copies.[144] The friendly press, as well as whitepapers of the "Economic Bureau" under the Russian general manager's control, gave Ostroumov the sole credit for the CER reforms.[145] As the US consul commented, the press campaign "led to a loss of face on the part of Doctor Wang" and transformed him and his Chinese colleagues into "non-entities."[146] Wang felt a lack of public acknowledgment and praise for his accomplishments.[147] He tried to modestly restore his prestige in the public, writing to one newspaper that although Ostroumov's energy and ability were "exceptional," the railway's improvements could not "be attributed to any one person," being instead a "common cooperation."[148] Certainly, Wang, who already harbored anti-Russian sentiments,[149] felt violated. His closest

143 NARA/CER, 14, 2739: Translation of Novosti Zhizni, August 31, 1922; 15: Albert W. Pontius to Jacob G. Schurman, September 8, 1922; 16, 2871: G. C. Hanson, Press Notes, October 30, 1922; 20, 3232: Dispatch No. 440, September 29, 1923; 20, 3274: The Far Eastern Times, "Gen. Ostroumoff, Director of C. E. R., Accorded Two Months' Leave," October 22, 1923; 23, 3524: G. C. Hanson to Jacob Gould Schurman, June 20, 1924. One of the purported successors was Fengtian-friendly Yuan Jinkai 袁金鎧 (1870–1947) (NARA/CER, 14: Consul General to Jacob Gould Schurman, September 8, 1922). According to the US consul in Harbin, the new Minister of Communications Gao Enhong 高恩洪 (1875–1943), not at best terms with his predecessor, Wang's friend Ye Gongchuo, was only "vainly fighting against his [Wang's] removal" (NARA/CER, 14: Consul General to Jacob Gould Schurman, September 8, 1922; 15: Albert W. Pontius to Jacob Gould Schurman, September 8, 1922).

144 Julean Arnold, *China: A Commercial and Industrial Handbook* (Washington, D. C.: Government Printing Office, 1926), 681.

145 NARA/CER, 17, 2988: M. de Bunting, "Chinese Exaction on C. E. R." The North China Herald, December 30, 1922; NARA/CER, 23, 3637: Far Eastern Times, "Russo-Manchurian Deadlock Unbroken," August 1, 1924. See Economical Bureau, C. E. R., *The Chinese Eastern Railway and Its Zone*, 24–32; I. Mihailoff, ed., *North Manuchria and the Chinese Eastern Railway* (Harbin: C. E. R. Printing Office, [1922] 1924).

146 NARA/CER, 19, 3104: G. C. Hanson to Secretary of State, May 7, 1923; 19, 3073: G. C. Hanson to Secretary of State, March 20, 1923.

147 NARA/CER, 19, 3075: G. C. Hanson to Secretary of State, March 21, 1923.

148 NARA/CER, 19, 3075: Wang, Jingchun and Danilevsky, S. I., "The Chinese Eastern Railway," The North China Herald, March 10, 1923.

149 Wang had previously warned that he had not forgotten about the 1900 Amur anti-Chinese pogroms at Blagoveshchensk, and that any aggression may lead to a similar

attempt at returning the public slander was in a press interview, in which he stated that although Ostroumov was "a man of ability and sagacity," he was "not able to vouch for his integrity."[150] Ostroumov's press campaign was so successful that history books relying on Russian sources reproduce it to this day.[151]

Beijing circles responded with an anti-Ostroumov campaign.[152] Their mouthpiece *Far Eastern Times* accused Ostroumov of having the railway "sucked dry and ruined" through bogus bookkeeping and of erecting a "railway dictatorship."[153] Its chief editor called for Ostroumov's police detention.[154] The *Harbin Dawning* joined the campaign, claiming Ostroumov to be "guilty of several hundred corrupt cases."[155] Allegedly, Ostroumov had systematically substituted retired and deceased Russian CER directors with acquaintances. Through the friendly BoD, Ostroumov allegedly attempted to secure corrupt schemes, for instance, having a befriended contractor buy railway stock for cheap and then sell it to the CER for an extortive price. Newspapers also criticized the diversion of the CER budget for constructing Ostroumov's new Harbin residence. Maybe the most discussed dubious affair was the signing of the Changchun Freight Agreement between Ostroumov and the South Manchurian Railway. Newspapers commented that the agreement contained unfavorable freight rates that lacked a rational explanation. Rumors that Japanese authorities had sealed the agreement with the help of a large cash payment

---

"massacre of all the Russians in the Railway Zone" by "enraged Chinese" (NARA/CER, 17, 2950: G. C. Hanson to Secretary of State, December 11, 1922. Also JACAE, 1 (3): Wang, Jingchun, Address at First General Meeting of the ACAE, Tianjin, October 6, 1920, 8–11: 9–10).

150 NARA/CER, 17, 2962: Translation from The International, "An Important Interview with Dr. C. C. Wang," December 10, 1922.

151 Urbansky's "global" history of the CER, shy of Chinese voices, never mentions Wang. Instead, the author calls Ostroumov a "master" whose name translates to "ingenuity," who received "hymns of praise" from the public, and who was the one responsible for the reforms of an era named after him (Urbansky, *Kolonialer Wettstreit*, 99–105). See also Leong, *Sino-Soviet Diplomatic Relations*, 107.

152 South Manchuria Railway Co., *Tō Shi tetsudō wo chūshin to suru Ro-Shi seiryoku no shōchō*, 949–53.

153 NARA/CER, 20, 3274: The Far Eastern Times, "The Latest Scandal on the Chinese Eastern Railway," October 22, 1923; Ostroumov sued the newspaper for slander. While the Beijing consular court ruled in favor of Ostroumov, the French court in Saigon eventually overruled the decision and acquitted the newspaper from all charges (NARA/CER, 21, 3325: The Far Eastern Times, "The Ostroumoff–Nachbaur Case," December 15, 1923).

154 NARA/CER, 16: Confidential Report No. 11 on the Changchun Conference by Lenox Simpson, October 11, 1922.

155 NARA/CER, 22, 3417: Translation from The Harbin Dawning, March 16, 1924.

never subsided.[156] Stirred against the Russian manager, the Chinese Chamber of Commerce declared a boycott of goods from Manzhouli and the Amur district.[157]

Apart from being uncooperative toward Wang Jingchun and bribing the press, nobody could detail the schemes with which Ostroumov allegedly tapped the CER's money.[158] However, a representative of the Russo-Asiatic Bank confirmed that Wang had uncovered how Ostroumov had—without approval—concealed ¥150,000 Japanese yen of railway money in a foreign bank account.[159] In a private discussion with the US consul, the representative further admitted that Ostroumov was "too autocratic and lacked tact."[160] In this respect, Ministry advisor John E. Baker compared Ostroumov to Oliver Cromwell (1599–1658) and Napoleon Bonaparte (1769–1821).[161] When the director of the Russo-Asiatic Bank, Jacques Raindre, sent one of his relatives to investigate, he likewise developed an anti-Ostroumov stance. He felt "discouraged and disgusted" by his experiences and, without success, joined Wang and the Harbin merchants in their calls for Ostroumov's disposal.[162]

Wang's frustrations reached their climax in the summer of 1923. Zhang Zuolin announced his intent to replace the CER land department with his own provincial one. On August 1, his forces attempted to seize the land department by force.[163] Zhang had only informed Wang of his plans two days earlier. Wang had first advised against the plans,[164] but, when faced with the facts, urged "all patriotic Chinese" to cooperate with the new bureau.[165] In reality, instead of playing an active role, he lay sick at Manzhouli Station for two months. During this time, Zhang's takeover attempt subsided.[166] The land department fiasco

---

156 NARA/CER, 16: Confidential Report No. 11 on the Changchun Conference by Lenox Simpson, October 11, 1922.

157 NARA/CER, 20, 3159: G. C. Hanson to Secretary of State, July 17, 1923.

158 NARA/CER, 22, 3429: G. C. Hanson to Jacob Gould Schurman, April 8, 1924.

159 NARA/CER, 19, 3073: G. C. Hanson to Secretary of State, March 20, 1923.

160 NARA/CER, 19, 3073: G. C. Hanson to Secretary of State, March 20, 1923.

161 NARA/CER, 20, 3160: John Earl Baker to William Cullen Dennis, June 8, 1923; 22, 3416: Confidential Dispatch No. 640, March 18, 1924.

162 NARA/CER, 21, 3382: Confidential Dispatch No. 602, February 28, 1924; 21, 3388: Confidential Dispatch No. 611, March 3, 1924; South Manchuria Railway Co., *Tō Shi tetsudō wo chūshin to suru Ro-Shi seiryoku no shōchō*, 946–49.

163 NARA/CER, 20, 3169: G. C. Hanson to Secretary of State, July 25, 1923; NARA/CER, 20, 3169: US, France, Great Britain, and Japan Ministers to Wellington Koo, August 11, 1923; NARA/CER, 21, 3348: Memorandum from B. V. Ostrooemov Concerning the "Violation of Rights of Russian Citizens," undated.

164 NARA/CER, 20, 3242: Confidential Dispatch No. 441, October 3, 1923.

165 Chiasson, *Administering the Colonizer*, 135–36.

166 NARA/CER, 20, 3195: Jacob G. Schurman to Secretary of State, September 18, 1923.

and his presidential impotency left Wang embarrassed. In late September, he transferred to a hospital in Harbin, and in early October, he moved to Beijing to receive additional treatment, planning to recover for at least six months. He cited asthma, a medical condition he had for "over ten years."[167] The Ministry of Communications ordered Wang to return to the CER, which he refused. As Wang's return seemed unlikely, resignation rumors reemerged.[168] The search for a successor proved unsuccessful: the position was still as unpopular as when it fell into Wang's hands two years earlier, with diplomat Wang Zhengting and retired military governor Zhu Qinglan 朱慶瀾 (1874–1941), who had already acted as the CER's president earlier, declining the offer.[169] In January 1924, after Zhang Zuolin reluctantly joined the call for his comeback, Wang Jingchun finally agreed to return to Harbin.[170]

Wang had just returned when Ostroumov added fuel to the fire. Under his supervision, the Railway Club (*tielu julebu* 鐵路俱樂部) performed an operetta that allegedly conveyed a sinophobic message. Word spread and Chinese residents staged a street demonstration against Ostroumov.[171] Although Wang conceded that the reaction toward the play was exaggerated and that agitators used it as a pretext to fuel their demands,[172] his railway police did not attempt to disperse the demonstrations.[173] For Stevens, the "spectacular" demonstrations against Ostroumov bore a striking resemblance to the anti-IATB protests of 1922. Again, US intelligence suspected them to be "staged and paid for" by Chinese officials.[174] Wang did not admit or reject allegations of personal involvement but stated that he admired the anti-Ostroumov protest as a "patriotic movement."[175] Eventually, the Ministry of Communications instructed Wang to calm the anti-Ostroumov movement in order not to upset the other

167 NARA/CER, 20, 3243: Translation from The International, "An Interview with Dr. Wang," September 29, 1923; 20, 3242: Confidential Dispatch No. 441, October 3, 1923; 20, 3248: Dispatch No. 449, October 18, 1923; Chiasson, *Administering the Colonizer*, 126, 135–36.

168 NARA/CER, 20, 3232: Dispatch No. 440, September 29, 1923.

169 NARA/CER, 20, 3267: Dispatch No. 471, November 14, 1923; 21, 3309: Translation from The Commerce, December 8, 1923.

170 *Dàgōng bào*, "Zhōngdōng lù zhī jìnxùn," January 17, 1924, 3.

171 NARA/CER, 21, 3371: Translation from Harbin Dawn, February 17, 1924.

172 *Dàgōng bào*, "Hābù zhī qū Wò wèntí," March 1, 1924, 3; NARA/CER, 21, 3397: Dispatch No. 622, March 8, 1924. Cf. NARA/CER, 21, 3382: Translation from Zarya, "The Secret Springs," February 28, 1924.

173 NARA/CER, 22, 3406: Translation from The Sungari Daily News, March 14, 1924.

174 NARA/CER, 22, 3414: John F. Stevens to Secretary of State, April 15, 1924.

175 NARA/CER, 21, 3391: Translation from The Sungari Daily News, "An Interview with Dr. C. C. Wang," March 2, 1924.

railway parties.[176] By that time, the protests against Ostroumov seemed to have worked to some degree. Mirroring Wang's reaction during the first press campaign that had favored the Russian general manager, by now Ostroumov felt "disgusted" by the lack of appreciation for "his hard work and efforts."[177] The Russian engineer stated he was "exceedingly tired" of his position and stated that he intended to resign with the next attack.[178]

The quarrels between Wang and Ostroumov demonstrate how a technocratic arrangement that would have allowed for a more efficient management of the CER wore itself out in a quest led by personal vanity and complex political meddling. What appeared to be the ground for a railroad management of experts, as Wang and his colleagues had called for, quickly became a political affair.

6 The End of Expert Management

No matter whom the newspapers rooted for, they liked to point out that officially, management of the CER had to be based on economic principles. But expert management of the CER never escaped its political agenda. As Dong Xianguang—formerly involved in river conservancy and now working as a journalist (chapter 3)—wrote, the 1920 agreement between Beijing and the Russo-Asiatic Bank had made the CER "purely commercial, not political."[179] Working in a time when proponents of economic management thought they were exploring a science based on natural laws, Wang Jingchun depicted the problems of the CER like a mathematical riddle that only needed application of the right approach to dissolve. In his interviews and government dispatches, Wang repeatedly wrote of his plans to "solve" (*jiejue* 解決) the "fundamental problems" (*genben wenti* 根本問題) of the CER.[180] Not just the language but

176 NARA/CER, 22, 3420: Translation from The Commerce, March 22, 1924; 22, 3429: Translation from The Commerce, April 1, 1924.

177 NARA/CER, 22, 3433: G. C. Hanson to Jacob Gould Schurman, April 19, 1924.

178 NARA/CER, 22, 3420: Translation from The Commerce, March 22, 1924.

179 NARA/CER, 13: Extract from The Weekly Review, April 25, 1922. Also NARA/CER, 12: Translation of The International, February 5, 1922; NARA/CER, 17, 3000: The Weekly Review, "The Legal Status of the Chinese Eastern Railway," January 20, 1923.

180 NARA/CER, 21, 3391: Translation from The Sungari Daily News, "An Interview with Dr. C. C. Wang," March 2, 1924; 22: Translation from Sungari Daily News, "Dr. Wang's Telegram," March 9, 1924; JSDG, 03-32-261-02-009: Wang Jingchun to Foreign Ministry, February 1923; 03-32-272-08-008: Wang Jingchun to Foreign Ministry, March 3, 1924; 03-32-268-03-018: Wang Jingchun, March 4, 1924; 03-32-272-06-001: Wang Jingchun, March 21, 1924; 03-32-272-06-002: Wang Jingchun to Sino-Russian Relations Office, March 21, 1924.

also the underlying ideology resembles technological solutionism, i.e., the mode of thought in which engineers believe they can solve any problem the world encounters through a rational, technology-driven solution. In this vein, Wang tried to counter a complex entanglement of political interests with efficient administration reforms. However, his looming realization as CER president must have been that this railway was not an issue that he could fix with the mindset of an economist.

The efficiency paradigm that guided the commercial operation was a pretext for political programs. Similarly, the Japanese South Manchurian Railway's president Kawamura Takeji 川村 竹治 (1871–1955) claimed that his railway was "only a commercial enterprise" when demanding the CER to convert its southern branch line from narrow gauge to standard gauge. This would have enabled through movement to Beijing and would have further integrated the South Manchurian line into the Chinese railway system. In this instance, the press quickly exposed the underlining political ambitions.[181] It is difficult to tell by what degree the propagators of the "purely commercial" myth believed their own framing. When opportune, Ostroumov rebuked the myth: the railway was not built as a commercial enterprise but to enable what he referred to as "colonization," including concessions, economic influence, rights of land, navigation rights, duty and tax exemptions, and the administration of railway towns.[182] However, once Ostroumov attempted to negotiate his future with the Soviets, he claimed he had "always tried to run the railway as a commercial enterprise" and that his past opposition toward Soviet demands had not been politically motivated.[183]

Wang Jingchun, as well as the leadership of the Beijing Republic, which included his old student friend and incumbent foreign minister Gu Weijun (chapter 1), displayed a guarded attitude toward the Soviets. In 1919, in a wave of early anti-imperialist enthusiasm, Moscow renounced all rights to the CER, only to retract the proclamation six months later.[184] Afterward, there was little will on Beijing's end to go for anything less.[185] However, after having experienced frustration applying the efficiency paradigm to the Chinese cause when so many other interpretations existed, Wang's position shifted. From early 1923 onwards, despite being an outspoken anti-Communist, Wang petitioned

181 NARA/CER, 23, 3490: Extract from Tribuna, May 18, 1924.

182 NARA/CER, 23.5, 3676: Memorandum from B. Ostroumov on the Agreement of May 31, 1924, undated.

183 NARA/CER, 22, 3416: Confidential Dispatch No. 640, March 18, 1924.

184 Chu, *V. K. Wellington Koo*, 101.

185 NARA/CER, 21, 3365: Translation from The International, "The Disagreeing Opinions," February 15, 1924.

the Beijing government to recognize the Soviet Union and negotiate a deal to replace the current CER arrangement.[186] Wang, who had left his position as chief of railways at the Ministry to assume the CER presidency, was evidently out of ideas or technical "solutions."

In August 1922, Soviet diplomat Adolph Abramovich Joffe (1883–1927) tried to negotiate a treaty with the Beijing Republic but met little interest. The attitude only changed in March 1923, after Joffe signed a similar agreement with Sun Yat-sen's rival government in the South instead.[187] In the aftermath of the "Lincheng Outrage"—an incident where rogue soldiers raided a Beijing-bound train on the Tianjin–Pukou line and took 300 passengers, including 25 westerners, hostage—foreign governments demanded to put the Chinese railway police under international control,[188] consequently increasing the Chinese willingness to reapproach the Soviets for the CER. In early 1924, a group of Beijing professors petitioned for the recognition of the Soviet Union. In the capital, influential militarist Wu Peifu echoed the calls, aiming at negotiations with the Soviets concerning the future of the CER.[189]

The Beijing government agreed to recognize the Soviet Union and to negotiate a new contract for the CER. Ostroumov reconciled with Moscow and was now in the same boat.[190] However, Zhang Zuolin, as leader of the autonomous government of Fengtian, refused to adhere to any Sino-Soviet agreement. He ordered Wang to fire Ostroumov, which Wang was unable to do, as he was under Beijing orders.[191] Zhang first placed Wang under house arrest and then made him leave for a long vacation to Switzerland to ensure he was unable to assist Beijing in the negotiations.[192] Instead of seeing Wang as a tragic figure,

186 NARA/CER, 18, 3002: Peking & Tientsin Times, "Dr. Wang's Memorial to President," January 16, 1923; 21, 3391: Translation from The Sungari Daily News, "An Interview with Dr. C. C. Wang," March 2, 1924; JSDG, 03-32-261-02-009: Wang Jingchun to Foreign Ministry, February 1923.

187 Chu, *V. K. Wellington Koo*, 103.

188 Chu, *V. K. Wellington Koo*, 83–88.

189 Chu, *V. K. Wellington Koo*, 107, 113; Craft, *V. K. Wellington Koo and the Emergence of Modern China*, 81.

190 NARA/CER, 19, 3073: G. C. Hanson to Secretary of State, March 20, 1923; NARA/CER, 21, 3393: Confidential Dispatch No. 619, March 7, 1924; 22, 3407: Confidential Dispatch No. 624, March 11, 1924.

191 NARA/CER, 22, 3429: Information, April 5, 1924; 23, 3637: Far Eastern Times, "Russo-Manchurian Deadlock Unbroken," August 1, 1924; 23, 3637: G. C. Hanson to Secretary of State, August 7, 1924.

192 NARA/CER, 23, 3531: Translation from The Sungari Daily News, June 21, 1924; 23, 3532: Dispatch No. 775, June 26, 1924; 23, 3532: Japan Adventure, "Chinese Railroad Head to See U. S." June 22, 1924.

the US mission ridiculed his departure, believing he had proactively shuffled out of responsibility.[193]

During the negotiations, Moscow and Beijing *de facto* abandoned the idea of running the railway as a purely commercial enterprise. The Sino–Soviet Agreement (*Zhong Su xieding* 中蘇協定) of May and the Fengtian–Russian Agreement (*Feng E xieding* 奉俄協定) of October 1924 transformed the former enterprise into a jointly managed institution.[194] Wang and Ostroumov were both relieved of their positions, with Ostroumov arrested and jailed on accounts of corruption.[195] The two engineers had fallen victim to the political meddling in which they inevitably participated. Instead of associating and cooperating under the premise of their common civil engineering profession, they competed against one another. For Wang, former leader of the world confederation of Cosmopolitan Clubs (chapter 1), internationalism presented itself with pro-American sympathies but did not extend to cooperating with the Russians who contested Chinese rule over the CER. The new agreement replaced Wang with Fengtian militarist Bao Guiqing, and Ostroumov with, as a US dispatch would have it, "an out and out communist, with no knowledge of engineering or business."[196] The new arrangement resembled the one before the Russian Civil War, as sycophancy triumphed over technical ability. With the professional engineers Wang and Ostroumov ousted, two years of attempted technocratic railway administration came to an end.

7 Conclusion

During the 1920s, civil engineers regularly lamented that, if fractured militarism was an ongoing concern, international investment was difficult to attract, and infrastructural development difficult to launch. Wang Jingchun stood out from his contemporaries in that he identified critical infrastructures as the

193 NARA/CER, 23, 3532: Dispatch No. 775, June 26, 1924; 23, 3531: G. C. Hanson to Jacob Gould Schurman, June 27, 1924.

194 NARA/CER, 23.5, 3676: B. Ostroumov to John F. Stevens, September 9, 1924; 23.5, 3676: B. Ostroumov on the Agreement of May 31, 1924, undated; 23.5: Dispatch No. 857, October 3, 1924; Chu, *V. K. Wellington Koo*, 108–16; Craft, *V. K. Wellington Koo and the Emergence of Modern China*, 79–84; Bruce Elleman, *Diplomacy and Deception: The Secret History of Sino–Soviet Diplomatic Relations, 1917–1927* (Armonk, London: M. E. Sharpe, 1997), 120–28; Líng, *Qīshí nián lái Dōng Qīng, Zhōng dōng, Zhōng Cháng tiělù zhī jīngguò*, 14–25.

195 NARA/CER, 23.5, 3704: Translation from The International, October 4, 1924; Gamsa, *Harbin*, 130–31, 168–72.

196 NARA/CER, 23.5, 3693: Dispatch No. 852, October 1, 1924.

motor for national unification and political consolidation, not vice-versa. In his theory, he argued that railroads could assist nation-building. However, his ambitions came down to earth when he inherited a dyed-in-the-wool colonial railway that was never meant to benefit a Chinese nation. His socio-technical imaginary of the railroad conflicted with the broader political entanglement, which left little scope for a technocratic administration and the nationalist utilization of the line. Large infrastructural endeavors required potent political consolidation, and political consolidation required enabling infrastructures. The reciprocal relationship resulted in an achievement of neither. Despite being a commercial enterprise on paper, Wang was unable to fully carry through his project of economic management. Instead, political interests stymied his project and intrigues hit his career. The frustration that accompanied the two years of attempted technocratic management of the Chinese Eastern between 1922 and 1924 helped expose the crux of civil engineers working in the era of dispersed political control in China. Wang was, by mindset, an engineer, not a politician, but found himself in a position where, despite the "purely commercial" myth, everything was political. The disappointment that followed this revelation would inform the pessimistic attitude that the once energetic civil engineers adopted over the limits of their profession.

In 1929, during the Sino-Soviet Conflict, the Chinese government attempted to gain unilateral control over the CER through military means. Writing from his US exile, Wang was critical of the government's actions. He explained that the Chinese forces had effectively seized the CER, which for Wang was an action "out of date."[197] He begged to respect the rulebook. His open critique may have signaled his exhaustion from political intrigues. In 1932, Wang met Mahatma Gandhi (1869–1948) at Cambridge University, where the two discussed contemporary Chinese politics.[198] Gandhi had earlier proclaimed, "Good travels at a snail's pace—it can, therefore, have little to do with the railways."[199] Given Wang's experience and disillusionment with the CER, he may well have agreed.[200]

197 Wang, "A Railroad Born in Sin," 64.

198 Ching-chun Wang and L. Pu, "A Discussion with Mr. Gandhi Concerning His Opinions of the Chinese National Government," *The China Critic* 5, no. 2 (1932). Wang gifted Gandhi a copy of Min-ch'ien Tuk Zung Tyau, *Two Years of Nationalist China* (Shanghai: Kelly and Walsh, 1930).

199 Mohandas Gandhi, *Hind Swaraj: And Other Writings* (Cambridge: Cambridge University Press, [1997] 2009), 47. Thanks to Tim Oakes for referring me to this quote.

200 For Wang's later career, see chapter 6 and the conclusion.

FIGURE 13 Graph of the ACAE members–workplace network (two-mode, undirected, unweighted, main component). Based on data from CERD, substituted with JACAE information about foreign members. Hydraulic engineers in the upper half, railroad engineers in the lower half.

CHAPTER 4

The Visible College

The Early Association of Chinese & American Engineers, 1919–1927

Concurrent with the return of US-educated civil engineers to China, the late 1910s saw an influx of foreign technicians assuming positions in the country. Americans worked on hydraulic projects such as the Grand Canal (chapter 2), as advisers to ministries, or inside commercial supply firms. Sharing a common educational upbringing, the two groups of US-educated Chinese and American engineers formed links of professional and social exchange. In sociology, the term "invisible college" describes epistemic communities that collaboratively identify areas of research, share ideas and ideals, and generate a "sense of commitment" and membership.[1] But in contrast to the concept of invisibility, a metaphor for opaque citation networks, the Sino-American engineering "college" was far from hidden. Instead, to establish civil engineering as a recognized profession and to promote the US–Chinese segment within it,[2] it sought publicity. The manifestation of this Sino-American "*visible* college" was the Association of Chinese & American Engineers (ACAE). Through the association, American and Chinese engineers attempted to establish a professional common ground, networked, and developed transnational friendships. Following this book's recurring motif of professional discontent, as a platform, the members used the ACAE to discuss their professional experiences and frustrations and came to identify with the problems they shared.

1 Formation and Organizational Makeup

Within the sociology of scientific knowledge, science is an "organized social activity."[3] Social scientific activities of scholarly exchange, such as conferences, editorial boards, and journal forums, manifest themselves in the form of associations that foster professionalization. Several factors worked against

1 Diana Crane, *Invisible Colleges: Diffusion of Knowledge in Scientific Communities* (Chicago, London: The University of Chicago Press, 1972), 35–6, 49–56, 72–6, 172.

2 See Bucher and Strauss, "Professions in Process," 334.

3 Norman Storer, *The Social System of Science* (New York: Holt, Rinehart and Winston, 1966), 3. Cited in Crane, *Invisible Colleges*, 4.

 | DOI:10.1163/9789004549555_006

the professionalization of civil engineers. Firstly, the conscious identification of civil engineers with their occupation was a young development, and professionalization usually takes time.[4] Secondly, while the public can easily develop an idea of what lawyers or physicians do, the esoteric work of engineers had a hard time arousing public interest or support.[5] Thirdly, already established institutions—the dominance of the bureaucracy, of kinship, and of native place—hindered the emergence of new organizational structures.[6] All these observations worked against the civil engineering profession in China. However, for the very reason that civil engineers were largely unable to build upon existing institutions to promote their goals, they also had an obvious reason to formally organize their own societies.

Overseas students had already started forming their networks before the emergence of a society that specifically catered to the profession. For instance, US returnees formed clubs in Guangzhou, Shanghai, Beijing, and Hankou that limited access to Euro-American graduates.[7] When they formed Phi Tau Phi, the first Greek letter organization in China, members of the US-based Phi Beta Kappa and Sigma Xi organizations were automatically eligible.[8] Christian organizations like the YMCA also kept the graduates together. Furthermore, former US students stayed updated through alumni magazines and sometimes returned to the States to visit old college friends.[9] In the case of prominent universities like Harvard and Yale, their graduates also founded local alumni organization "exclaves" in China. But all these clubs catered to returned students at large, not to a specific profession. This changed when in 1919, Chinese and American engineers formed their own professional organization, the Association of Chinese & American Engineers.

The ACAE was not the first engineering association in China, nor did it intend to act as the representative society of the broader profession. In China, the railway engineers Zhan Tianyou, Yan Deqing 顏德慶 (1878–1942), and Xu Wenjiong 徐文炯 had founded three early engineering societies in 1912, which had consolidated into the Chinese Institute of Engineering (*Zhonghua*

4 Jackson, "Professions and Professionalization," 8.

5 Freidson, *Professionalism*, 168.

6 C. Turner and M. Hodge, "Occupations and Professions," in Jackson, *Professions and Professionalization*, 20.

7 Brooks, *American Exodus*, 58–59.

8 JACAE, 5 (8): 31–32.

9 *The Daily Illini*, "Dr. C. C. Wang Makes Visit to Friends at Alma Mater," October 8, 1912, 19; YARO, 830:723, Young, In: Correspondence with Ethel A. Young, May 15, 1949.

gongchengshi hui 中華工程師會, CIE) in 1913.[10] Apart from the CIE, in 1918, Chinese engineering students in the US had formed the China Engineering Society (*Zhongguo gongcheng xuehui* 中國工程學會, CES), which hosted its first domestic congress in 1923, after its founding members had returned to China.[11] Given that the CES was the product of engineering students who wanted to connect during their study in the US, the CES bore American influence, including in its organizational structure.[12] As demonstrated by the CES' merger with the CIE in 1931, the two Chinese societies had more in common with each other than with the transnational ACAE, whose unique characteristic remained the bilateral exchange.

The CIE and CES were founded in the tradition of study societies: their journals and annual meetings served as forums to discuss new engineering developments. In their function as professional societies, they furthermore aimed to represent the engineering scene. Through membership requirements, they introduced certification to the field: not everyone working in a technical environment met the requirements to become a full member of society and, consequently, the profession. Through entry requirements such as college degrees and recommendation letters from peers, the members introduced an autonomous means to legitimize themselves as engineers.

As reflected in its name, the primary function of the ACAE differed from that of these regular professional societies. The ACAE was not a "study society" (*xuehui* 學會) but an "association" (*xiehui* 協會). Its primary aim was not to advance engineering research or to represent the profession as a whole but to represent the specific segment of the Sino-American engineers within the broader field. The ACAE specifically targeted engineers with a US educational background. In contrast to the CES, which catered only to *Chinese* students and graduates, the ACAE—as the name suggests—also welcomed members from other nationalities, especially from the US. Consequently, the ACAE worked as a complement to the larger umbrella organizations and as a more exclusive, condensed club for practitioners associated with the Sino-American civil engineering culture.

ACAE membership overlapped with that of other professional organizations, including the CIE. 37 of 44 CIE leaders had studied abroad, including in

10 Le, "The Struggle for Autonomy from the State," 96; Fáng, *Jìndài gōngchéngshī qúntǐ de 'mínjiān lǐngxiù'*, 41.

11 Fáng, *Jìndài gōngchéngshī qúntǐ de 'mínjiān lǐngxiù'*, 52–53.

12 For US cultural influence on the CES, see Fáng, *Jìndài gōngchéngshī qúntǐ de 'mínjiān lǐngxiù'*, 53, 93.

the US, Japan, and Europe,[13] and seven of the ones with US backgrounds also assumed leading positions in the ACAE (table 2, table 3). In the early years, Yan Deqing simultaneously acted as ACAE president and as CIE vice-president. Sun Duoyu 孫多鈺 (Taoyuh C. Sun, 1882–1951), Sa Fujun, and Wang Chongyou 王寵佑 (Chung-yu Wang, 1881–1958) also, at different times, led both the CIE[14] and the ACAE. CIE President Shen Qi 沈琪 (1871–1930) was also an honorary member of the ACAE. Later examples include Li Shutian 李書田 (Shu-tien Li, 1900–1988) and Ling Hongxun (chapters 1, 7), who were eminent figures in both organizations. Furthermore, while the CIE did not accept American engineers as members, they welcomed ACAE members to their conferences and featured directors like Murray Sullivan (1881–1941) as guest speakers.[15]

The international membership was not the only difference between the CIE and the CES on the one side and the ACAE on the other. The ACAE concentrated on civil engineers, while the Chinese societies catered to all engineering specializations. In 1918, 38% of CES members had a background in civil engineering,[16] which closely reflected the 41% of civil engineers within the larger engineering pool at that time.[17] In contrast, the ACAE's emphasis lay more clearly on civil engineering, which 69% of Chinese members had studied.[18] Even when members had a background in other engineering disciplines, they at least shared an interest in civil engineering. For instance, one of the association's later presidents, Samuel M. Dean, held a degree in mechanical engineering but had worked in architectural bureaus, where he consulted on questions of structural design, thereby working in proximity to civil engineering regardless of his original major.[19]

In pure numbers, the ACAE was not a large organization. At its inception in 1919, the association counted 53 Chinese members.[20] In comparison, in the year of the ACAE's founding, the CES counted 132 members and the CIE 435.[21]

13 Fáng, *Jìndài gōngchéngshī qúntǐ de 'mínjiān lǐngxiù'*, 108–9.
14 Fáng, *Jìndài gōngchéngshī qúntǐ de 'mínjiān lǐngxiù'*, 104–8.
15 JACAE, 1, no. 2 (1920): "Annual Meeting of the Chinese Institute of Engineers," 39; JACAE, 7, no. 7 (1926): Notes, 34.
16 Fáng, *Jìndài gōngchéngshī qúntǐ de 'mínjiān lǐngxiù'*, 50.
17 Pelzer et al., CERD, [{"t":"10"}, {"linked_to:940":[{"t":"58"}, {"f:9":"<=1918"}, {"f:997":" 9483"}]}].
18 Pelzer et al., CERD, [{"t":"10"}, {"f:1000":"9547"}, {"linked_to:940":[{"t":"58"}, {"f:997":" 9483"}]}].
19 JACAE, 16, no. 3 (1935): "S. M. Dean, Scholastic and Professional Record," 123.
20 JACAE, 3, no. 8 (1922), 1.
21 Fáng, *Jìndài gōngchéngshī qúntǐ de 'mínjiān lǐngxiù'*, 44, 52.

At any given time, less than 150 Chinese engineers had a running membership with the ACAE.[22] To put the number in comparison, over 880 Chinese engineers graduated in the US before the war.[23] Instead of being a mass organization, many ACAE members worked in senior positions, were highly connected, and led influential engineering projects.

In the CERD sample, the overall ratio between railway and hydraulic engineers is about 2:1.[24] Inside the ACAE, however, the two sub-disciplines formed two equally important groups. As the network graph, which links the association members through their shared places of employment, demonstrates, the center of the network neatly divides between the two clusters (Figure 13).[25] Smaller, less integrated groups were those of educators and merchant-engineers. The importance of hydraulic engineers within the ACAE was a result of the ACAE's Grand Canal heritage and member recruitment among the engineers' later conservancy workplaces. Personal continuities show the close association with the Grand Canal Board (chapter 2) as an informal predecessor of the association (Figure 13). Charter members included Grand Canal figures Yang Baoling (chapters 1, 2) and Joseph Ripley (chapter 2).[26] The first General Meeting featured an address by the Canal's former director Xiong Xiling,[27] and former President Li Yuanhong, acquainted with the American engineers since the Grand Canal project, likewise supported the new association.[28] Although the railroad and hydraulic engineers socialized with each other, they had little occupational links running between them. The two exceptions were the China International Famine Relief Commission (CIFRC, chapter 5) and the Chinese Engineering and Development Company (*Dachang* 大昌), which were of central importance for the network in that

22 JACAE, 8, no. 2 (1927): "List of Members," 31–44.

23 Pelzer et al., CERD, [{"t":"10"}, {"linked_to:940":[{"t":"58"}, {"f:9":"<=1937"}, {"linked_to:964":[{"t":"57"}, {"linked_to:238":[{"t":"12"}, {"f:26":"119"}]}]}]}]; [{"t":"10"}, {"linked_to:58:940":[{"t":"58"}, {"f:9":"<=1937"}, {"f:997":"9483"}]}].

24 This ratio is only a rough estimate based on string matching. Pelzer et al., CERD, [{"t":"10"}, {"linked_to:942": [{"t":"61"}, {"linked_to:21": [{"t":"4"}, {"any": [{"f:1":"鐵道"}, {"f:1":"鐵路"}]}]}]}]; Pelzer et al., CERD, [{"t":"10"}, {"linked_to:942": [{"t":"61"}, {"linked_to:21": [{"t":"4"}, {"any": [{"f:1":"水"}, {"f:1":"河工程"}]}]}]}].

25 See also Wu, *Měiguórén wèijìng de Zhōngguó mèng*, 284–86.

26 JACAE, 1, no. 1 (1920): Notes, 35; 5, no. 1 (1924): Minutes of the First Meeting of the Association, November 22, 1919, 3.

27 JACAE, 1, no. 2 (1920): Addresses at First General Meeting of the ACAE, 22–24.

28 JACAE, 3, no.5 (1922), 1.

TABLE 2 National ratio of the ACAE members

	Chinese	Foreign
Honorary members	5 (45.5%)	6 (54.5%)
Members	71 (58.2%)	51 (41.8%)
Plus, absent	—	37
Associate members	46 (70.7%)	19 (29.3%)
Plus, absent	—	11
Total including absentees	122 (49.6%)	124 (50.4%)

they formed occupational links between the otherwise weakly connected groups (Figure 13).

Despite the lack of a constitutional quota to enforce a balance between Chinese and American members, the ACAE achieved a fifty-fifty ratio (table 2). Subtracting those foreign engineers who joined the association when in China but had since returned to their home country, the Chinese engineers were in the majority. The balance extended to the board of directors and leadership positions, as senior members from China and the US regularly held offices (table 3). This national balance was in stark contrast to the Sino-British Engineering and Architectural Society. There, the presidency remained white during the first half of the 20th century,[29] hinting at a colonial hierarchy of UK technicians which, if it existed, remained invisible in the Sino-American engineering world.

Requirements for membership in a professional association indirectly define the profession.[30] The more exclusive the requirements, the more prestigious the promoted version of the profession: in the US, the American Society of Civil Engineers (ASCE) did not accept anyone in military service, and full members had to work in high ranks.[31] Like the ASCE, apart from honorary members, the ACAE was strictly professional. This makes it different from,

29 Michiko Kobayashi and Hideo Izumida, "Kindai chūgoku no gijutsusha kyōkai no kikanshi ni tsuite: Kindai chūgoku no gijutsusha kyōkai ni kan suru kenkyū sono 1," *Journal of Architecture and Planning* (*Transactions of AIJ*) 76, no. 669 (2011); Wu, "Foreign Engineers' Activities in China and the Process of China's Internationalization," 381–384.

30 Layton Jr., *The Revolt of the Engineers*, 26–28. Also Davis, *Thinking Like an Engineer*, 164–65.

31 Layton Jr., *The Revolt of the Engineers*, 109–10.

TABLE 3 Timeline of ACAE executives, 1920–1928

<table>
<tr><th></th><th>President</th><th>First Vice-President</th><th>Second Vice-President</th><th>Treasurer</th><th>Secretary</th></tr>
<tr><td>1920</td><td rowspan="2">Kuang Jingyang</td><td rowspan="2">Murray Sullivan</td><td rowspan="2">Yang Baoling</td><td rowspan="2">Shi Zhaoxiang</td><td rowspan="2">Cai Guangji</td></tr>
<tr><td>1921</td></tr>
<tr><td rowspan="2">1922</td><td rowspan="2">Yan Deqing</td><td rowspan="2">Yang Baoling</td><td rowspan="2"></td><td>Tan Zhen</td><td>Tan Zhen</td></tr>
<tr><td>Thomas Meloy</td><td>Thomas Meloy</td></tr>
<tr><td rowspan="2">1923</td><td rowspan="2">Sun Duoyu</td><td rowspan="5">C. J. Carroll</td><td rowspan="2">Sa Fujun</td><td>H. C. Faxon</td><td>H. C. Faxon</td></tr>
<tr><td rowspan="7">Harry A. Raider</td><td>Joseph H. Ehlers</td></tr>
<tr><td>1924</td><td rowspan="2">Sa Fujun</td><td rowspan="2">Wang Chongyou</td><td rowspan="2">Gu Zhen</td></tr>
<tr><td>1925</td></tr>
<tr><td rowspan="2">1926</td><td>Frank H. Clark</td><td>Kuang Jingyang</td><td>Jin Tao</td></tr>
<tr><td>Shi Zhaoxiang</td><td>Kuang Jingyang</td><td>John A. Ely</td><td>Wu Zhuopan</td></tr>
<tr><td>1927</td><td rowspan="2">Gu Zhen</td><td rowspan="2">John A. Ely</td><td rowspan="2">Oliver J. Todd</td><td rowspan="2">Zhong Guidian</td></tr>
<tr><td>1928</td></tr>
</table>

for instance, the geological "Earth Studies Society" (*dixue hui* 地學會), founded in 1909, which allowed all kinds of literati in its ranks.[32]

There was no outspoken restriction of the sexes, but in addressing its members, for instance, through the frequent reminders to bring one's wife to excursions,[33] the ACAE made clear it was an exclusively male, heteronormative society. This pattern reflected the factual distribution of the sexes. The first batch of female engineering students only graduated in the 1930s,[34] meaning they could not have gained the experience necessary to be eligible for membership until shortly before the ACAE's demise.

Through the years, the ACAE gradually eased its requirements for membership. In 1921, the directors allowed graduates from Chinese universities to become full members,[35] and later, the ACAE started to accept members from other nationalities, e.g., Sigurd Eliassen (1884–1960) from Norway, although

32 Shen, *Unearthing the Nation*, 77.

33 OJTP, 13A.2: "Annual Convention. Association of Chinese and American Engineers," April 6–7, 1935, Beiping National Library.

34 Pelzer et al., CERD, t:10 f:20:415.

35 JACAE, 2, no. 4 (1921): Notes, 8–9.

they could not assume executive positions. Consequently, through the decade, membership numbers rose steadily. Memberships quintupled from 53 charter members in 1919 to 247 in 1922,[36] and then stagnated throughout the decade.[37] For reference, the CIE also experienced stagnation through the decade, with membership numbers never exceeding the high 400s, even shrinking temporarily.[38]

Within the already privileged ACAE, the same handfuls of individuals rotated seats back and forth in the various presidency and directorship positions (table 3). Similarly, the pool of authors for conference papers and journal articles consisted of a fraction of the total membership. Members were aware of the existence of an inner circle within the already exclusive ACAE.[39] That the ACAE failed to assemble all US-educated engineers, let alone have all of them actively participate, may have strengthened the ties of those involved but also left the association more vulnerable (chapter 6).

2 Functions of the Association and Its *Journal*

The ACAE's objectives date September 1, 1920. Printed in the association's constitution and its journal's preamble, they defined the "fostering of the spirit of co-operation," the advancing of knowledge and the maintenance of standards as the association's three *raisons d'être*.[40] For these purposes, the association hosted conventions and printed a journal. The organizing committee tried to convene the annual meetings in a different city each year.[41] Locations included Beijing, Hankou, Shanghai, Mukden, and Qingdao. Official functions of the conventions included socializing, delivering speeches, presenting papers, holding business meetings, conducting elections, and excursions to technical facilities.[42] The socializing aspect would help foster a group identity. In this context, the association released golden badges with blue enamel,[43] proudly worn during conference group photos to signal group affiliation (Figure 14).

---

36 JACAE, 3, no. 8 (1922): "The Second Anniversary," 1.

37 JACAE, 8, no. 2 (1927): "List of Members," 31–44.

38 Fáng, *Jìndài gōngchéngshī qúntǐ de 'mínjiān lǐngxiù'*, 48.

39 PJRF, 129: Yang Baoling to Freeman, April 26, 1928.

40 JACAE, 1, no. 1 (1920): Preamble, ii; JACAE, 3, no. 8 (1922): "Association of Chinese and American Engineers. Constitution," 10, Art I.3.

41 JACAE, 3, no. 4 (1922): Deqing Yan, Address of the President of the Association, Shanghai, April 4, 1922, 4.

42 JACAE, 3, no. 4 (1922): "Spring Convention," 1–2.

43 JACAE, 3, no. 1 (1922): Notes, 31.

FIGURE 14 Group photo of the annual ACAE fall meeting, Hankou, October 5, 1921 (JACAE, 2.10, 1921).

When a segment promotes itself, its members propagate their vision of the profession to the public.[44] Accordingly, the ACAE invited prominent politicians to the conferences to help the profession gain a "better standing."[45] Most importantly, the ACAE reached out to the Ministry of Communications, whose head *ex officio* became an honorary member of the association for the period of their term and regularly addressed the members at the annual conventions. In the upcoming years, the ACAE developed an indirect dependency on state railways as advertisers in its journal (chapter 6), but this did not make the ACAE a state organization. While the government welcomed and endorsed professional societies, it did not nationalize them as official institutions. The professional organizations received their legitimacy through the public reputation of its members and acknowledgment of the broader professional community, not through state endorsement.[46]

After the overseas student clubs had acted as a training ground for democratic practices (chapter 1), the ACAE provided another stage to perform elective procedures. Letters between the committee members and the board of directors show contestations of authority, with reference to the interpretation of the ACAE constitution. The inner democracy was not perfect. For

44 Bucher and Strauss, "Professions in Process," 334.

45 OJTP, 33.06: Todd to Tan Zhen, March 30, 1935.

46 Shen, *Unearthing the Nation*, 96–97. See also Wang, "Saving China through Science," 319.

instance, the ACAE copied the concept of US engineering associations to have nomination committees preselect candidates for the offices,[47] meaning that while elections were free, there was not much of a choice. The board of directors approved the list before the members had a choice—that is, if there was a choice, as especially in early years, there was at times only a single option to select.[48] Consequently, in 1922, there was a protest with reference to the violation of perceived democratic principles, after which elections became more meaningful.[49] In another attempt to progress toward democratic ideals, in this case transparency of the board's work, the ACAE started to publish the board's meeting minutes.[50]

The ACAE required a printed medium to print its business and to record its history,[51] to spread news concerning the profession, and to formalize and concentrate social and professional exchange. As Communications Minister Wu Yulin 吳毓麟 (1871–1944) noted, it would be through associations like the ACAE that techno-scientific communities in China finally obtained a platform to transmit, exchange, and conserve their ideas.[52] The platform was the *Journal of the Association of Chinese and American Engineers* (*Zhong Mei gongchengshi xiehui yuekan* 中美工程師協會月刊), published between 1920 and 1941. The ACAE usually issued ten numbers per year, although some variation occurred. Temporarily supertitled *The Oriental Engineer* between 1926 and 1927, it was an English-language publication that informed its members and other private and institutional subscribers of recent developments in the field of engineering. It was primarily a platform for its members to publish articles about their current projects. Other regular sections included book reviews, news about new job assignments of its members, tenders, and career opportunities. The journal editors further offered to print job searches for its members, to recommend job-seeking engineers to its advertisers, and to provide interested members with information about ways to visit or study in the United States.[53] Finally, like the conferences, the journal aimed to establish the ACAE's reputation among outside readers,[54] who could subscribe to it without ACAE

47 Akin, *Technocracy and the American Dream*, 15; Layton Jr., *The Revolt of the Engineers*, 15.
48 JACAE, 3, no. 7 (1922): "Minutes of the Board of Directors," August 8, 1922, 39.
49 OJTP, 13A.2: Wu Zhuopan to Thomas Meloy, August 18, 1922; 13A.2: Todd to Wu Zhuopan, August 22, 1922.
50 OJTP, 13A.2: Todd to Shao, March 16, 1936.
51 JACAE, 19, no. 1 (1938): Oliver J. Todd, "Engineering Experience in China during the Last Two Decades," 28). See also Layton Jr., *The Revolt of the Engineers*, 16.
52 JACAE, 5, no. 10 (1924): Yuling Wu, ACAE Convention Address, 15.
53 JACAE, 4, no. 6 (1923): Backend; JACAE, 4, no. 9 (1923), 38.
54 OJTP, 13A.2: Todd to Yang Baoling, October 20, 1929.

membership. The journal was volunteer work but seen as a serious duty. As secretary of the ACAE and editor of the *Journal*, every Friday, Joseph Henry Ehlers (1892–1988) traveled from Beidaihe to Beijing to work on association matters over the weekend.[55]

In 1920, when the ACAE published its first *Journal*, the publication of the CIE, the *Zhonghua gongchengshi hui huikan* 中華工程師會會刊 ("Periodical of the Chinese Engineers' Society"), had already been running for seven years. Like the ACAE magazine, the CIE journal featured discussions, proposals, and research.[56] However, like the way the associations differed from one another in their membership makeup, the ACAE *Journal* offered a more focused editorial direction informed by US civil engineering, including articles from Chinese members influenced by this perspective. Many articles echoed contemporary US engineering discourse. For instance, ACAE member Walter C. Lowdermilk (1888–1974) led the discussion on reforestation,[57] then an important point of discussion in the United States.[58]

While the basic functions of the journal remained, the following years saw some conceptual experiments. Originally, the editors expressed their vision for the magazine as a platform to catch up with the current projects of colleagues. The journal was explicitly not to publish articles of "a highly scientific nature."[59] In 1925, the directors decided to make the journal "more popular and extend its circulation."[60] During this period, the journal featured lightweight articles about ancient architectural marvels in China and, for a short time, ran under the new title *The Oriental Engineer*. Later, articles of a more technical nature returned. Despite some highly active American authors and a steady stream of Chinese-authored articles, the number of original articles never sufficed to fill the journal. The publication committees had to fill the *Journal* with reprints of selected articles from American engineering publications.

55 Ehlers, *Far Horizons*, 50.

56 Fáng, *Jìndài gōngchéngshī qúntǐ de 'mínjiān lǐngxiù'*, 8–10, 46.

57 JACAE, 6, no. 4 (1925): Walter C. Lowdermilk, "Some Practical Possibilities in Forestry for China," 30–36. Also JACAE, 1, no. 2 (1920): Gongchuo Ye, Address at First General Meeting of the ACAE, 24; *The Conference Daily*, "Freeman Sees Great Future for China," September 11, 1923; *The Conference Daily*, "Waddell Proposes Concrete Program for China Regeneration," September 12, 1923.

58 Hays, *Conservation and the Gospel of Efficiency*, 27–48.

59 JACAE, 5, no. 12 (1924), 2.

60 JACAE, 7, no. 1: Jin Tao, "The November Meeting of the Association," October 28, 1925, 8.

ACAE Secretary Oliver J. Todd (chapters 2, 5, 6) explained that the *Journal* was in English so that foreign libraries would subscribe to it.[61] The English language of the *Journal* had two outcomes. Taking part in ACAE activities, including the authoring and reading of articles in the *Journal*, was a symbolic performance that established one's place within the English-speaking international academic community.[62] Joining an annual conference was not exclusively about socializing, and authoring an article in the *Journal* was not about compensation. Rather, to publish in the *Journal*, which was sent to libraries worldwide, meant to set oneself off against the pool of domestic engineers and to make the claim of belonging to the privileged group of Sino-American professionals. Seen from a different angle, the language choice revealed the power hierarchy between the American and Chinese members. While members of both countries assumed positions equally without an explicit quota, and the sources of their friendly relations left no trace of racial discrimination, American engineers still expected their Chinese counterparts to read English, not vice-versa. It is, therefore, not surprising that a few energetic foreign members, such as J. A. L. Waddell (chapters 2, 6) and Todd, regularly delivered articles. They generated an output that Chinese writers could not catch up to, even though Chinese members consistently sent in articles (Figure 15).

On the one hand, the *Journal*'s practical value must have been higher for the foreign readership, as those members able to read Chinese would already have known of recent engineering developments, job offers, and tender postings through newspapers and publications of the Chinese engineering associations, such as the CIE *Periodical*. On the other hand, the *Journal* gave Chinese readers convenient, selected access to US engineering theory. Articles discussing engineering ethics or the ideal setup of engineering schools were often reprints from American professional associations, which the US members may already have read via other channels. Although at least 26 Chinese engineers eventually became overseas members of the ASCE,[63] and at least some of them sporadically read the ASCE's influential *Proceedings*,[64] their involvement remained passive. Therefore, reprints of these articles would have

61 JACAE, 19, no. 1 (1938): Oliver. Todd, "Engineering Experience in China during the Last Two Decades," 28.

62 Shen observes this for the geologists: Shen, *Unearthing the Nation*, 91–92.

63 JACAE, 4, no. 10 (1924): J. A. L. Waddell, "The Functions of Both Pure and Applied Science in the Future Development of China," 17. By 1940, Jin Tao, Tan Zhen, Xue Zhuobin, and Yang Baoling had become members of the ASCE (OJTP, 31.03: Todd to Zhang Jichun, May 3, 1940).

64 OJTP, 31.03: Zhang Jichun to Todd, February 20, 1940.

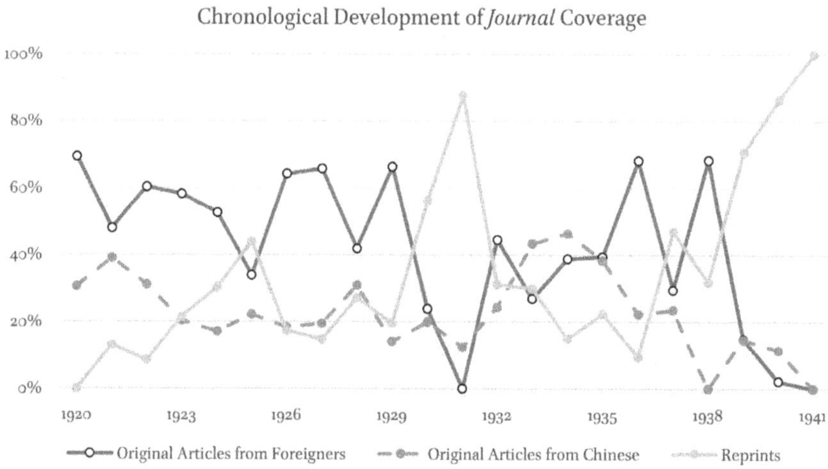

FIGURE 15 Chronological development of *Journal* coverage (own survey), by page count. Articles without named authors excluded.

catered more to the Chinese readership. In any case, both the *Journal* portions catering to foreign readers and the portions catering to Chinese readers display the same hierarchical structure of sending and receiving: The editors selected the engineering news because US engineers could profit from them, and they included the reprints as a form of readers' digest through which they could make the Chinese audience aware of professional ideals and theories popular in the US. In other words, while the *Journal*'s two-sided approach in content catered to the binational ACAE membership base, the underlying agency was primarily American.

3 Ethics and Socialization

Professionalization theories see the functions of professional associations in spreading scientific knowledge, molding networks, and forming rules of conduct.[65] As to the latter, professions subscribe to ethics as a collective, but the codes focus on the normative behavior of the professional individual. To describe forms of disciplined conduct, Foucault reintroduced the term "care

65 A classic editorial volume to this regard is John Jackson, ed., *Professions and Professionalization,* Sociological Studies 3 (Cambridge: Cambridge University Press, 1970).

of the self."[66] He defined the underlying techniques as "intentional and voluntary actions by which men not only set themselves rules of conduct, but also seek to transform themselves, to change themselves in their singular being, and to make their life into an *oeuvre*."[67] Modern self-care, though now spread through media instead of independently found through meditation, has conserved some of its qualities: teacher–disciple relations transmit ideals within an institutionalized environment. For the ACAE *Journal* authors, self-care was a group-marker of the professional—a regulated path that the ideal engineer was ought to follow. The ACAE's obsession with the "care of the self" highlights the entangled relationship between the group and individual conscience that leads to the formation of the professional subject.[68]

A broad range of articles imagined the main professional duty of engineers to be service to the people, the country, or the "nation."[69] This would require engineers to voluntarily dismiss personal, pecuniary benefits. While the engineer's work was material, their rewards were supposed to be immaterial, idealistically manifested in the country's wealth and welfare.[70] In the development of the country, one of the engineers' main duties was to use their inventions to turn poverty into prosperity.[71]

---

66 Michel Foucault, *The Care of the Self*, The History of Sexuality 3 (New York: Pantheon Books, [1984] 1986), 43–68.

67 Michel Foucault, *The Use of Pleasure*, The History of Sexuality 2 (New York: Vintage, [1984] 1990), 10–12.

68 The analysis abstracts positions voiced by various individuals to construct a common ideology of the ACAE. Naturally, the disadvantage of this method is that it generalizes nuanced positions. However, discordance between the authors was extremely rare. There was also no noticeable change in ideology as the two decades progressed. Therefore, the synoptic analysis emulates the ensemble of sentiments members received when reading the *Journal* in its entirety over the years of its publication.

69 JACAE, 1, no. 2 (1920): Jingyang Kuang, Address at First General Meeting of the ACAE, 3; 1, no. 2 (1920): Xiling Xiong, Address at First General Meeting of the ACAE, 22; 2, no. 4 (1921): Jingyang Kuang, Address at ACAE Annual Spring Meeting, April 6, 1921, 11–12; 2, no. 5 (1921): Gongchuo Ye, Opening Speech at Ninth Through Traffic Conference, May 2, 1921, 27; 3, no. 4 (1922): Deqing Yan, Address of the President of the Association, Shanghai, April 4, 1922, 4; 4, no. 1 (1923): Enhong Gao, "How to Improve the Engineering Works of China," ACAE Convention, Beijing, November 1, 1922, 11; 5, no. 2 (1924): Gongchuo Ye, "The Outlook of Communications in China," 5–7, 19; 5, no. 11 (1924): Julean Arnold, "Railways—China's Greatest Need," 61–67; 18, no. 5 (1937): Oliver J. Todd, "Future Work for the Engineer in China," 301; 19, no. 3 (1938): Robert Ridgway, "The Engineer in Public Life," 204–205; 22, no. 3 (1941): Daniel W. Mead, "Standards of Professional Relations and Conduct," 104.

70 JACAE, 2, no. 1 (1921): Philip Walter Henry, "A Few Remarks on Cooperation," 4; 6, no. 10 (1925): John F. Stevens, Address at Macky Auditorium, May 22, 1925, 87.

71 JACAE, 2, no. 10 (1921): Jingyang Kuang, Address at ACAE Annual Spring Meeting, April 6, 1921, 12; 2, no. 10 (1921): Xiling Xiong, Address at ACAE Fall Meeting, Hankou, October

Apart from the duty to serve the development of the country, Chinese and American authors were eager to stress the cosmopolitan worldview of the profession.[72] The maxim "Above all national interest is service to humanity," which Minister of Communications Zhang Zhitan 张志潭 (1883–1946) voiced in an ACAE address, closely mirrors the Cosmopolitan Club motto (chapter 1) "Above all nation there is humanity."[73] In this context, one of the recurring themes in normative ACAE articles was that of cooperation. This cooperation specifically hinted at transnational cooperation between American and Chinese engineers,[74] but it also transported the broader idea of a team spirit in professional work life and a "loyalty" to the profession.[75]

Professional behavior was not limited to one's job. It was an approach to life, a "care of the self" which required strict control over one's actions and hygiene of the body. There was no emic translation of the term, but the idea transpired in ideals of conduct, discipline, and exercise. Waddell reminded the Chinese readership to "limit liquor and tobacco" and the extent of vacations.[76] Likewise, a US reprint on the "standards of professional relations and conduct" included the demand to stay away from gossip and the "minor pleasures of life" in order to "budget one's time."[77] Another reprint further explained to the readers that qualities such as initiative, judgment, dependability, self-assurance,

5, 1921, 9; 3, no. 4 (1922): John R. Freeman to John A. Ely, March 31, 1922, 7; 5, no. 6 (1924): John F. Stevens, "The Problem of China's Transportation," 2; 6, no. 10 (1925): William J. Showalter, "The Challenge of the Future to the Scientist and the Engineer," 67; 9, no. 1 (1928): H. Chatley, "Engineering and Economics," 22; 12, no. 6 (1931): J. A. L. Waddell, Presentation of First Clausen Award, March 11, 1931, 13; 17, no. 1 (1936): Earle B. Norris, "Research as Applied to Engineering," 62; 18, no. 5 (1937): Oliver J. Todd, "Future Work for the Engineer in China," 300.

72 JACAE, 4, no. 10 (1923): J. A. L. Waddell, "The Functions of Both Pure and Applied Science in the Future Development of China," 21; 18, no. 4 (1937): Middleton C. A. Smith, "Engineering Development in China," 257, 260.

73 JACAE, 2, no. 10 (1921): Zhitan Zhang, Address at ACAE Fall Meeting, Hankou, October 5, 1921, 8.

74 JACAE, 1, no. 2 (1920): Jingyang Kuang, Address at First General Meeting of the ACAE, 3. See also OJTP, 30.03: Arthur M. Shaw to Todd, January 27, 1931.

75 JACAE, 2, no. 4 (1921): Xiling Xiong, Address at ACAE Annual Spring Meeting, April 5, 1921, 34; 2, no. 10 (1921): Xiling Xiong, Address at ACAE Fall Meeting, Hankou, October 5, 1921, 10; 7, no. 7 (1926): Oliver J. Todd, "Modern Highways in China," Chinese Engineering Society, Beijing, August 24, 1926, 22.

76 JACAE, 6, no. 3 (1925): J. A. L. Waddell, "The Necessity for the Development and Application of the Prin-ciples of Economics in All Lines of Engineering Work," 64.

77 JACAE, 22, no. 3 (1941): Daniel W. Mead, "Standards of Professional Relations and Conduct," 99, 111–113.

personality, good health, and appearance were important.[78] Despite the engineers' unhappiness with manual workers, physical endurance was another quality of the ideal engineer.[79] Waddell recommended temporarily accepting "unworthy" positions of the "lower grade of the laboring class" to gain practical experience.[80] That young Chinese engineers paid manual laborers to carry their instruments and climb trees baffled Todd.[81] He urged his peers to adopt "a policy of 'coats off and hard work'" to "prove our right to be called professional men."[82] In this respect, he later recounted how he had climbed a tree to demonstrate to young engineers that they needed to use their muscles.[83]

These disciplined self-care values are mirrored in the *Journal*'s descriptions of "remarkable men." Although (in)visible colleges are a collective enterprise and serve to attract new members into the profession, they also utilize reputable, monumental "leaders" as guiding, archetypal figures and as network brokers. In their position, these figures have the personality and the reputation that people require to mold a profession's paradigms.[84] In network analysis terms, they are the nodes that tied the Sino-American engineers to the wider invisible college of global civil engineering. Honorary members such as John Ripley Freeman (chapter 2), John Frank Stevens and Wang Jingchun (both chapter 4), or government consultant J. A. L. Waddell (chapters 2, 6) may not have had the strongest ties within the Sino-American engineering community on the ground (Figure 13). Nonetheless, they were eminent figures of more recent engineering history and recognized in US institutions like the American Society of Civil Engineers (ASCE).

The descriptions of these individuals give clues to the engineering ideals of the association. In an early 1920 address, Yang Baoling praised Wang Jingchun as "upright, capable, energetic and progressive, and most faithful and efficient."[85] Likewise, another article called Waddell "the prototype of the ideal

---

78 JACAE, 18, no. 5 (1937): Harrison P. Eddy, "What an Employer Looks for in a Young Engineer," 333–334.

79 JACAE, 2, no. 10 (1921): Jingyang Kuang, Address at ACAE Fall Meeting, Hankou, October 5, 1921, 6; 19, no. 1 (1938): J. A. L. Waddell, "China's Needs for Proper Development," 1, 19.

80 JACAE, 12, no. 7 (1931): J. A. L. Waddell, "Engineering," 4.

81 JACAE, 19, no. 1 (1938): Oliver J. Todd, "Engineering Experience in China during the Last Two Decades," 31. See also Will, "The Emergence of the Modern Civil Engineer in China, 1900–1940," 3:105–7; Köll, "The Making of the Civil Engineer in China," 157.

82 OJTP, 29.13: Todd to J. A. L. Waddell, April 4, 1929.

83 Todd, *The China That I Knew*, 17–18.

84 Crane, *Invisible Colleges*, 88–89. Honorary members are briefly discussed in Wu, *Měiguórén wèijìng de Zhōngguó mèng*, 257–58.

85 JACAE, 1, no. 2 (1920): Baoling Yang, Chair at First General Meeting of the ACAE, 19–20.

engineer," similarly citing his generous, noble, and cooperative character.[86] Freeman was self-aware of his status when he sent Yang a portrait of himself for the office, claiming that it showed "the personality of the engineer."[87] After his passing, he received a memorial issue (the only one in the *Journal*'s history). Following a dedicated poem from former Chinese Prime Minister Xiong Xiling, five Chinese and American authors stressed the qualities they found in Freeman and thought to be exemplary. Apart from personal characteristics such as kindness and goodwill, they focused on workplace morals. They described Freeman as someone who "worked without pause," as "a dynamo of energy," someone who "hardly stopped to eat and sleep," and who worked from day to night without observing holidays.[88] These descriptions illustrate the level of work devotion the engineers found commendable.

The ACAE members continued the distinction strategy seen a decade earlier on American campuses (chapter 1): The aspiring profession made use of what Bourdieu described as an effort to distance oneself from perceived lower groups through behavior and taste.[89] Despite claims that "all classes of people" ultimately appreciated the engineers' work (chapters 2, 6),[90] portrayal of the working class and non-technical elites was across-the-board negative. Waddell demonstrated his disdain for the common people when claiming that altruism and unselfishness, values he found desirable for engineers, usually belonged "to the highly-educated class."[91] Peasants were deemed "ignorant" for opposing new engineering projects.[92] Ultimately, the authors believed themselves to be the avant-garde of a new scientific mode of truth, with all opposition stemming from either conspiracy or ignorance. In their mission to end the latter, they envisioned the engineer as a "high priest before the altar of progress,"[93]

86 JACAE, 19, no. 1 (1938): C. Calor Mota, "Dr. J. A. K. Waddell," 60.
87 PJRF, 128: Freeman to Yang Baoling, January 16, 1922.
88 JACAE, 14, no. 1 (1933): Xiling Xiong, Poem, ii; JACAE, 14, no. 1 (1933): Oliver J. Todd, "In Memoriam," 1, 3; 14, no. 1 (1933): Zhen Tan, "Memories of John R. Freeman," 10; 14, no. 1 (1933): Zhuobin Xue, "A Source of Inspiration," 13.
89 Pierre Bourdieu, *Distinction: A Social Critique of the Judgement of Taste* (Cambridge: Harvard University Press, [1979] 1984), 169–225, 260–317.
90 JACAE, 2, no. 4 (1921): Xiling Xiong, Address at ACAE Annual Spring Meeting, April 5, 1921, 34.
91 JACAE, 4, no. 10 (1923): J. A. L. Waddell, "The Functions of Both Pure and Applied Science in the Future Development of China," 21.
92 JACAE, 2, no. 4 (1921): Xiling Xiong, Address at ACAE Annual Spring Meeting, April 5, 1921, 34.
93 JACAE, 19, no. 4 (1938): Dewitt R. Jordan, "The Engineer," 234.

and imagined themselves as "missionaries of the cause" who must "preach" and "spread the gospel of strict honesty" to the public.[94]

Engineers not only distinguished themselves from artisans and workers, but also positioned themselves as new members of the upper class. ACAE President Samuel M. Dean's memories of graduation displayed the contradictory relationship between practical experience and the desire to attain an upper-class, gentlemanly standing. On the one hand, he presented a nostalgic memory of the "untamed, rough and tough" engineering student in the US that "soiled his hands" and "bent his back." On the other hand, though regretfully the new generation of engineering students abstained from physical exercise, it had attained a "clean" and "aristocratic" bearing that helped its professional standing.[95] Journal articles attempted to portray the ideal engineer as "noble" and as a "gentleman."[96] In one of several Herbert Hoover reprints, the future US president lamented that the public still contrasted between engineer and gentleman.[97] Alluding to the issue, articles promoted the study of history and politics,[98] and an ACAE member that worked as an educator argued that engineers needed to receive an education in classical Chinese literature in order to gain acceptance from established elites.[99]

The problem ran deeper and extended to the question of whether the young professionals were men at all. As the ACAE's first president, the eminent railway engineer Kuang Jingyang (chapter 2) remarked, within Chinese society, the engineer was a "new creature."[100] Accordingly, the authors first had to humanize the engineer as an individual. The scientific method appears to have given civil engineers the public image of a "technical machine,"[101] leading to the challenge to create professionals who were convincingly "*men* as well as

94 JACAE, 5, no. 11 (1924): Chongyou Wang, "Industrial Research," 8; OJTP, 29.13: Todd to J. A. L. Waddell, April 4, 1929.

95 JACAE, 19, no. 5 (1938): Sam M. Dean, "Rip Van Winkle Returns to an American Technical College," 272.

96 JACAE, 4, no. 1 (1923): Enhong Gao, "How to Improve the Engineering Works of China," ACAE Convention, Beijing, November 1, 1922, 11; 8, no. 9 (1927): Herbert Hoover, "A Tribute to Engineers," 22.

97 JACAE, 8, no. 9 (1927): Herbert Hoover, "A Tribute to Engineers," 22.

98 JACAE, 2, no. 2 (1921): Ernest P. Goodrich, "Internationalism in Engineering," 5.

99 JACAE, 4, no. 1 (1923): John A. Ely, "The Education and Training on Chinese Engineers," 19.

100 JACAE, 2, no. 4 (1921): Jingyang Kuang, Address at ACAE Annual Spring Meeting, April 6, 1921, 10.

101 JACAE, 2, no. 1 (1921): Philip Walter Henry, "A Few Remarks on Cooperation," 4.

engineers."[102] The problem was the engineer's *habitus* of quantifying, working efficiently, and materializing scientific "truth."[103]

Through the interwar years, the engineering profession in the US was in a deep crisis. Instead of establishing their role as the bearers of progress and welfare, the public increasingly saw engineers as the creators of the maleficent weaponry of the First World War and as the culprits of job cuts as machines started replacing human labor.[104] Despite China being only at the periphery of the war, and industrialization still in its infancy, the discussion spread to China through the ACAE.[105] Accidents, such as the fatal Jiaozhou–Jinan Bridge Collapse of 1923 or the collapse of the Gansu–Haizhou Railway Lingbao tunnel in 1926, may have added to the perception that engineers held grave responsibilities.[106]

A code of ethics is what makes an occupation become a profession: it creates a paradigm, a contract with society, and a service ideal between the professionals.[107] However, historically, engineering professionals did not attach high importance to a written-out code: the ASCE did not adopt a code of ethics until 1914, and the code remained largely symbolic and irrelevant.[108] In an effort to emulate American engineering society, in 1927, President Li Shutian called to prepare an ACAE "Code of Ethics."[109] Although such a codex apparently never materialized, the call shows that the ACAE, including its Chinese leadership, attached importance to engineering ethics. The CIE eventually published the earliest written professional code in 1932.[110] "The Engineers' Creed" had its

102 JACAE, 4, no. 1 (1923): John A. Ely, "The Education and Training on Chinese Engineers," 20.

103 JACAE, 6, no. 10 (1925): John F. Stevens, Address at Macky Auditorium, May 22, 1925, 82; JACAE, 8, no. 9 (1927): Herbert Hoover, "A Tribute to Engineers," 22–23.

104 Akin, *Technocracy and the American Dream*, 156–63; Layton Jr., *The Revolt of the Engineers*, 228–29.

105 JACAE, 5, no. 10 (1924): John H. McCracken, "The Education of the Engineer," 49; 5, no. 11 (1924): Chongyou Wang, "Industrial Research," 3; 11, no. 10 (1930): J. F. Coleman, "Reflections on the Status of the Engineer," 24; 18, no. 4 (1937): J. K. Finch, "The Social Responsibilities of the Engineer," 278, 280.

106 JACAE, 4, no. 2 (1923): P. H. Chen, "The Kiaochao-Tsinan Railway Brid[g]e Accident," 26–28; 11, no. 11 (1930): Notes, 42. See also JACAE, 12, no. 6 (1931): Frank W. Skinner, "Service Brings Reward," 19–24.

107 Davis, *Thinking Like an Engineer*, 49–50.

108 Layton Jr., *The Revolt of the Engineers*, 109–14.

109 JACAE, 8, no. 4 (1927): Shutian Li, "Message from the President," 233.

110 Fáng, *Jìndài gōngchéngshī qúntǐ de 'mínjiān lǐngxiù'*, 76–79; Nanyuan Cao, Junbin Su, and Mingyan Hu, "Ethical Awareness in Chinese Professional Engineering Societies: Textual Research on Constitutions of Chinese Engineering Organizations," in *Philosophy and Engineering: Reflections on Practice, Principles and Process*, ed. Diane Michelfelder,

members identify as a profession that serves a national cause and that internalizes work ethics transcending the interests of a profit-seeking occupation.[111]

As for responsibilities when planning and handling a project, the failure of the Grand Canal improvement led to a rethinking of the scope of projects engineers should take on in China. Instead of pursuing "white elephants"—vast infrastructural projects of symbolic character but without sufficient economic foundation—eminent figures like Minister of Communications Ye Gongchuo (chapter 3) and Waddell urged to start small and plan realistically.[112] The ideal of economic efficiency, waste reduction, and, in their own words, "scientific management,"[113] was a reoccurring ideal, especially promoted by the American members aware of similar discussions in the US.[114] However, Chinese members also used the paradigm of secure and efficient management, for instance, when they argued that engineers had to know about the situation on the ground they planned to work on. This meant that Chinese engineers would slowly replace foreign engineers, but it also meant that Chinese engineers from the metropolitan areas were to get accustomed to the country's interior (chapters 5, 7).[115]

As for the goals and ideals of the technocracy movement, a US movement that promoted political governance based on techno-scientific rationales,[116]

Natasha McCarthy and David Goldberg (Dordrecht, Heidelberg, New York, London: Springer, 2013), 208–9; Köll, "The Making of the Civil Engineer in China," 170.

111 Zhōngguó gōngchéngshī xuéhuì, ed., *Zhōngguó gōngchéngshī xuéhuì sānshízhōu jìniànkān: Sānshí niánlái zhī Zhōngguó gōngchéng*, 2nd ed. (Nanjing: Guó gōngchéng chūbǎn gōngsī, [1946] 1948), 1.

112 JACAE, 2, no. 11 (1921): Gongchou Ye, Message for ACAE Fall Meeting, Hankou, October 5, 1921, 26; 3, no. 7 (1922): J. A. L. Waddell, "Report upon a Project for Bridging the Yangtze River between Cities of Nanking and Pukow," 2; 4, no. 10 (1923): J. A. L. Waddell, "The Functions of Both Pure and Applied Science in the Future Development of China," 17–18; 18, no. 4 (1937): J. K. Finch, "The Social Responsibilities of the Engineer," 279–283.

113 JACAE, 2, no. 10 (1921): Zhitan Zhang, Address at ACAE Fall Meeting, Hankou, October 5, 1921, 8; 2, no. 11 (1921): Shichang Xu, Address at ACAE Fall Meeting, Hankou, October 5, 1921, 24.

114 JACAE, 2, no. 3 (1921): E. J. Corkery, "Modern Manufacturing Methods in China," 12; 4, no. 10 (1923): Oliver J. Todd, "China's Call to the Engineer," 24; 6, no. 3 (1925): J. A. L. Waddell, "The Necessity for the Development and Application of the Principles of Economics in All Lines of Engineering Work," 64, 66; 6, no. 9 (1925): John F. Stevens, "The Relation of the Engineer to the Development of New Lines of Railways," 2; 9, no. 1 (1928): H. Chatley, "Engineering and Economics," 20; 18, no. 4 (1937): J. K. Finch, "The Social Responsibilities of the Engineer," 282.

115 JACAE, 1, no. 2 (1920): Gongchuo Ye, Address at First General Meeting of the ACAE, 24; 2, no. 10 (1921): Jingyang Kuang, Address at ACAE Fall Meeting, Hankou, October 5, 1921, 6.

116 See Akin, *Technocracy and the American Dream*; Layton Jr., *The Revolt of the Engineers*.

the ACAE members supported the call on principle.[117] However, ACAE President Kuang was careful to emphasize that although engineers worked for the benefit of the country, they should not take part in its governance.[118] A main source of discontent was the experience with nonexpert authorities managing and administering engineering facilities. Todd likened the situation to doctors turning hospitals over to nonmedical staff.[119] The frustration about bureaucratic management led to the notion that society underappreciated the engineer's role. The engineers understood that "unadulterated glory" was not enough of a return when a low salary appeared to signal a lack of appreciation from the employer, which in the case of costly infrastructural projects was the government.[120] In some contradiction to the ideal of laboring for the sake of the country and not for financial returns, there was a sense of bitterness when articles claimed, for instance, "the world owes a debt of gratitude it never can pay."[121] In this vein, Stevens lamented that society had not recognized the engineer's "importance which it justly deserves,"[122] and Todd looked forward to a time in which the engineering profession would receive "the standing it deserves."[123] Treatment was especially unfair because, as the authors reassured each other, the engineer had "on his shoulders the greatest burden of all"[124] and helped the world to keep "moving in the right direction,"[125] while an ignorant society did not know the difference between "a cow and a kilowatt."[126]

Sharing one's discontent in the *Journal* was not about reaching out to the public and asking for appreciation. The repeated assertion that engineers shoulder a heavy responsibility, or that few people properly acknowledge

117 JACAE, 5, no. 10 (1924): J. A. L. Waddell, "The Engineer of the Future," October 3, 1924, 45; 12, no. 6 (1931): J. A. L. Waddell, Presentation of First Clausen Award, March 11, 1931, 13; 19, no. 1 (1938): J. A. L. Waddell, "China's Needs for Proper Development," 1. See also JACAE, 18, no. 4 (1937): Middleton C. A. Smith, "Engineering Development in China," 257.

118 JACAE, 2, no. 4 (1921): Jingyang Kuang, Address at ACAE Annual Spring Meeting, April 6, 1921, 11.

119 JACAE, 19, no. 1 (1938): Oliver J. Todd, "Engineering Experience in China during the Last Two Decades," 36.

120 OJTP, 37.02: Huang Rongde to Todd, February 7, 1928; 30.02: Todd to D. W. Edwards, November 19, 1932; 30.05: Todd to Murray Sullivan, November 7, 1933.

121 JACAE, 6, no. 10 (1925): William J. Showalter, "The Challenge of the Future to the Scientist and the Engineer," 67.

122 JACAE, 7, no. 1 (1926): John F. Stevens, "The Locating of Railways," 12. Similar in JACAE, 2, no. 10 (1921): Jingyang Kuang, Address at ACAE Fall Meeting, Hankou, October 5, 1921, 6; 11, no. 10 (1930): J. F. Coleman, "Reflections on the Status of the Engineer," 23.

123 JACAE, 17, no. 1 (1936): Oliver J. Todd, "As an Engineer Sees America in 1935," 50–51.

124 JACAE, 18, no. 5 (1937): Oliver J. Todd, "Future Work for the Engineer in China," 301.

125 JACAE, 17, no. 1 (1936): Oliver J. Todd, "As an Engineer Sees America in 1935," 50–51.

126 JACAE, 8, no. 9 (1927): Herbert Hoover, "A Tribute to Engineers," 22–23.

their (at first highly remunerated) "service" to the public, had little potential to motor change when kept inside the profession's echo chamber. Instead, the primary function of the articles appears to have been the creation of a shared professional consciousness. This included the promotion of techniques of the self, such as discipline and sacrifice, and the shared identification with a tragic figure who devotes himself to the public without receiving the recognition he or she deserves.

4 Career Opportunities

Already in his student days, Wang Jingchun had stressed the hierarchical relationship between young intellectuals who received their education in the US on the one hand, and in China or in Japan on the other, as he described US-returned students to be "more competent for, and better adapted to work; and they are found more sober."[127] Applicants who graduated in civil engineering from the US between 1905 and 1918 often assumed positions at railroad companies (70%), ministries and national commissions (52%), schools and colleges (33%), hydraulic works (30%), companies (24%), and provincial and municipal offices (21%).[128] The graduates did not confine their workplace to their native place but moved within a broader territory that opened up spaces west of and in between the original hubs of Guangzhou, Hankou, Shanghai/Nanjing, and Beijing/Tianjin.[129]

Civil engineers from US colleges were much more likely to assume high ranks than those graduating from European or Chinese institutions (table 4). For instance, compared to domestic degree holders, they were five times more likely to become directors general or chief engineers. In contrast, domestically trained engineers often worked in the capacity of assistant engineers (*fu gongchengsi* 副工程司, *bang gongchengsi* 幫工程司). Civil engineering graduates with a US degree also attained professorships almost three times more often than their Europe- and China-trained counterparts. Supported by US textbooks and inspired by US curricula, English remained the language of instruction for

127 CSM, 3.4 (1908): Jingchun Wang, "How Should the Refunded Indemnity Be Used?," 148.

128 Pelzer et al., CERD, [{"t":"10"}, {"linked_to:940": [{"t":"58"}, {"f:9":"<=1918"}, {"f:9":">=1905"}, {"f:997":"9483"}, {"linked_to:964": [{"t":"57"}, {"linked_to:238": [{"t":"12"}, {"f:26":"119"}]}]}]}, {"linked_to:942": [{"t":"61"}, {"linked_to:21": [{"t":"4"}, {"any": [{"f:1":"[學 / 鐵道|鐵路 / 委員會|交通部 / 河工程|水 / 建設廳|市 / 公司]" }]}]}]}].

129 Pelzer et al., CERD, [{"t":"12"}, {"f:26":"109"}, {"linkedfrom:134": [{"t":"4"}, {"linkedfrom:21" :[{"t":"61"}, {"linkedfrom:942": [{"t":"10"}, {"linked_to:940": [{"t":"58,59"}, {"f:9":"<=1918"}, {"linked_to:964": [{"t":"57"}, {"linked_to:238": [{"t":"12"}, {"f:26":"119"}]}]}]}]}]}]}].

TABLE 4 Comparison of the most prominent positions held

Job description appearing in CV	US degree	European degree	Domestic degree
Department Head (*chuzhang* 處長)	35%	21%	5%
High-Ranking Technician (*jizheng* 技正)	32%	21%	10%
Intern (*shixi* 實習)	31%	21%	8%
Director General (*juzhang* 局長)	23%	14%	5%
Professor (*jiaoshou* 教授)	21%	7%	5%
Section Chief (*kezhang* 科長)	21%	7%	7%
Chief Engineer (*zong gongchengshi* 總工程師)	18%	7%	4%

prospective engineers well throughout the 1920s.[130] The appointment of foreign professors became unpopular amidst a growing "anti-imperial" discourse, and foreign staff gradually left China due to insufficient salaries (chapter 6). Consequently, US-trained engineers were the natural choice to gradually replace foreign academic staff and continue their legacy. Most of them did not remain in academia for their lifetime but used college appointments as career entries or as a stopgap between productive work assignments. Conversely, domestic degree holders often remained lower-ranking lecturers (*jiaoyuan* 教員).[131] In summary, the positions ACAE members assumed during their career confirms the notion of the Sino-American segment as a privileged group within the civil engineering profession.

130 Hóngxūn Líng and Zǔkāng Zhào, "Gōngchéng jiàoyù diàochá tǒngjì zhī yánjiū," *Gōngchéng: Zhōngguó gōngchéng xuéhuì huìkān* 2, no. 4 (1926): 248–49; Shūfāng Wèn and Yǒuguāng Huò, "Yè Gōngchuò duì Jiāotōng dàxué de gǎigé jí yǐngxiǎng," *Jiāotōng gāojiào yánjiū* 3 (2004): 22–23.

131 Pelzer et al., CERD, [{"t":"10"}, {"linked_to:940":[{"t":"58"}, {"f:9":"<=1918"}, {"f:9":">=1905"}, {"f:997":"9483"}]}, {"linked_to:942":[{"t":"61"}, {"any":[{"f:983":"總工程師|處長 / 副工程司|幫工程司 / 教授 / 教員"}]}]}].

5 Network for Collaboration

While the general privilege of US-trained engineers can be explained by greater professional reputation and language expertise, the ACAE functioned as a network to increase job opportunities further. For workplaces that employed ACAE members, the share of ACAE civil engineers within the workforce was more than twice as high (2.7%) as the ACAE share among the total pool of Republican-era civil engineers with documented employments (1.3%).[132] In other words, the members formed clusters based on their mutual educational upbringing and professional organization.

ACAE members had shared career paths since before the association's inauguration. For instance, Harry A. Raider (b. 1880), the ACAE's long-term treasurer, had come to China at the request of the association's first president Kuang in 1909, when they started working together on the Guangzhou–Hankou Railway. Later, the future ACAE presidents Sa Fujun and Wang Jinzhi 王金職 (Jick G. Wong, 1892–1939) worked under Raider, and in 1921 Raider, together with ACAE executives Yang Baoling, Shi Zhaoxiang 施肇祥 (S. C. Thomas Sze, 1881–1961), and Murray Sullivan, formed the Chinese Engineering & Development Company.[133] Likewise, Tan Zhen, Oliver J. Todd, and their colleague Zhou Yichun 周詒春 (Y. T. Tsur, 1883–1958) attempted to jointly enter the peppermint oil industry.[134] In another example of patronage, in 1927, the support of ACAE honorary member Xiong Xiling enabled Yang Baoling to improve the Yongding River, which the Beiyang government had previously failed to upkeep.[135]

Similar to how traditional Chinese gentry tried to preserve their status through generations of civil examination candidates, parts of the US-educated elite soon forged their own cliques and lineages. Chinese students urged their peers to send their "sons & grandsons."[136] Early examples of such lineages include the Yale electrical engineering student and later ACAE secretary Zhong

132 Pelzer et al., CERD, [{"t":"10"}, {"linked_to:940": [{"t":"58"}, {"f:997":"9483"}]}, {"linked_to:942": [{"t":"61"}, {"linked_to:21": [{"t":"4"}, {"linkedfrom:21": [{"t":"61"}, {"linkedfrom:942": [{"t":"10"}, {"f:1000":"9547"}]}]}]}]}].

133 JACAE, 6, no. 7 (1925), 1; PJRF, 125: Yang Baoling to Freeman, May 18, 1921 Sullivan eventually left the ACAE in disarray following discrepancies in his accounting (PJRF, 129: Yang Baoling to Freeman, August 30, 1922).

134 OJTP, 02.01: Todd to H. B. Elliston, December 28, 1927; 02.03: C. F. Chou to Todd, August 27, 1929; 33.06: Todd to Tan Zhen, November 22, 1927; 33.06: Todd to Tan Zhen, December 2, 1932; 31.03: Todd to Zhang Jichun, January 10, 1938.

135 PJRF, 129: Wei Yi to Freeman, February 17, 1927; Jonathan Spence, *To Change China: Western Advisers in China* (New York: Penguin, [1969] 1980), 212.

136 HCSR, September 29, 1909.

Guidan 鍾桂丹 (Daniel McClean Chung, 1889–1956), who was a son of Yale alumnus and railway administrator Zhong Ziyuan 鍾紫垣 (Mun-yew Chung, 1861–1945), and a nephew of Yale alumnus and legendary railway engineer Zhan Tianyou.[137]

The above-discussed senior, honorable members proved especially helpful in securing employment. For instance, when the ACAE honorary member Waddell acted as a consultant for the Ministry of Communications, he had ACAE members Ling Hongxun and Thomas Meloy assist him.[138] A decade later, he invited Ling to join the Ministry of Railways (chapter 7).[139] Similar to Waddell's support, Freeman secured Xue Zhuobin's (chapter 2) employment at the Wusong Port Development Administration Commission (*Wusong shangbu ju* 武松商埠局).[140] Xue and Tan Zhen (chapter 2) then mediated their own friends and assistants to Freeman.[141] Freeman invited them to study at his office in Providence, prepared detailed travel plans, and introduced them to his various engineering contacts in the US and Europe.[142] Tan also introduced his student Zhang Jichun 張季春 (C. C. Chang, 1902–1946) to ACAE secretary Todd, who immediately hired him (chapter 5).[143] The ACAE may sometimes have acted as an alternative to bureaucratic coterie. A junior member of the association, Liu Rusong 劉如松 (J. S. Liu, b. 1901), lamented that he had "nobody to help [him] from the top"—and instead asked Todd to help him secure a job in Ling Hongxun's railway office.[144] Although a junior–senior mentality existed, benefits were not unidirectional. Todd functioned as a patron for others but profited from Waddell who guided his admittance into the prestigious Institute of American Consulting Engineers and helped him get appointments with Minister of Railways Sun Ke 孫科 (1891–1973, chapter 6).[145] Another patron was Charles Richard Crane, who became an ACAE honorary member, not through professional prestige but *ex officio* as minister to Beijing. Crane, the son of an engineer, remained a financier of US–Chinese engineering work throughout

137 YARO, D. M. Chung.

138 JACAE, 2, no. 6 (1921): Notes, 8; Líng, *Qīshí zìshù*, 36.

139 Líng, *Qīshí zìshù*, 47.

140 PJRF, 125: Freeman to L. J. Buckner, November 22, 1920.

141 PJRF, 130: Xue Zhuobin to Freeman, November 7, 1922.

142 PJRF, 128: H. T. Wang to Freeman, May 10, 1921; 129: H. T. Wang to Freeman, May 15, 1922; 129: Freeman to Shi Zhaoji, July 3, 1923; 130: Freeman to M. T. Yu, June 27, 1923; 130: Letters from John R. Freeman, dating April 9, 1923, and June 27–28, 1923.

143 Todd, *The China That I Knew*, 59.

144 OJTP, 30.04: Liu Rusong to Todd, September 31, 1932.

145 OJTP, 29.13: J. A. L. Waddell to Todd, March 31, 1929; 31.03: Todd to Zhang Jichun, January 10, 1938.

his life. He sponsored irrigation work that ACAE members conducted in Yan Xishan's 閻錫山 (1883–1960) Shanxi domain, paid Todd's salary when the latter investigated the Yellow River floods in 1935, and also paid for the printing costs of Todd's and Sigurd Eliassen's lengthy "Yellow River Problem" article in the ASCE's *Proceedings*.[146]

The most important employer for ACAE members was the China International Famine Relief Commission (*Zhongguo Hua Yang yizhen jiuzai zonghui* 中國華洋義賑救災總會, CIFRC) (Figure 13).[147] There were other big employers, such as the Ministry of Communications, but while just a fraction of Ministry employees were members of the ACAE, the vast majority of CIFRC engineers were part of the association. Many ACAE members worked for the CIFRC, including Sigurd Eliassen, Huang Rongde 黃榮德 (Wing Tak Wong, d. 1932, chapter 5), Shao Fuwu 邵福旿 (b. 1897, chapter 5), Tan Zhen, Oliver J. Todd, Peng Binglu 彭炳祿 (Ben Luke Pond, b. 1891, chapter 5), Xue Zhuobin, Yang Baoling, and Zhang Jichun. The ACAE's merchant members, otherwise less involved with the association's activities, also profited from the network, as the CIFRC purchased survey instruments from Andersen, Meyer & Co. (*Shanghai shen chang yanghang* 上海慎昌洋行), where many members worked.[148] The circumstance that so many members of the inner circle depended on the CIFRC would prove a factor in the ACAE's demise (chapter 6).

6 Transnational Friendship

Apart from what could be described as a professional iteration of *guanxi*, the engineers grew private ties and cared for each other outside the realm of deal-making. For instance, ACAE secretary and *Journal* editor Ehlers described Waddell as a friend,[149] and Freeman composed his letters to Todd in a private language. Friendliness and intimacy were not limited to connections between white, American engineers. Waddell called Ling Hongxun a "highly-valued" friend and Kuang Jingyang among the "most lovable men."[150] Ehlers, in his memoirs, remembered his Chinese peers as "clever and brilliant, friendly and

146 JACAE, 20, no. 3 (1939): Oliver J. Todd, "A Friend of China–Hon. Chas. R. Crane," 101; Oliver Todd and Sigurd Eliassen, "The Yellow River Problem," *Transactions of the American Society of Civil Engineers* 105 (1940); Fisher, "Service to China," 37, 142.

147 See also Wu, *Měiguórén wèijìng de Zhōngguó mèng*, 288.

148 OJTP, 01.12: Todd to Clement Noble Mylne, December 18, 1926.

149 Ehlers, *Far Horizons*, 2.

150 JACAE, 11, no. 11 (1930): J. A. L. Waddell to the ACAE Editor, October 7, 1930, 48.

with a keen sense of humor."[151] After Freeman returned to Providence, he continued to provide his Chinese contacts with information about his health, hinting at a relatively intimate nature of acquaintanceship.[152] He became anxious when his former Chinese assistants stopped writing and inquired about their well-being.[153] Freeman described Yang Baoling as a "very likable fellow," as honest, and as having "an ear to the ground, an eye to the future."[154] Freeman's office displayed a photograph of him with Yang, together with the Canal's contractor William Francis Carey (1878–1951) and superior Pan Fu (chapter 2).[155] He found the Chinese engineers "full of zeal toward absorbing western scientific knowledge."[156] His continued letter exchanges with his former Chinese employees long after the end of his career show that Freeman grew comfortable and developed a personal relationship with his former staff.

An example of how the Sino–American acquaintanceships transgressed professional boundaries is Tan Zhen's social environment, especially the support it rendered during hardships. Tan was Freeman's favorite student and assistant, and later became an intimate friend. Normally, Freeman and his Chinese students addressed each other formally even after forming a friendly relationship, but Tan referred to all the Freeman family members by their first names.[157] Already in 1920, Tan wrote to Freeman that their relationship was "just as father is to son,"[158] and he later added that Freeman had treated him "kindly like a teacher and more than a teacher."[159] Tan's parents gifted Freeman an old mandarin garment,[160] and Tan sent him lacquer boxes with jujubes prepared by his wife.[161] He claimed Freeman taught him "things which cannot be obtained in school or any where," [*sic*] once again demonstrating the attention to the practical component of engineering education.[162] Freeman explained that Tan was "one of the brightest, nicest young men [he] have ever had."[163] In

151 Ehlers, *Far Horizons*, 39.

152 PJRF, 125: Freeman to Lin Quancheng, June 4, 1921; Freeman to Yang Baoling, June 11, 1921.

153 PJRF, 125: Freeman to Xue Zhuobin, June 10, 1921.

154 PJRF, 129: Freeman to Charles K. Edmunds, June 28, 1922.

155 PJRF, 130: Freeman to Yang Baoling, December 1, 1921.

156 PJRF, 129: Record of the Testimonial Dinner to Dr. Charles K. Edmunds, March 2, 1922; 129: Freeman to President and Director New England, Waterworks Association, Boston, June 13, 1923.

157 PJRF, 128: Tan Zhen to Freeman, January 8, 1920.

158 PJRF, 128: Tan Zhen to Freeman, January 8, 1920.

159 PJRF, 130: Tan Zhen to Freeman, October 13, 1922.

160 PJRF, 125: Freeman to Tan Zhen, January 4, 1922.

161 PJRF, 129: Tan Zhen to Freeman, March 14, 1926.

162 PJRF, 128: Tan Zhen to Freeman, January 8, 1920.

163 PJRF, 130: Freeman to Xiong Xiling, January 23, 1924.

1929, looking back at a decade of acquaintanceship, Tan wrote of Freeman as "a great man whom I love and respect."[164]

In 1921, Tan Zhen became engaged to the daughter of CEM veteran Kuang Jingyang.[165] Forging a link between the railway engineer of the 1880s and the new generation of US-educated engineers, the engagement was a modern love marriage,[166] something still unusual outside the realm of the new cosmopolitan class. Tan Zhen's hardships started around the same time. Government positions did not appear any more secure than private enterprise. Tan's father-in-law Kuang quit the railway service out of frustration. As the changing political regimes brought with them frequently changing administrations, Kuang began running a wool factory in Beijing instead. Likewise, his son-in-law Tan became an engineer at Lam, Glines & Co. (*Yun Yuan shiye gongsi* 允元實業公司) in Beijing and Tianjin, a private enterprise dealing with reinforced concrete.[167] He eventually started his own factory, the Tangent Manufacturing Works (*Tan Zhen zhizao gongchang* 譚真製造工廠), in 1923.[168] Tan's company constructed the Continental Bank Building but, while gaining a reputation, suffered financial losses from the project.[169] By 1927, Tan's company had run into serious trouble. He blamed his risky management "during such an abnormal time." In a desperate attempt to escape the financial troubles, he had bid exceptionally low on a tender for the Tianjin arsenal works. Using the project's budget to fill a financial gap in another project, his mismanagement led to a delay in the completion of the important Tianjin military building. Having incurred the wrath of Zhang Zuolin's government, he found himself in debtor's prison and awaited execution in the coming weeks.[170] Having gained knowledge of Tan's existential financial trouble, Freeman inquired about details and offered his help.[171] Todd, who later in his memoirs remarked that Tan "had the brain of a philosopher and an inventor, the energy of an acrobat and the loyalty of a true friend,"[172] was the first to help. He immediately sent Tan $1,500, leaving him "practically out of cash."[173] Their colleague Zhou

164 PJRF, 129: Tan Zhen to Freeman, undated, received January 13, 1929.

165 OJTP, 33.06: Tan Zhen to Todd, June 6, 1921.

166 PJRF, 125: Tan Zhen to Freeman, January 1, 1921.

167 PJRF, 130: Tan Zhen to Freeman, October 13, 1922.

168 PJRF, 129: Tan Zhen to Freeman, March 14, 1926.

169 PJRF, 129: Yang Baoling to Freeman, February 22, 1927.

170 OJTP, 02.01: Todd to H. B. Elliston, December 28, 1927; 33.06: Tan Zhen to Todd, November 19, 1927; 33.06: Tan Zhen to Todd, November 17, 1927; PJRF, 129: Yang Baoling to Freeman, April 26, 1928; Todd, *The China That I Knew*, 58.

171 PJRF, 129: Freeman to Tan Zhen, August 31, 1928.

172 Todd, *The China That I Knew*, 56, 59.

173 OJTP, 02.01: Todd to John Hendry, November 25, 1927; Todd, *The China That I Knew*, 59.

Yichun contributed another $500.[174] The last-minute aid of his friends allowed Tan to make up for the deficit, hence successfully evading capital punishment. It took Tan several more years to overcome his "very miserable" condition in which he found "no ladder [to] lead [him] out of the hole."[175] During this time, Tan continued to receive favors from his peers. For instance, at Tan's request, Todd (unsuccessfully) intervened with Sun Ke to have him join Ling Hongxun on his trip to the International Highway Congress in Washington.[176] Later, when Tan was ill in the hospital with a seemingly terminal lung disease, Todd proactively offered his help.[177] The professional acquaintanceships had clearly evolved into cross-cultural friendships.

While Tan's case of financial ruin remained singular in its severity, it was not an isolated instance of mutual aid. When ACAE member Zhang Jichun was on the brink of selling his home farm, Todd loaned another $1000 at 1% monthly interest.[178] He also tried to secure a scholarship for Zhang to study in the US, arguing that he felt "responsible" for him.[179] A similar patronizing element showed when the US-educated civil engineer Peng Binglu was assigned to the southwest interior, and Todd urged foreigners to support him so that he would not "fall back [...] into provincial ways."[180]

In summary, the intimate contacts between civil engineers of different nationalities show the association's potential to better intercultural understanding. The contacts were not pecuniary, but, as Todd put it, demonstrated "justice, duty, loyalty and charity."[181] They also demonstrate that nationalist historiography, which tends to paint the Republican era as a period of foreign capitalist-imperialist oppression and includes the US Open Door policy in the "century of humiliation" paradigm, was not the lived experience of privileged Chinese Americophiles. The relations involved intimacy and mutual respect. As Todd put it, the Americans became aware that the Chinese "do not all stand on their heads or walk on their hands."[182] Rather than bilateral distinction among the engineers, the Sino-American segment distanced

174 OJTP, 33.06: Todd to Tan Zhen, November 21, 1927; 33.06: Todd to Tan Zhen, November 22, 1927.

175 PJRF, 129: Tan Zhen to Freeman, undated, received January 13, 1929.

176 OJTP, 33.06: Tan Zhen to Todd, July 26, 1930; 33.06: Todd to Sun Ke, August 4, 1930; 30.01: J. Usang Ly to Todd, August 15, 1930.

177 OJTP, 02.04: Todd to Yang Baoling, January 29, 1934; Todd, *The China That I Knew*, 60.

178 OJTP, 31.03: Todd to Zhang Jichun, April 20, 1931; 31.03: Todd to Zhang Jichun, January 10, 1938.

179 OJTP, 31.03: Todd to Geo. T. Seabury, June 3, 1937; also 31.03: Todd to Paul Monroe, June 14, 1937.

180 OJTP, 01.12: Todd to R. F. Lankester, December 18, 1926.

181 OJTP, 02.04: Todd to L. O. Lee, August 11, 1931.

182 OJTP, 01.12: Todd to Willis T. Ryan, April 13, 1926.

themselves together against nonprofessionals. Within the realm of those they deemed respectable, nationality was an afterthought.

7 Conclusion

Throughout the 1920s, the Association of Chinese & American Engineers attracted the who-is-who of the US–Chinese engineering community. The annual conferences and the *Journal* fulfilled the functions of a professional organization, namely, to produce, diffuse, and discuss knowledge, and to shape an understanding of the professional self. However, more specifically, the association represented the US–Chinese civil engineers as a segment. As a formal organization, the ACAE promoted its segment to the public and created distinction from its professional colleagues. Their formal organization showed that their shared characteristics of time, place, and field of study do not just make them a retrospective analytical category, but that they identified as a common group during their lifetime, as well. The ACAE network demonstrates that the experiences gained during engineering assignments, such as the ones analyzed in the other chapters of this book, led to more than a collection of isolated voices of discontent. The individuals behind these voices interacted with one another on a professional and personal level, even when they did not always work on the same assignments.

As a privileged circle with restricted admission, the members also helped each other get into occupational positions. However, the sense of community transgressed the realms of professional interaction. Their intimate communications show that beneath the institutional surface of late-stage infrastructural imperialism, corporate interests, and global capitalism, there was a personal side to US–Chinese civil engineering, informed by professional exchange. In close cooperation with their American colleagues, the Chinese professionals exchanged their professional values and beliefs. Through contact with US members and through *Journal* reprints, they took part in American engineering discourse long after they had graduated and left the country. As a young profession, the ideals of civil engineering were in flux and its ethical code remained unwritten. During this period, the ACAE functioned as an arena for professional ideals to ripen. For instance, officials seen as selfish, inefficient, or idle provided a counter-image to the professional civil engineer, who regulated himself through a disciplined lifestyle in service to the public and the imagined nation. Therefore, shared frustrations and professional discontent were not only a reaction to disillusionment but also a source for ideals to form.

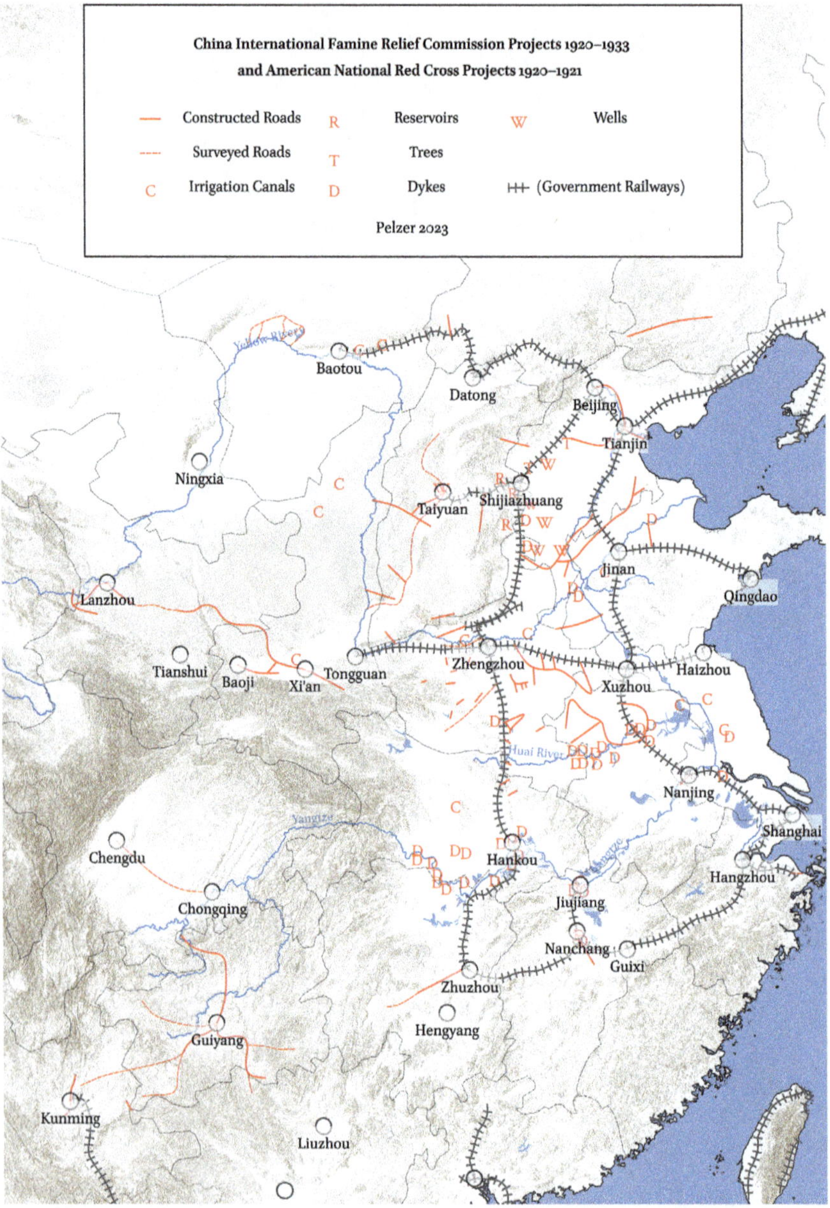

FIGURE 16 Map of ANRC and CIFRC Projects, 1920–1933 (based on CIFRC 1934, *Accomplishments*).

CHAPTER 5

Cohesion and Exclusion

The Relief Commission Paves the Provinces, 1926–1934

As seen in the previous chapters, civil engineers claimed that their service benefited the nation and that their infrastructures aided the cohesion of its people. But despite grand, national ambitions, not all construction projects were of national proportions and played out on the national stage. The China International Famine Relief Commission (CIFRC), a major employer for the core members of the Association of Chinese & American Engineers (ACAE, chapters 4, 6), focused its construction work on smaller infrastructures in the provinces, including the so-called Chinese interior. For example, the commission started highway projects in the remote regions of Yunnan and Guizhou (1926–1928), and in the Shaanxi–Gansu border region (1933–1934). These projects relied on social exclusion, a strategy within the broader process of professional distinction.[1] They demonstrate that civil engineers, apart from distinguishing themselves from non-experts in government and bureaucracy, also took care to distance themselves from the laboring class. The engineers exercised rigid control over common people through conscription and labor relief programs. Evidently, the professional work ethics allowed for a flexible reinterpretation of nationalist and participatory ideals when it suited the outcome of the work. It remained questionable who really benefited from the projects. In his monumental volume on the history of famine relief, Communist propagandist Deng Yunte 鄧雲特 (1912–1966) had little to say about the CIFRC engineering department, being careful not to give its infrastructural projects too much credit.[2] Given the international nature of the program, the reluctance of nationalist historiographers to acknowledge the work of the CIFRC is understandable. However, on closer examination, later nationalized institutions took up the specific mode of construction championed in the Sino-American environment.

1 Bourdieau, *Distinction*, 169–225, 260–317; Siegrist, "Professionalization as a Process," 177; Torstendahl, "Essential Properties, Strategic Aims and Historical Development," 52–59.

2 Yunte Deng, *The History of Famine Relief in China* (Cambridge, New York: Cambridge University Press, [1937] 2020), 2, 80, 369–377.

 | DOI:10.1163/9789004549555_007

1 "Good Roads" and Labor Relief

In 1920s China, despite the government's notoriously cash-strapped treasury, road construction was in full swing: the last six years of the Beijing era saw highways expanding from less than 2,000 to almost 30,000 km.[3] A major factor that enabled this development was the utilization of (un)free human resources to supplement the lack of monetary resources. These labor forces consisted of famine sufferers, schoolchildren, and soldiers.

As for the group of famine sufferers, their exploitation had its modern-day origins in 1917, when the American National Red Cross (ANRC) and the Chinese government cooperated in a labor relief program to construct a highway from Beijing to Tongzhou.[4] The idea of labor relief (*yi gong dai zhen* 以工代賑) was that instead of providing free relief (*zhenji* 賑濟)—i.e., using relief funds to provide the famine-struck population with soup kitchens—constructive enterprises would offer food provisions in return for manual labor.[5] The infrastructures built during the interventions aimed at preventing future famines, e.g., through dikes that prevented future floods or through highways that allowed for the interregional transportation of food.

The idea of labor relief had historical predecessors,[6] but Chinese reformers themselves understood it as a new development that started with the ANRC.[7] The new institutions put labor relief into a modern context of social science by fusing "self-help and efficiency," often intertwining poverty "with punishment and compulsory labor."[8] Accordingly, the 1920s opened a "new discourse" that "reinforced current trends among modernizers and state builders in China toward the use of coercive and punitive work regimes."[9]

---

3 Eastman, *The Abortive Revolution*, 271–72; Geographical Section of the Naval Intelligence Division, ed., *China Proper: Volume III. Economic Geography, Ports and Communications*, Geographical Handbook Series (Cambridge: Naval Intelligence Division, 1945), 432–33.

4 JACAE, 2, no. 4 (1921): Xiling Xiong, Address at ACAE Annual Spring Meeting, April 5, 1921, 32.

5 For the history and method of *zhenji*, see Pierre-Étienne Will, *Bureaucratie et Famine en Chine au 18e Siècle*, Civilisations et Sociétés 66 (Paris, The Hague, New York: De Gruyter Mouton, 1980), 93–134.

6 Yanet Chen, *Guilty of Indigence: The Urban Poor in China, 1900–1953* (New Jersey: Princeton University Press, 2012), 7, 32; Pierre Fuller, *Famine Relief in Warlord China* (Cambridge, London: Harvard University Press, 2019), 230–31.

7 Fuller, *Famine Relief in Warlord China*, 229.

8 Chen, *Guilty of Indigence*, 7, 32, 43, 74.

9 Fuller, *Famine Relief in Warlord China*, 231, 233. For international observers critical of the absence of labor relief in China, see Kathryn Edgerton-Tarpley, *Tears from Iron: Cultural Responses to Famine in Nineteenth-century China* (Berkeley: University of California Press, 2008), 118–19.

The institutionalization of modern labor relief set in after the Great Huabei Flood of 1920–1921.[10] In the wake of the resulting famine, the ACAE fall convention of 1920 passed a position paper that supported replacing free relief with labor relief. Headed by hydraulic engineer Yang Baoling (chapter 2), an ACAE workgroup submitted the proposal as a memorial to Chinese president Xu Shichang.[11] Chinese government advisor John Earl Baker supervised a charity drive that collected $1,214,000 for famine and labor relief programs. The primary donor was the ANRC, which returned to China to administer the majority of the donations, $750,000,[12] to construct highways through a labor relief program.[13] In total, the ANRC supervised eight constructions: roads in Shandong, in Pingding (Shanxi), from Tianjin to Baoding, from Tianjin to Tongzhou, from Jinghua to Qiyuan, and from Handan to Daming, plus wells in Ding County (Zhili).[14] In total, the ANRC directed approximately 900,000 people to construct the roads, which would expedite the interregional delivery of goods and thereby reduce the extent of future famines.[15]

In the aftermath of the Huabei Flood, in August 1921, the Peking United International Relief Committee (PUIRC) joined with the North China International Society of Famine Relief, the Chinese Foreign Famine Relief Committee, and local famine relief committees in Shandong, Shanxi, Hankou, and Hunan to form the CIFRC.[16] The CIFRC assumed "quasi-state functions."[17] In the beginning, traditional famine relief and rural credit schemes, inspired by Raiffeisen cooperatives, made up the bulk of the commission's work.[18] Two years later, in 1923, executive secretary Walter Hampton Mallory (1892–1980) started the CIFRC's version of labor relief. Directly referring to Baker's work with the Red

10 See Peking United International Famine Relief Committee, *The Famine of 1920–21 in North China* (Beijing: s. n., 1922); Fuller, *Famine Relief in Warlord China*.

11 JACAE, 1, no. 2 (1920): Baoling Yang, Thomas S. C. Sze, T. H. Wiggin, H. B. Merrick, R. D. Goodrich, Oliver J. Todd, H. A. Raider, Memorial Submitted to the President of the Republic of China, 40–43; Fisher, "Service to China," 47.

12 JACAE, 8, no. 8 (1927): Hardy Jowett, "Communications of China [cont.]," 19.

13 JACAE, 2, no. 11 (1921): Oliver J. Todd, "Famine Relief and Road Building in Shantung," 28.

14 JACAE, 2, no. 3 (1921), 20–21; JACAE, 5, no. 4 (1924): John Earl Baker, "An Outline of the Permanent Results of the Red Cross Famine Relief Construction Work," 5.

15 J. Clements, *China: Automobile Conditions and the Good Roads Movement* (Washington, D. C.: Government Printing Office, 1922), 4. For the ANRC in 1920–21, see Fuller, *Famine Relief in Warlord China*, 228–33.

16 Hong Fu and Calum Turvey, *The Evolution of Agricultural Credit during China's Republican Era, 1912–1949* (London: Palgrave Macmillan, 2018), 241; Andrea Janku, "The Internationalization of Disaster Relief in Early Twentieth-century China," *Berliner China-Hefte* 43 (2013): 12; Fuller, *Famine Relief in Warlord China*, 246–48.

17 Janku, "The Internationalization of Disaster Relief in Early Twentieth-century China," 9.

18 Fu and Turvey, *The Evolution of Agricultural Credit during China's Republican Era, 1912–1949*, 233–56.

Cross, the goal of the new engineering department was to continue the ANRC project of engaging famine-struck workforces for productive enterprises.[19]

The CIFRC received its budget from the Chinese government, through donations from the Chinese people, and from the American National Red Cross. Annual proportions between the three pillars fluctuated but were overall balanced.[20] The commission executives had received their education in the US but were not engineers. The civil servant Liang Ruhao 梁如浩 (M. T. Liang, 1863–1941), diplomat Yan Huiqing, and educationist Zhou Yichun led the commission. From 1921 to 1935, the CIFRC spent $50,000,000 on relief work, of which it used $12,355,000 on constructive engineering activities, including the construction of 7,481 km of highways, 1,400 km of dikes, 460 km of canals, and 5,050 wells (Figure 16).[21]

Apart from the direct influence of the ANRC labor relief model, the CIFRC engineering department also subscribed to the goals of the contemporary "Good Roads Movement." Instead of employing famine-struck populations, this movement primarily aimed at conscripting soldiers of the various provincial armies. A coalition of militarists and industrialists, including "Christian General" Feng Yuxiang 馮玉祥 (1882–1948), Shanxi's "Model Governor" Yan Xishan,[22] and Nantong entrepreneur Zhang Jian (chapter 2), supported the movement and its political goal to downsize provincial armies by conscripting soldiers to roadwork. Providing soldiers with new employment would prevent them from turning into bandits (Figure 17). For this purpose, in 1921, the two Christian politicians Wang Zhengting and Wu Shan 吳山 (1884–1936) organized the National Good Roads Association (*Zhonghua quanguo daolu jianshe xiehui* 中華全國道路建設協會, NGRA) in Shanghai.[23] The association was an indirect offspring of the American movement by the same name, in which cyclists

19 See Andrew Nathan, *A History of the China International Famine Relief Commission* (Cambridge: Harvard University Press, 1965), 11–26, 40–56.

20 Wéndé Huáng, "1921 zhì 1935 nián qíjiān fēizhèngfǔzǔzhī guójì císhàn mùkuǎn huódòng: Yǐ Zhōngguó Huá Yáng yìzhèn jiùzāi zǒnghuì wéi lì zhī yánjiū," in *Jìndài Zhōngguó wàijiāo de xīn shìdài guāndiǎn*, ed. Mǐnshū Liào (Taipei: Zhèngdà chūbǎn shè, 2018), 84.

21 "China International Famine Relief Commission Urgently Appeals for Help," *The China Reconstruction & Engineering Review* 7, no. 3 (1935): 107; Huáng, "1921 zhì 1935 nián qíjiān fēizhèngfǔzǔzhī guójì císhàn mùkuǎn huódòng," 72.

22 In the mid-1920s, Yan Xishan pursued a 1,000-mile road project in Shanxi, for which he used his military troops as laborers. (JACAE, 8, no. 7 (1927): Hardy Jowett, "Communications of China [cont.]," 6). Yan and Todd met and stayed in contact.

23 Shān Wú, "Běn huì sì dà jìhuà zhī jìnxíng shū," *Dàolù yuèkān* 4, no. 2 (1923): 2; JACAE, 2, no. 5 (1921), 41.

and car manufacturers pushed for paved roads.[24] The Chinese version started as a Yangzi Delta interest group lobbying for motor roads between Shanghai, Hangzhou, and Nanjing, but soon developed into a wider movement.[25] By 1923, the association had 70,000 members whose membership fees were added to the charity drive. Its fourteen branches had repaired between 20,000 and 25,000 km of roads and completed new ones, including a network between Shanghai and its neighboring municipalities of Longhua, Taicang, and Zhelin.[26]

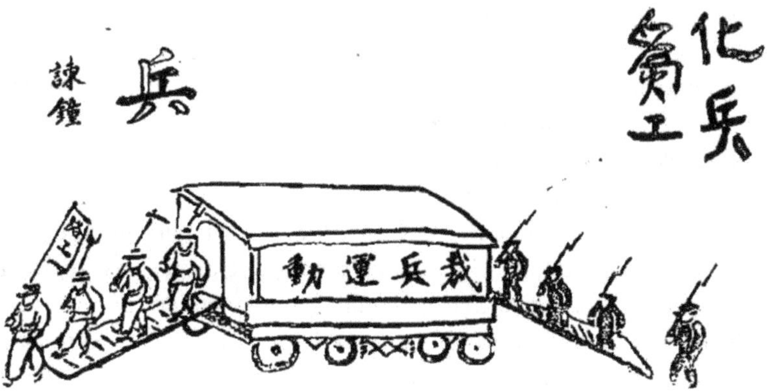

FIGURE 17 "Transforming Soldiers into Workers" for roads through an "army reduction campaign." The Chinese character for "soldier" transforms into the character for "work." Caricature by Jian Zhong (*Daolu yuekan* 2.1, 1922).

Like the NGRA, the ACAE promoted the construction of "good roads" through conscription labor. In 1923, the association established a road committee for the cause. Oliver J. Todd (chapters 2, 4, 6), who led the committee, proposed using prisoners and soldiers, as well as enforcing road and poll taxes payable in manual labor.[27] Ye Gongchuo, the on-and-off Minister of Communications with close ties to the Sino-American engineering world, proposed a "compulsory labor law" in the ACAE *Journal*. To build highways quickly and cheaply, he

24 Carlton Reid, *Roads Were Not Built for Cars: How Cyclists Were the First to Push for Good Roads & Became the Pioneers of Motoring* (Washington, D. C., Covelo, London: Island Press, 2015), 143–58.

25 Clements, *China*, 9–10; "The Road Building Program in China," *The Oriental Motor* 3, no. 1 (1921): 7.

26 JACAE, 3, no. 9 (1922), 24; JACAE, 4, no. 8 (1923), 35.

27 JACAE, 3, no. 10 (1922): Oliver J. Todd, et al., "Progress Report of Roads Committee," 9.

argued, the state needed the option to mandate the public, including retired soldiers and prisoners, to unpaid labor.[28] Already earlier, the association's honorary member Wang Jingchun (chapters 1, 3) had proposed "making use of the sufferers" to build railways and open mines.[29] The Good Roads Movement and the ACAE exchanged their ideas. For instance, in 1929, the association invited NGRA founder Wang Zhengting to discuss his movement at the ACAE spring meeting with Minister of Railways Sun Ke and ACAE honorary member J. A. L. Waddell.[30]

The common link between the ACAE, the ANRC, the NGRA, and the CIFRC engineering department was Oliver J. Todd.[31] Despite being specialized in hydraulic engineering, he had overseen the construction of the 1921 ANRC roads in Shandong and Shanxi. Accordingly, he was the ideal candidate when the CIFRC looked for someone to lead its engineering department. Todd, as a graduate from Michigan, the heartland of US automobile manufacturing, was enthusiastic about cars and the Good Roads Movement he knew from the US. After he was stranded in Suiyuan during one of his surveys, he had a Ford shipped into China so that he could "be independent and get things done Michigan fashion."[32] However, modern roads to use the car were scarce. To construct them, he had to rely on premodern, manual labor. Through massive cheap labor mobilization, Todd's projects would unite "technology and the coolie."[33] Highway construction projects in Yunnan, Guizhou, and the Shaanxi–Gansu border region would put the model to the test.

2 Contested Authority in Yunnan

While work relief and conscripted labor meant little cost for the workforce, highway construction still required a budget to work with. In Yunnan, opportunity arrived after the customs service (*haiguan* 海關) had imposed

28 JACAE, 5, no. 2 (1924): Gongchuo Ye, "The Outlook of Communications in China," 14.

29 Wang, "The Effect of the Revolution upon the Relations between China and the United States," 269.

30 OJTP, 29.13: Todd to Roy D. Chapin, March 11, 1929.

31 See also Thomas Campanella, "'The Civilising Road': American Influence on the Development of Highways and Motoring in China, 1900–1949," *The Journal of Transport History* 26, no. 1 (2005): 85–86.

32 OJTP, 29.08: Todd to Han Ngan, August 23, 1925.

33 Spence, *To Change China*, 208.

a famine-relief surtax for four winter months between 1925 and 1926.[34] The local CIFRC committee in Yunnan received $300,000 silver dollars of the 3-million-dollar income. Of this portion, it had to spend 70% on productive labor relief. Consequently, governor Tang Jiyao 唐繼堯 (1883–1927) ordered the construction of a motor road leading from the capital Kunming to Shuifu, located at the border to Sichuan.[35] The road would follow the course of the larger Yunnan–Sichuan Railway. US engineers had surveyed the line between 1909 and 1911, but it had remained unconstructed.[36] The relief funds would suffice for the first 10% of the road.[37]

Todd put Peng Binglu (chapter 4, Figure 18) in local charge of the roadwork. Peng held degrees in civil engineering from the universities of Missouri and Cornell. He had the typical cosmopolitan biography of a US-returned student: during his college days, he acted as the local officer of the Chinese Student Christian Association,[38] joined the Cornell Cosmopolitan Club, and was a member of the Chinese Students' Club.[39] Being of the later student generation that put a stronger emphasis on practical training (chapter 1), from 1923–1926, he was part of a Chinese student group working at the Ford Motor Corporation in Dearborn, Detroit, to gain practical experience in automobile engineering.[40]

When in 1919, John R. Freeman had recruited Todd for the Grand Canal project (chapter 2), it had been the unwritten rule that, at least in projects with

---

34 "Hǎiguān zhènzāi fùshuì zhī jīngguò," *Jiùzāi yuèkān* 4, no. 1 (1926): 1.; China International Famine Relief Commission, ed., *Custom Surtax of 1925 and Plan for Its Utilization,* B 15 (Beijing: China International Famine Relief Commission, 1925). For a comparable 1924 surtax, see Walter Mallory, *China: Land of Famine* (New York: American Geographical Society, 1926), 70–71.

35 OJTP, 29.10: CIFRC Yunnan Committee to Yunnan Government, October 18, 1927.

36 OJTP, 29.10: Peng Lubing to the CIFRC Yunnan Committee, June 9, 1927. See also Elisabeth Kaske, "Sichuan as a Pivot: Provincial Politics and Gentry Power in Late Qing Railway Projects in Southwestern China," in *Southwest China in a Regional and Global Perspective (c. 1600–1911)*, ed. Ulrich Theobald and Cao Jin (Leiden: Brill, 2018), 409–12.

37 JACAE, 8, no. 8 (1927): "Engineering Projects Costing Forty-two Million Dollars Considered by China International Famine Relief Commission," 32.

38 "Officers and Committees of Chinese Student Christian Association in North America: 1918–1919," *The Chinese Students' Christian Journal* 5, no. 1 (1918): 73. See also "Personal, Local and Home News," *The Chinese Students' Christian Journal* 6, no. 6 (1920): 351.

39 C. Sherwood, ed., *The 1920 Cornellian,* The Cornellian 52 (Ithaca: The Cornell Annuals, 1920), 193.

40 *Ford News,* "Ford Trade School Prepares Chinese Students for Career," June 22, 1923, 3. Cited in Chelsea Zuzindlak, "'Tell 'Em You're from Detroit': Chinese Americans in the Model City," in *Asian Americans in Michigan: Voices from the Midwest,* ed. Sook Wilkinson and Victor Jew (Detroit: Wayne State University Press, 2015), 57.

FIGURE 18
Peng Binglu as a student (*The Cornellian* 52, 1920).

international investment, senior engineers were foreigners. In contrast, half a decade later, Todd started installing Chinese engineers like Peng in leading positions. In theory, Chinese civil engineers in responsible positions would help foster a capable young generation of engineers and empower the new domestic technical elite. Apart from his Michigan sympathies, Todd likely asked Peng to lead the construction because the latter was working at the local Yunnan University, meaning, at first glance, he had better connections with the residents. However, Peng was originally from Guangdong,[41] which was a distant place in the locals' perception of spatial scale. For the residents of Yunnan, the foreign-educated, Christian Cantonese technician Peng was barely less foreign than Todd. Therefore, while hierarchies had become more permeable as domestic professionals gradually ascended the ranks, the engineers in leading positions were still not truly local in the perception of the residents. This created problems for the urban Chinese engineers that resembled the resentment against foreign engineers.

Just as the Yunnan locals could not identify with Peng, conversely, the latter also eagerly distinguished himself from the locals, especially displaying his disdain for the workforce. When overseeing the road in Yunnan, he repeatedly called his workers "inferior natives" incapable of physically keeping up with the "daily routine" of engineers. He ridiculed the locals' fear of rain and cold

41 Michael Feighan, *81st Congress, First Session, Report No. 153: Ben Luke Pond, Shao Hung Pond, and David Yat Wei Pond* (Washington, D. C.: United States Government Printing Office, 1949), 1–4.

temperatures, found at least a part of them to be disobedient and incompetent, and suspected foremen of squeezing money before dividing pay among the workers.[42] Parts of the workforce tried to extort a higher salary by threatening the use of violence, while others attempted burglary in Peng's office. During an uprising, Peng "drew a pistol and fired into the air" and had some of the workers arrested and sentenced to jail.[43]

The lack of "harmony" or cohesion between the local population and Chinese engineers, who held foreign degrees and came from big cities, was an ongoing issue shared across the various CIFRC engagements. For instance, Grand Canal veteran Xue Zhuobin (chapter 2) complained about locals obstructing his irrigation work around the Marco Polo Bridge near Beijing: to him, landowners had been excessive, wasteful, and disruptive to engineers exerting their right-of-way, causing a "serious delay" of half a year.[44] In a different case, the CIFRC's resident engineer for the southern section of the Nanchang–Fuzhou (Jiangxi) motor road resigned, lamenting about local women "crying, scolding, grab[b]ing" surveyors and splashing the engineers' faces with natural fertilizer.[45] In another episode, one of the CIFRC engineers left for Tianjin after being subjected to, in Todd's words, "groundless attacks [...] by worthless *paitous* [village heads]."[46] A CIFRC irrigation project in Shishou, Hubei, had two Chinese engineers quit their job after local soldiers physically abused them.[47] When working on dikes in the same province, Todd saw his Chinese engineers quit after a little over a week, after they had been unsuccessful in making the local people follow their instructions.[48] These experiences with the interior population demonstrate that a Western/Chinese dichotomy is too simplistic. Regarding the perception and treatment by and of nonprofessional locals, the cosmopolitan Chinese engineers joined the ranks of the foreign engineers as an "other."

---

42 OJTP, 29.10: Peng Lubing to Todd, ca. January 1927.

43 OJTP, 29.10: Peng Lubing to Todd, April 9, 1927.

44 JACAE, 8, no. 7 (1927): Yuanshan Zhang, "Farmer Versus Engineer," 14; 9, no. 4 (1928): Zhuobin Xue, "Shih Lu Irrigation Work," 4; OJTP, 29.10: Zhang Yuanshan to Todd, May 31, 1927; "Shih Lu Project Delayed by Local Problems," *Jiùzāi yuèkān* 4, no. 6 (1927): 31–32.

45 OJTP, 29.11: Todd to Mr. Sung, May 1, 1928; 29.11: K. H. Sung to Todd, March 13, 1928; 37.01: Todd to Xue Zhuobin, October 19, 1928.

46 OJTP, 02.04: Todd to Dwight W. Edwards, June 11, 1930.

47 OJTP, 01.11: Zhuo Kangcheng to Todd, September 22, 1925.

48 OJTP, 29.19: Todd to Zhang Yuanshan, May 15, 1927; 29.10: Todd to Zhang Yuanshan, June 17, 1927; 29.10: Todd to CIFRC Hubei Committee.

In addition to the troubles with the common workforce, Peng encountered opposition from influential landowners and bureaucrat officials. Different from two decades earlier, when Zhan Tianyou had overseen the construction of the Beijing–Kalgan Railway and had encountered opposition rooted in geomantic beliefs,[49] this time, the resistance had clear economic motives. The battleground was Dabanqiao, a rural area just north of the capital. Peng, following the efficiency paradigm, ordered the construction of a road straight through the Dabanqiao rice fields (Figure 19). The enraged landowners started a campaign to halt the construction. To avoid the expropriation of their farmlands, the group petitioned to run the road across the nearby mountains instead. Peng, acting in service to the provincial government, attempted to enforce his right-of-way. However, he soon realized that within the local CIFRC committee that administered his funds, his opponents held a comfortable majority. Peng complained about "damaging posters" on Kunming walls stirring opinion against him,[50] and that an unauthorized engineer, contracted by the CIFRC Yunnan committee, had encouraged the locals to plow and plant where the road bureau had already reserved its right-of-way.[51]

Peng reported having received a "threatening letter" in which the Dabanqiao village head demanded the change of the road's location in exchange for a bribe.[52] He did not give in to the demands. In return, the village head, supported by local CIFRC chairperson You Yunlong 由雲龍 (1877–1961), charged Peng with arbitrary alignment, attempted extortion, and illegal arrest.[53] Peng pointed out that the main accusation—attempting to extort money to change the course of the road—was implausible, as it was not within the engineer's authority to make late alterations to the road location.[54] He argued that the proposed alternative route through the mountains was not safe, too lengthy, too expensive, and subjected to damage by mountain torrents.[55] Peng built his argument around efficiency, ignoring or disdaining the personal interests of the local population. For chairperson You, this rationalist logic was "high

49 Le, "The Struggle for Autonomy from the State," 62–64.
50 OJTP, 29.10: Peng Lubing to CIFRC Yunnan Committee, July 2, 1927.
51 OJTP, 29.10: Peng Lubing to CIFRC Yunnan Committee, July 12, 1927.
52 OJTP, 29.10: H. C. Page to You Yunlong, April 15, 1927.
53 OJTP, 29.10: Peng Lubing to CIFRC Yunnan Committee, June 9, 1927.
54 OJTP, 29.10: Peng Lubing to CIFRC Yunnan Committee, July 2, 1927.
55 OJTP, 29.10: Peng Lubing to CIFRC Yunnan Committee, July 12, 1927.

FIGURE 19 Road inspection at Dabanqiao fields, October 1927 (CIFRC 1927, *Accomplishments*).

handed" in that it disregarded "custom, usage, and public opinion."[56] The situation appeared deadlocked, with the locals unreceptive to arguments of technical efficiency, and Peng unwilling to give in to municipal politics. With the Dabanqiao conflict between engineers and locals ongoing, Peng felt increasingly unsafe. As a last resort, Peng requested Todd to buy and bring him shotgun cartridges, as he suspected a conspiracy of "some big men" against him and speculated that without the support of foreign committee members, local officials would have jailed or killed him already.[57] Peng and Todd felt let down by the Yunnan governor, who had originally ordered the road construction.

The situation was resolved when news broke that You Yunlong, the CIFRC chairperson who had prominently sided with the Dabanqiao landowners, was himself an owner of the affected farmland. The revelation tainted his claim of impartiality and led him to resign from office.[58] The Yunnan government renewed its support of Peng's position against gentry opposition.[59] However, by then, the road project had already been artificially delayed.

56 OJTP, 29.10: Copy of Chairman You Yunlong's Resignation, June 28, 1927.

57 OJTP, 29.10: Peng Lubing to Todd, July 8, 1927.

58 OJTP, 29.10: Copy of Chairman You Yunlong's Resignation, June 28, 1927.

59 OJTP, 01.12: Peng Lubing to Todd, September 26, 1926; 01.12: Todd to Mr. Petterson, November 10, 1926.

A military coup in Yunnan, which ultimately ousted Governor Tang in the spring of 1927, led to the suspension of the work.[60] It became too dangerous for Yunnanese troops to guard Peng, stationed just 24 km outside Tang's capital. At the time of termination, the CIFRC had surveyed a 100 km track from the provincial capital to Yanggai (Songming) and finished construction of an even shorter part, a 20 km highway between the center of the capital and the nearby fields of Dabanqiao (Figure 16).[61] Thereafter, to Todd's frustration, even this small stretch of roads was "not properly maintained and little used."[62] The provincial government's abeyance, its lack of direction, and lack of immediate support against the revolting landowners had left a negative impression on the involved engineers.

3 Guizhou—The Strong State

While the Yunnan project demonstrated how discontent emerged out of opposition from local landowners and hesitance of the local leadership, the second case led the CIFRC to the neighboring province of Guizhou, where these obstacles did not surface. There, a strong local state enabled the commission to materialize its project. On the surface, social cohesion led the roadbuilding to success. However, a closer look reveals the deep chasm between the engineers involved and the local population.

Governor Zhou Xicheng 周西成 (1893–1929) followed an ambitious scheme: he envisioned a star-shaped highway system around the capital Guiyang, consisting of five trunk routes and seven branch lines (Figure 16).[63] The system would serve two functions simultaneously: it would cement the capital's status as center of the province, and establish links to the neighboring provinces of Sichuan, Yunnan, and Guangxi. This included a north-south axis from the Guangxi border of Libo via Anshun to the capital of Guiyang, and further to Zunyi, Tongzi, and the Sichuan–Yunnan border town of Chishui. The young governor planned to cooperate with the CIFRC to spend $300,000 silver, which the province, like Yunnan, had received from the famine-surtax. The famine-surtax would primarily be used for the first 64 km of the 269 km long

60 "Todd Returns from South-west," *Jiùzāi yuèkān* 5, no. 12 (1927): 11.

61 Addie Smith and Anselm Chuh, *Motor Roads in China* (Washington, D. C.: United States Government Printing Office, 1931), 129–30.

62 JACAE, 9, no. 1 (1928): Oliver J. Todd, "Extending Motor Roads in the South," 8.

63 See Smith and Chuh, *Motor Roads in China*, 90–95.

portion north of the capital. When completed, the project promised to cut down travel time between Guiyang and Chishui from 15 to three days.[64]

As in Yunnan, Todd limited his role to that of a supervisor. He delegated the civil engineer Huang Rongde (chapter 4) from Beijing to the province. Huang was a civil engineering graduate from Ohio Northern University,[65] who, just like Peng in Yunnan, had spent three years at the Ford Motor Works in Detroit.[66] Governor Zhou's highway bureau sent eight division engineers and 81 students to assist Huang. Each student took charge of seven kilometers of the road, and the division engineers oversaw the road in equal parts.

In contrast to Yunnan, which he found to be full of reluctant, corrupt, and opium-smoking bureaucrats,[67] Todd was impressed by the Guizhou governor's efforts and energy. The governor may have bolstered this favorable impression by hosting Todd like a "royal."[68] During his first visit in December 1926, he quartered him in an empty museum, lent him his private servant, and awarded him a gold medal.[69] More impressive than these personal favors, however, was Zhou's political control and the efficiency of construction this enabled. Todd remarked frenetically that the governor was "practically a dictator."[70] That is, a benevolent, selfless dictator known for his "zeal on behalf of the people."[71]

According to Huang, the local gentry resented engineers from Beijing, as relatives had lost their positions in the highway department against the educated newcomers from the capital. Huang took the officials to be jealous of his "fat pay envelope," even though he still felt underpaid by a hundredfold. The large discrepancy between the local committee's and the engineers' salary expectations led to the local committee refusing payment and asking the provincial government to take over.[72] In a situation paralleling the events of Yunnan, Huang and Todd sought an alliance with the provincial government, not with the local branch of the CIFRC.

---

64 JACAE, 8, no. 8 (1927): "Engineering Projects Costing Forty-two Million Dollars Considered by China International Famine Relief Commission," 32.

65 "Tóngrén xiāoxī," *Jiùzāi yuèkān* 4, no. 3 (1927): 14.

66 OJTP, 29.10: Todd to Guizhou CIFRC Committee, March 11, 1927.

67 OJTP, 01.12: Todd to Doc [?], December 9, 1926.

68 OJTP, 01.12: Todd to Fuller, December 12, 1926.

69 OJTP, 01.12: Todd to Geo. L. Davis, December 1, 1926; 01.12: Zhou Xicheng to Todd, December 20, 1926.

70 OJTP, 01.12: Todd to Doc [?], December 9, 1926.

71 OJTP, 18.03: Oliver J. Todd, Article Draft "Kweichow's Progressive Leader—Chao Hsi Tsun," May 1927.

72 OJTP, 29.11: Huang Rongde to Todd, May 16, 1928.

The decisive support from the Guizhou government institutions allowed resident engineer Huang to ignore the objections of Zunyi residents when building the road directly through their city.[73] Zhou discharged and jailed any magistrate whose district did not finish the road on time. Todd held the governor's disregard for private land rights in high esteem, venerating that the governor's policies allowed for "no sacrifice in efficiency," as nobody's land was "sacred"—a view that was "the only view to take." For Huang, the level of political control was a "pleasure."[74] He was full of praise for the governor, as he found his jailing of inefficient magistrates "so efficient." For Huang, the governor's "military way" established "a record to challenge the whole world."[75] Like Peng's experience with the landowners of Dabanqiao, Huang also encountered occasional local opposition. However, different from Yunnan, the decisive support of the Guizhou government institutions allowed him to ignore these objections. Radically following the efficiency paradigm, Huang surveyed the shortest route, cutting right through fertile farmland and destroying it.[76]

4 Guizhou—Mass Mobilization

Governor Zhou's first pillar was his army, which had to work on the roads. Just as he emulated General Feng Yuxiang by outlawing tobacco and enforcing the prohibition of opium, he also followed Feng's example of constructing "good roads" through conscript labor. But Zhou utilized other forms of compelled labor, as well. With a workforce of over 200,000 people,[77] the Guizhou roadwork is a prime example of how CIFRC engineers used the ideas of the Good Roads Movement and stretched the definition of labor relief to its limits to achieve their goals.

As a second pillar of compelled labor, the Guizhou governor mobilized schoolboys and schoolgirls over fourteen years of age (Figure 20, Figure 21). There was a weekly rotation to enforce sex division and enable the children to visit school every other week. The government forced them to perform manual labor, although their days were shorter than those of the enlisted adults.[78] In

73 OJTP, 37.01: Huang Rongde to Todd, July 16, 1927.
74 OJTP, 29.10: Huang Rongde to Todd, December 15, 1927.
75 OJTP, 37.02: Huang Rongde to Todd, February 7, 1928.
76 OJTP, 37.01: Todd to Howard Page, August 24, 1927.
77 "Motor Road Penetrates Center of Kweichow," *Jiùzāi yuèkān* 5, no. 4 (1928): 25; Oliver Todd, "Engineering and Altruism in China," *The Peking Leader*, November 7, 1929, 4.
78 OJTP, 37.01: "Transportation and the Food Problem in South China," manuscript, ca. 1927.

the 1920s, awareness of child labor was on the rise, and business owners in the international community of Shanghai started self-enforcing restrictions in their factories in this regard.[79] Foreign and foreign-trained engineers did not share the concerns of their urban compatriots. For instance, John E. Baker, adviser to the Beijing Ministry of Communications, defended industrial labor conditions against the accusation of "foreign exploitation" by asking: "Hours are long; but what else have these laborers to do? Child labor is revolting, but would they prefer infanticide?"[80] Todd followed Baker's pragmatism. In his diary, he wrote of drinking his afternoon tea on the road while watching schoolgirls digging dirt.[81] He described the pupils commandeered off school "in good humor."[82] Resident engineer Huang did not like working with the students, but not for ethical reasons. Rather, he found them inefficient. In his perception, instructing them made him "waste lots of time," and their bodies, not used to strenuous physical work, caught sickness easily.[83]

FIGURE 20 Middle school students digging earth in Guizhou, 1926 (OJTP 67.3).

79 Robin Porter, *Industrial Reformers in Republican China* (Armonk, London: M. E. Sharpe, 1994), 75, 98-121, 157.
80 John Baker, *Explaining China* (London: A. M. Philpot, 1927), 161.
81 OJTP, 8.16: Diary, December 21, 1926.
82 JACAE, 8, no. 5 (1927): Oliver J. Todd, "Motor Roads for South China," 9.
83 OJTP, 37.01: Huang Rongde to Todd, July 16, 1927.

As the use of soldiers and middle school pupils proved insufficient, the government introduced a third pillar, which obligated residents to commit to the roadbuilding process. Residents did not necessarily need to handle the workload themselves. To fulfill their allotted work, wealthy merchants hired substitutes, whom they, according to Todd, paid "little more than food."[84] The government reimbursed the gentry for the costs of one daily wage of manual labor in the form of road bonds. In return, the property owners, in proportion to their wealth, had to delegate up to three laborers to join the roadwork.[85] The compensation in road bonds was, for the time being, worthless, and its eventual value uncertain.

Whether described as "compulsory,"[86] or "coercive and punitive,"[87] on an analytical level, the work arrangements marked forms of compelled labor.[88] Historically, however, the International Labour Organization (ILO), which the Chinese Republic co-founded in 1919, defined "forced labor" only in 1930. Even within this framework, some forms of work relief and conscripted labor remained legal, as the international convention excluded "normal civic obligations" and "service exacted in cases of emergency" from the definition.[89] Further, only 18 states ratified the convention through the decade, including Japan, the UK, and France, but excluding China.[90] In other words, while subjective moral objections to forced labor existed, the arrangements were legal during the time. Lawmakers legitimized them as a means to prevent famines, achieve social cohesion, or promoted them as patriotic duty.

Despite his political views and his constant fight for famine prevention on the macro level, Todd never questioned the local labor conditions and interpreted them as a form of social cohesion. In Todd's interpretation, everything about the labor recruitment was "fair play." The schoolchildren had received "a course in road building" to "develop *esprit de corps* and a keener sense

84 JACAE, 9, no. 1 (1928): Oliver J. Todd, "Extending Motor Roads in the South," 4.

85 OJTP, 29.11: Huang Rongde to Todd, May 16, 1928.

86 JACAE, 5, no. 2 (1924): Gongchuo Ye, "The Outlook of Communications in China," 14; Chen, *Guilty of Indigence*, 74.

87 Fuller, *Famine Relief in Warlord China*, 231.

88 See Michael Bush, *Servitude in Modern Times* (Cambridge: Polity, 2000), 5.

89 International Labour Organization, "C029: Forced Labour Convention, 1930 (No. 29)," accessed June 25, 2021, https://www.ilo.org/dyn/normlex/en/f?p=1000:12100:0::NO::P12100_ILO_CODE:C029, §2.

90 International Labour Organization, "Ratifications of C029: Forced Labour Convention, 1930 (No. 29)," accessed June 25, 2021, https://www.ilo.org/dyn/normlex/en/f?p=NORMLEXPUB:11300:0::NO::P11300_INSTRUMENT_ID:312174.

of community interest,"[91] while the adult workers had "paid their road tax in labor."[92] He blathered about a "united spirit on the part of all classes" to complete the motor roads,[93] and of an "intense enthusiasm" shared by the governor and "the people generally."[94] In contrast, resident engineer Huang saw commoners cooperating simply because "nobody dar[ed to] disobey orders."[95] As Todd's biographer Fisher—who blames Governor Zhou and vindicates the involved engineers—writes, the common people "were building the roads out of fear, not out of eagerness."[96] Labor questions were primarily a question of recruitment, not a question of labor rights, adequate payment, or dignified handling.

Politically and in socialization, Todd was a progressive. He had sympathies for the liberal pastor Harry Emerson Fosdick (1878–1969) and believed in a Jeffersonian democracy against the "accumulation of wealth at the cost of the efforts of others."[97] Furthermore, in the upcoming years, he would become friends with the leaders of the left wing of the Chinese Nationalist Party (chapter 6). While Todd supported the authoritarian labor tax scheme, the US Good Roads Movement, which he aimed to bring to China, supported the abolition of statute labor. Deemed ineffective and antiquated, by the turn of the century, US states had begun to explore new ways to recruit labor for the construction of public highways.[98] That a progressive mind like Todd supported labor schemes in China that civil engineers in the US had already moved on from hints at a double-sided interface of Sino-American transnationality. Not only were Chinese engineers influenced by US engineering discourse, but American engineers also modified their professional ideals under the impression of the Chinese *status quo*.

The idea that technological advancement and, in particular, infrastructure could help national cohesion or bring "harmony" between its citizens was a vision shared by theoreticians and practitioners across the globe.[99] Around the

91 OJTP, 18.03: Todd, O. J., Article Draft "Kweichow's Progressive Leader—Chao Hsi Tsun," May 1927.

92 Todd, "Engineering and Altruism in China," 5.

93 OJTP, 37.01: "Transportation and the Food Problem in South China," manuscript, ca. 1927.

94 Oliver Todd, "Construction Enterprises in South-west China," *The Far Eastern Review* 24, no. 9 (1928): 419.

95 OJTP, 37.01: Huang Rongde to Zhang Yuanshan, November 27, 1928.

96 Fisher, "Service to China," 94.

97 OJTP, 02.01: Todd to Margaret Smith, March 13, 1927; 02.01: Todd to Margaret Smith, June 5, 1927.

98 Alex Lichtenstein, "Good Roads and Chain Gangs in the Progressive South: 'The Negro Convict is a Slave'," *The Journal of Southern History* 59, no. 1 (1993): 86–88.

99 van Laak, *Weiße Elefanten*, 33, 87.

turn of the century, progressivist engineers in the US argued that infrastructures and the rational, scientific management of their work would help end class conflict.[100] The Taylorites, the prime propagators of scientific management, argued that while democratic institutions interfered with their work, efficient scientific management would, in its own right, help create an equal society.[101] Likewise, pioneers of the technocracy movement, for instance, the consulting engineer Walter Rautenstrauch (1880–1951), argued that his profession could bridge the gap between pure science and the values of society, aiding the creation of "harmony" between industrial production and distribution.[102] The engineering progressivism that promised to end class conflict through technical progress eventually transitioned into a technocratic movement with totalitarian overtones.[103] The proponents found their quest for social harmony and abundance justified the exploitation of land and people, including the mass conscription of unfree labor. The engineers in Guizhou followed their rationale.

FIGURE 21 Governor Zhou Xicheng lifting a Chinese hoe, surrounded by military and enlisted schoolchildren, 1926. Photo taken by Oliver J. Todd (JACAE 8.5, 1927).

100 Akin, *Technocracy and the American Dream*, 4–10; Layton Jr., *The Revolt of the Engineers*, 67–68. See also Camprubí, *Engineers and the Making of the Francoist Regime*, 164; Jarausch, *The Unfree Professions*, 125.

101 Layton Jr., *The Revolt of the Engineers*, 145–49.

102 Akin, *Technocracy and the American Dream*, 54–59.

103 Layton Jr., *The Revolt of the Engineers*, 226–28. Also Akin, *Technocracy and the American Dream*, 62–63, 131; van Laak, *Weiße Elefanten*, 44, 240–242.

The Guizhou engineers appeared to forget about the original aim of the CIFRC engineering department, i.e., the construction of famine-preventing infrastructures through labor relief. Instead, they gave priority to the efficient completion of the road through various forms of labor exploitation. As an example of how secondary famine relief became during the project, Todd complained that the CIFRC had employed too many sick, "unfit," and "useless" people for the roads and suggested they be cared for through other means instead.[104] He disregarded the fact that his construction project limited the free relief funds for those unfit for work. Instead, he proposed to further reallocate the remaining funds—designated for immediate, traditional famine relief—to fill financial holes in his construction work.[105] This became necessary as three years of good harvest had reduced the number of famine sufferers to an amount that it was difficult to find enough workers willing to finish the road for a mere subsistence wage.[106]

As the locals were no longer famine sufferers, who depended on work offered through the construction project, the roadbuilding was no longer a genuine form of "labor relief." The integrity of the CIFRC's superordinate cause came under scrutiny, as a direct connection between the construction efforts and famine relief became less and less apparent. Although Todd pointed out that the new roads in underdeveloped Guizhou would help medical personnel

FIGURE 22 New road in Guiyang, September 1927 (CIFRC 1927, *Accomplishments*).

104 OJTP, 01.12: Todd to CIFRC Guizhou Committee, December 20, 1926.
105 OJTP, 37.01: Todd to CIFRC Guizhou Committee, September 3, 1927.
106 OJTP, 37.01: Hu Pi Hsiang to CIFRC Executive Secretary, October 22, 1928.

travel faster in future times of need,[107] he also promoted the roads as a way to enable the exploitation of mineral resources and build health resorts.[108] Observers started pointing out that even the past famines had not been of natural cause, but largely self-inflicted through "bandits, civil war and exorbitant taxation."[109] What followed was an academic discussion about the definition of a "famine," culminating in a wrecking report by a Red Cross investigation in 1929, which claimed that the CIFRC's long-term infrastructural projects were a fundamentally unsuitable method of famine relief.[110]

As funds proved insufficient to pay the locals what they could earn by returning to their farms, the roadwork ended with a low-budget version of the original plans. Huang decided to have a ferry cross the Wu River to transport cars between Zunyi and Chishui, as there was no money to build a bridge.[111] He further opted for zigzag roads along the mountains to achieve an 8% grade without heavy cutting of the rocks. In contrast, around Zhou's powerbase of Guiyang, workers had already macadamized roads six meters wide and built sidewalks of 1.5 meters width (Figure 22).[112] Having used up all surtax funds, Huang left Guizhou in the autumn of 1928.[113]

Despite the budget cuts, the project was a success in engineering terms: Huang had overseen the completion of the Anshun–Guiyang–Chishui motor road, where a bus service regularly plied the 250 km between Anshun and Zunyi. Afterward, he went on to survey an extension of the road 300 km south to the Guangxi border town of Libo.[114] The Guizhou highway work confirmed the engineers' suspicion that an unconsolidated central state had been the

107 OJTP, 01.12: Todd to CIFRC Guizhou Committee, December 20, 1926.

108 OJTP, 37.01: "Transportation and the Food Problem in South China," manuscript, ca. 1927.

109 OJTP, 29.10: E. F. Stanton to Todd, December 30, 1927. Todd did not always deny this, see OJTP, 18.03: Todd, O. J., Article Draft "Kweichow's Progressive Leader—Chao Hsi Tsun," May 1927.

110 American Red Cross and China International Famine Relief Commission, *Famine in China's Northwest: American Red Cross Commission's Findings and Rejoinders Thereto* (Beijing: China International Famine Relief Commission, 1930), 1–33; Nathan, *A History of the China International Famine Relief Commission*, 16–22, 45–48; Fisher, "Service to China," 118.

111 OJTP, 29.10: Huang Rongde to Todd, December 15, 1927; 37.01: Huang Rongde to Zhang Yuanshan, November 27, 1928. This may well be the reason why the Tongzi–Chishui section ran out of popularity in the 1930s. The former Tongzi–Songkan branch road extended to Chongqing, while Chishui became a part of the extended Guiyang–Bijie branch road.

112 JACAE, 9, no. 1 (1928): Oliver J. Todd, "Extending Motor Roads in the South," 6.

113 OJTP, 37.01: Todd to Xue Zhuobin, October 19, 1928.

114 "Motor Road Penetrates Center of Kweichow," 25; Wing Tak Wong, "Kweiyang to Kwangsi by Motor Before Next Xmas," *Jiùzāi yuèkān* 5, no. 5 (1928): 34.

root of their discontent: after all, Zhou's authoritarian rule over the province had finally allowed them to see a project to its end. There is no hint that the involved engineers sympathized with the large workforce, which labored under the strong state that allowed the engineers to thrive.

5 Guizhou—Beneficiaries and Burden Bearers

The turn to a cybernetic mélange of progressivism and frigid technocratic project management did not leave room for the humanization of the workforce. The distinction from the laboring class went hand in hand with the professionalization of civil engineering: professionalization requires the maintenance of social inequality and the stabilization of a differentiated occupational stratification.[115] The concept of professionalism requires hierarchy and is, therefore, no matter how progressive the supposed goals, "always inequal."[116] As civil engineers could not build on professional autonomy to establish their group as doctors and barristers were able to, their contractual dependency impelled the drive with which they tried to distinguish themselves from common laborers.[117] In other words, the civil engineers, regardless of the corporeality of their work and their limited contractual autonomy, defended their entitlements and felt distinct from the working class.

While the CIFRC liked to portray the mass mobilization of labor as an expression of social cohesion, workers profited much less than the upper strata. The avant-garde nature of the motor roads demonstrates that Todd's "united spirit on the part of all classes" was a fiction the civil engineers chose to believe. Existing gaps widened as landowners eventually profited from rising land value along the roads. Already in 1917, when the Chinese government and the ANRC had cooperated to employ flood refugees on construction work in Beijing, wealthy families had made use of the opportunity and built summer resorts along the road.[118]

Even in more developed provinces, cars were mostly found in urban settlements. As ACAE secretary Ehlers later remembered, when he had become lost on his surveys through Shandong and ended up in a remote "adobe mud hut,"

115 Burrage, Jarausch and Siegrist, "An Actor-Based Framework for the Study of the Professions," 222. Also Murphy, "Proletarianization or Bureaucratization," 79–80; Noble, *America by Design*, 39–40.

116 Layton Jr., *The Revolt of the Engineers*, 63–65.

117 Murphy, "Proletarianization or Bureaucratization," 74–79.

118 JACAE, 2, no. 4 (1921): Xiling Xiong, Address at ACAE Annual Spring Meeting, April 5, 1921, 33.

villagers had surrounded and curiously inspected his Ford, as they had never seen a car before.[119] In 1924, of 8,510 automobiles in China, Yunnan counted one registration and Guizhou none.[120] In September 1927, that number had grown to one. Manual laborers had disassembled the car and carried it through the mountains to reach Zhou Xicheng's domain.[121] The governor was so proud that the coins he had minted showed his car instead of his face.[122] During a time when the province was so bankrupt that it had to delay indefinitely the proper pavement of the roads outside the vicinities of the capital,[123] the governor ordered seven more cars.[124]

The laboring locals had never seen a motorcar in real life, let alone ride one. As they could not imagine the vehicle for which they built the avant-garde roads, they constructed 30% grades instead of the maximum 8%. A mule could surmount this slope, but not a Model T.[125] While the sight (Figure 23) caused some amusement among the engineers, it is perhaps the clearest demonstration of how far away the usefulness of an exclusive motor road was from the reality of the locals. What ran as a charitable labor relief project was, at the same time, also the vanity project of a provincial governor.

The CIFRC engineers recommended the governments to forbid commoners from using their traditional wheelbarrows on the new roads,[126] as the weight distribution on the narrow wheels would gradually carry off the new road surface. From the viewpoint of the engineers, rational arguments justified the exclusion: the ANRC roads in Shandong had started decaying just two years after their 1921 construction,[127] and the "good roads" of Feng Yuxiang and Yan Xishan faded a few years after their completion, as well. Although the highways succeeded as short-time labor relief, because they faded quickly, they did not contribute to long-term economic development.[128] Consequently, the US

119 Ehlers, *Far Horizons*, 47.

120 JACAE, 5, no. 3 (1924), 37.

121 OJTP, 37.01: Todd to Lynn W. Franklin, September 4, 1927.

122 Paul Bevan, "Zhou Xicheng's 'Guizhou Auto Dollar': Commemorating the Building of Roads for Famine Relief," *Journal of the Royal Asiatic Society* 29, no. 2 (2019): 354–55.

123 OJTP, 37.01: Huang Rongde to Todd, July 16, 1927.

124 OJTP, 37.01: Todd to Howard Page, August 24, 1927.

125 OJTP, 37.01: Todd to CIFRC Guizhou Committee, September 17, 1927; 37.01: Todd to Zhou Xicheng, September 19, 1927; JACAE, 9, no. 1 (1928): Oliver J. Todd, "Extending Motor Roads in the South," 4.

126 See also Eastman, *The Abortive Revolution*, 211.

127 Ehlers, *Far Horizons*, 46.

128 Donald Gillin, *Warlord: Yen Hsi-shan in Shansi Province 1911–1949* (Princeton: Princeton University Press, 1967), 91–92; OJTP, 30.04: Todd to the CIFRC Executive Committee through the Executive Secretary, June 7, 1932.

FIGURE 23 Bad grade near Zunyi, Guizhou, 1927 (CIFRC 1927, *Accomplishments*).

Bureau of Foreign and Domestic Commerce cautioned American motorcar companies against investing in Chinese enterprises, citing the highways' poor quality, lack of maintenance, and especially the destruction put forth by the heavy narrow-tired wheelbarrows.[129] To keep traditional carts off the roads, apart from excluding them through adequately cut-out culverts,[130] government advisor John E. Baker proposed invoking higher tolls or selling licenses for narrow-tired carts and introducing a complete prohibition of them after five to seven years.[131] Todd agreed, proposing scrapping bonuses as an additional means to keep the roads in good condition.[132] The prohibition, as well as the proposed solutions, excluded most of the laborers from the roads they had constructed.

The civil engineers did not express regret at the exclusion of commoners from the roads. Their emphasis on practical experience in the field (chapter 4) meant that civil engineers had to strategically distinguish themselves from vocational artisans and unskilled laborers in order to be recognized as part of society's upper, gentlemanly echelon. As students in the United States, they had already learned to differentiate themselves from Chinese laundrymen (chapter 1). Writings of the Sino-American circle reiterate this distancing through a cold and calculating treatment of the working force. For instance,

129 Clements, *China*, 21.
130 JACAE, 8, no. 8 (1927): Hardy Jowett, "Communications of China [cont.]," 15.
131 JACAE, 2, no. 5 (1921): John Earl Baker, "Broad Tires the Next Step towards Good Roads," 35.
132 JACAE, 2, no. 11 (1921): Oliver J. Todd, "Famine Relief and Road Building in Shantung," 39.

John E. Baker of ANRC fame compared the required feeding and physical limits of humans and pack animals and contrasted their cost efficiency.[133] Yang Baoling, who at that time controlled 7,000 workers on the improvement of a dike, evaluated different gangs according to their native place. He described how laborers from one village "were the best" because they carried 45 kg instead of the average 34 kg but named as a disadvantage that they "had to be fed with wheat rice."[134] There is little giveaway that Yang's naturalizing statistics describe human workers from different villages and not different breeds of cattle.

In Guizhou, common interests subordinated under the ambitions of the military leader in command. When Governor Zhou Xicheng prepared his army to march along the roads and attack Communist strongholds in neighboring Hunan,[135] it became evident that the new civil motor roads benefited militarist expeditions. The civil engineers, vocal against the political instability that "militarism" had been causing, tried to remain ignorant about the misuse of their work. Todd, two months after learning about Zhou's military plans, continued to praise the motor roads for "opening up" Guizhou's economy toward Hunan and the Yangzi Valley.[136] A few months later, Governor Zhou fell in one of his battles. Legend has it that he died in his automobile. He planned to spearhead his army in his lonely motorcar but drove too far ahead and into the enemy.[137] This legend could act as a metaphor for how the Sino-American engineers, despite having been allowed to construct their motor roads, did not reach far in their professional service to the country, as the broader population could not upkeep the avant-garde infrastructure nor profit from it.

6 Continuities between Xi'an and Lanzhou

The methods explored through the Sino-American civil engineering ventures subsisted in later institutions. The Xi'an–Lanzhou Road was a transitional project in that it started as part of the labor relief program of the CIFRC and was completed by a new national commission.

133 Baker, *Explaining China*, 190–95.

134 JACAE, 8, no. 2 (1927): Baoling Yang, "A Story of Repair Works on the Yung Ting Ho," 8.

135 OJTP, 37.01: Sia O. Ling to Todd, June 9, 1927.

136 OJTP, 37.01: Todd to CIFRC Guizhou Committee, September 17, 1927; 37.01: Todd to Zhou Xicheng, September 19, 1927.

137 *The China Weekly Review*, "'Motor-Car' Dollar Which Brought Bad Luck to Kweichow General," November 11, 1933, 450. Cited in Bevan, "Zhou Xicheng's 'Guizhou Auto Dollar'," 358.

The highway from Shaanxi to Gansu was originally a response to the Northwestern famine of 1928. Feng Yuxiang, the "Christian General" and short-lived vice-premier of the new Republic, urged the CIFRC to erect a road from Lingbao, the temporary terminal of the Gansu–Haizhou Railway located near Xi'an (chapter 7), to the capital of Gansu. Sun Ke cabled the request to Todd, who was currently fundraising in the US.[138] The Xi'an–Lanzhou Highway project utilized American donations totaling $350,000 silver and a $200,000 silver budget from the national Chinese government. When completed, the motor road would shorten travel time from 18 to three days.[139]

The CIFRC engineers Shao Fuwu and Zhang Jichun (chapters 4, 6), both ACAE members, oversaw the construction work. Zhang is a unique example of how the Sino-American civil engineering segment influenced and permeated the larger professional body in China. Zhang had never visited the States, but through the training he received in the Beiyang University classes of ACAE member Tan Zhen (chapter 2), and through his job assignments from Todd, who was introduced to Zhang via Tan, the young engineer integrated into the group's cosmos and joined the association. He worked with other association members on the Xi'an–Lanzhou Road and on the Salaqi irrigation project (chapter 6). It was not until his mid-30s, in the middle of his career, that he unsuccessfully applied for scholarships to study in the US.[140] Despite being a member of the ACAE and an author of its journal, he never left China throughout his life.

The mountainous Shaanxi–Gansu border region made for a difficult project. While resident engineers Zhang and Shao were able to survey the region close to the provincial capitals by car, they had to rely on mules for the most tiresome and most scenic section of the road, Mt. Liupan. Situated west of the border town of Pingliang and east of the Gansu counties of Jingning and Longde, cutting a 7% grade across the 2,700 km high elevation required lengthy manual labor and the largest portion of the project's budget.[141]

---

138 OJTP, 29.12: Sun Ke through Li Jinlun to Todd, October 26, 1928. Feng had earlier built roads in the region, including a decayed "sky line" from Anding to Jingning: JACAE, 4, no. 8 (1923), 37; 8, no. 8 (1927): Hardy Jowett, "Communications of China [cont.]," 18; OJTP, 30.04: Todd to the CIFRC Executive Committee through the Executive Secretary, June 7, 1932.

139 Oliver Todd, "The Si–Lan Road," *The China Critic* 7, no. 44 (1934): 1074; JACAE, 19, no. 1 (1938): Oliver J. Todd, "Engineering Experience in China during the Last Two Decades," 35.

140 OJTP, 31.03: Todd to Geo. T. Seabury, June 3, 1937.

141 OJTP, 30.05: Zhang, Jichun, "Investigation Report on Silan Road," October 12, 1933; China International Famine Relief Commission, ed., *Engineering Accomplishments: Famine Prevention and Relief Projects, 1932–33*, A 38 (Beijing: s. n., 1934), 5–6.

FIGURE 24 Manual cutting of rocks at Tingkou, West Shaanxi, July 1933 (CIFRC 1934, *Accomplishments*).

The highway was the final project for which the CIFRC followed its proven concept. Utilizing famine labor meant that the undertaking also encountered similar issues as the earlier projects. Labor relations remained impersonal, with Zhang's and Shao's contact with the workforce limited to a chosen few foremen. The region bordered the Northern Shaanxi Mountains, a largely lawless region that would become the destination of the communists' Long March in 1934. Making use of the weak state, highwaymen robbed the workers' pay before they could exchange it for food. The enlisted famine sufferers remained scapegoats for delays, with Zhang lamenting that stone cutting (Figure 24) was slow because the "local people" were "stupid for the work."[142] Like the other CIFRC highways, once completed, narrow-wheeled carts and donkeys would be forbidden, thereby excluding the peasant population that was building the road.[143] Nonetheless, during his visit to China, the Dean of Canterbury Hewlett Johnson (1874–1966), communist sympathizer and engineer by profession, was amazed by the progress made and called for missionaries deployed in the famine area to abandon their soup kitchens and start recruiting for the road

142 OJTP, 30.05: Zhang, Jichun, "Investigation Report on Silan Road," October 12, 1933.

143 OJTP, 30.05: Zhang, Jichun, "Investigation Report on Silan Road," October 12, 1933.

instead.[144] In summary, the Xi'an–Lanzhou Road was a continuation of both the labor relief campaign's defects and the ignorance with which the engineers celebrated its advantages.

The prohibition of narrow-wheeled carts did not prevent the highway's rapid decay. After a mountain flood in 1933, the unsurfaced road already required repair. In response, the CIFRC, which had run out of famine funds to administer, prepared plans for the newly established National Economic Commission (*Quanguo jingji weiyuanhui* 全國經濟委員會, NEC, chapter 6) to macadamize a particularly vulnerable 113 km stretch of the road.[145] The state institution inherited the project in March 1934 and opened the highway in May 1935.[146]

As exemplified by this transfer of responsibility, the CIFRC's compulsory labor programs, promoted by the ACAE and by the Good Roads Movement, acted as a blueprint for road building under the Nanjing regime. The Nationalists adopted the system of forced, unpaid labor and the seizure of land.[147] As Generalissimo Jiang Jieshi 蔣介石 (Chiang Kai-shek, 1887–1975) explained, his 1935 "People's Economic Reconstruction Movement" (*Guomin jingji jianshe yundong* 國民經濟建設運動), the material complement to his spiritual "New Life Movement" (*Xin shenghuo yundong* 新生活運動), aimed at creating "harmony between men and object" by "harmoniously" co-relating all human endeavors with factors for production.[148] In other words, of the two reasonings the CIFRC had delivered—the idea of relieving famine sufferers, and the idea that compulsory labor prepared for prosperity under the impression of a "harmony" across classes—the latter "consensus frame" now officially sufficed.[149] Jiang's agenda to "turn soldiers into laborers,"[150] and his expression that "people from all walks of life […] must work together, each class along its respective line, toward [a] common goal,"[151] had their precursor in the Good Roads

---

144 OJTP, 30.04: Todd to the CIFRC Executive Committee through the Executive Secretary, June 7, 1932.

145 China International Famine Relief Commission, *Engineering Accomplishments*, 6, 27. See also Zhihong Chen, "Romance with the Periphery: Chinese Travel Writing about the 'Frontier' during the Nanjing Decade (1927–1937)," *Journal of Tourism History* 8, no. 3 (2016): 231–32.

146 Feng Chin, "National Economic Council," in *The Chinese Year Book: 1936–37*, ed. Chao-ying Shih and Chi-hsien Chang, 2nd ed. (Shanghai: Commercial Press, 1936), 367.

147 Eastman, *The Abortive Revolution*, 208–10.

148 Kai-shek Chiang, "Development of the People's Economic Reconstruction Movement," *The China Reconstruction & Engineering Review* 8, no. 1 (1935): 7.

149 Borrowing the concept from William Gamson, *Talking Politics* (Cambridge, New York: Cambridge University Press, [1992] 1996), 260.

150 Chiang, "Development of the People's Economic Reconstruction Movement," 8–9.

151 Chiang, "Development of the People's Economic Reconstruction Movement," 10.

Movement and the CIFRC's work relief campaigns. Jiang ordered compulsory labor in the winter months in support of water conservancies, afforestation, and road construction. Government employees, including soldiers, teachers, and students, had to work ten days each.[152] For the construction of the Yunnan–Burma Highway, ACAE member Ling Hongxun later recalled, the government forced large numbers of "illiterate aborigines and Tibetans," the "aged, women, and even children," to take part in the construction.[153] As China historian George E. Taylor observed in 1936 when lecturing in the country, the nationalist Reconstruction Movement, including the unfree labor campaigns, led to a "widening gap" between "the intelligentsia and the peasantry."[154]

7 Conclusion

The highway projects of the CIFRC reflect the social distinction processes of professionalization. Resident engineer Peng Binglu met resistance from established groups in Yunnan, which climaxed in a public campaign, threats, and a police report against him. He had to leave his project uncompleted, leading to frustration and professional discontent. In contrast, in Guizhou, resident engineer Huang Rongde's project enjoyed the consistent cooperation from a strong local state. The authoritarian governor had the means to enforce a mass mobilization of labor, which enabled the involved engineers to see their project through to the end. In comparison to Peng's professional discontent, Huang's story appears to be, at least in professional terms, a success. However, when looking at the beneficiaries and burden bearers, their roadwork paints a less idealized picture of a more differentiated reality. The involved professionals did not contest that their infrastructures, apart from potentially aiding famine relief and economic development in the long run, were erected under conditions already illegal in the US environment they admired.

The young civil engineers wanted to "save," reconstruct, and shape the Chinese "nation," but this did not imply that all subjects of the nation would benefit equally. In their attempt to gain recognition as reputable, professional

152 "Nationwide Drive for Compulsory Labour to Facilitate Reconstruction, Dyke-Building and Afforestation," *The China Reconstruction & Engineering Review* 7, no. 3 (1935): 116.

153 Hóngxūn Líng, "China's Epic Struggle in Developing Its Overland Transportation System During the Sino-Japanese War," in *Nationalist China During the Sino-Japanese War, 1937–1945*, ed. Paul Sih (Hicksville: Exposition Press, 1977), 258.

154 George Taylor, "The Reconstruction Movement in China," in *Problems of the Pacific, 1936: Aims and Results of Social and Economic Policies in Pacific Countries*, ed. W. Holland and Kate Mitchell (Illinois: University of Chicago Press, 1936), 402–5.

gentlemen, the young, specialized men made sure to distance themselves from the laboring class. The content of the profession was the discontent of others. The engineers' content was primarily a question of appreciation and potency, while the infrastructure's effect on the national citizens remained an afterthought. The avant-garde nature of the highway in Guizhou, catering to a single registered car in the province, demonstrates this aptly. When the Chinese state took over the highway project, it continued its exploitative labor policies. The idea of labor relief, institutionalized by the Sino-American commission, became a state policy. The CIFRC mass conscription policies combined professional distinction with the professional paradigm of efficiency. As a transitional project, the Xi'an–Lanzhou Road, the last of the CIFRC's ventures, demonstrates the reach and impact of the Sino-American segment's practices upon the wider civil engineering culture in China. While the Guomindang state eventually nationalized much of the larger civil engineering work, features of Sino-American heritage remained. As personified in resident engineer Zhang, who emerged as an industrious member of the Sino-American engineering community despite never having been to the States, Sino-American engineering culture pervaded the profession and left its imprint.

CHAPTER 6

Demise without Exhaustion

The Withdrawal of US Engineers, 1928–1941

After the Nationalists had consolidated their power in 1928, the new Nanjing government sped up infrastructural progress through ambitious development plans, supported by increased treasury income, potent state commissions, and the rise of a new bureaucratic technostructure. Nanjing nationalization politics did not abruptly end the influence of the Sino-American engineering segment, but the latter's dominant position slowly faded. The ACAE as an institution suffered from financial strains and structural hurdles, mirroring the political and economic difficulties engineers faced in their specific projects. The association slowly unraveled under the impression of world affairs, expedited through a crisis of reputation originating in the failure of the Salaqi irrigation project (1929–1932).

1 Early Troubles of the ACAE

The ACAE encountered problems years before the commencement of Nationalist rule. These problems mirrored the financial and infrastructural constraints of the war-ridden twenties. The miserable financial situation of the 1920s Beijing government had journal advertisers—most of them state-owned railway lines—fail to pay their fees.[1] Members employed by the government could no longer pay membership dues, either.[2] Due to civil warfare and the resulting suspensions of train traffic, members from entire regions were unable to join the annual conventions. Frequently, the organizers had to cancel excursions, and in 1923, 1924, and 1927, called off the convention altogether.[3] For the involved members, the experiences with the ACAE served as another reminder of dependence on a consolidated state. When attendance rates did

1 JACAE, 5, no. 10 (1924): T. C. Sun, ACAE Convention Address, 15–16.

2 PJRF, 129: Yang Baoling to Freeman, February 22, 1927.

3 JACAE, 3, no. 4 (1922): "Spring Convention," 2; 4, no. 5 (1923): Notes, 34; 5, no. 10 (1924): Sun Duoyu, ACAE Convention Address, 16; 7, no. 1 (1926): Jin Tao, "The November Meeting of the Association," October 28, 1925, 7; 7, no. 10 (1926): "Association News," 35; 8, no. 11 (1927): "Announcement," 8.

 | DOI:10.1163/9789004549555_008

not reach the quorum of the constitution, and even the ACAE president and vice presidents were unable to join the meetings,[4] the unconsolidated state of the republic limited the members in organizing association affairs and in exercising its internal democratic control.

Ironically, it was the Guomindang's Northern Expedition (1926–1928)—which in nationalist historiography marks the stabilization of the country's affairs—that almost phased out the ACAE and its patriotically motivated members.[5] At the height of the Nationalist conquest, the ACAE still printed a portrait of China's Northern dictator, the Guomindang opponent Zhang Zuolin, and hosted its last convention in his powerbase Mukden. The association and its journal depended on advertisement income from Beijing's Ministry of Communications and its railway lines. The ACAE had no official contacts with the rival Nationalist government, which established its own Ministry of Communications in Nanjing in April 1927, and later moved authority over the railway lines to a newly established Ministry of Railways. On the contrary, in late 1927, the association still elected Beijing–Mukden Railway director Gu Zhen 顧振 (C. "Dan" Ku, b. 1894) as its president, after the latter promised to see personally to his line paying its outstanding advertisement dues.[6] Gu's ascent in leadership only diminished the ACAE's political outlook further. In June 1928, the Japanese army assassinated Zhang Zuolin in a bombing of his train. His son Zhang Xueliang 張學良 (1901–2001) dismissed and temporarily arrested ACAE president Gu, who was in charge of the railway on which the bombing had occurred.[7] After Gu's arrest, the ACAE remained leaderless for over a year and suspended the publication of its journal. It effectively ceased operations.

2 Support for the Command Economy

The ACAE's support of the Beijing government had been strategic and financial. However, its members had become frustrated through failed projects such as the Canal Board (chapter 2), which led to growing disillusionment throughout

---

4 JACAE, 3, no. 5 (1922): "The Directors Meeting, June 11, 1922," 38; 7, no. 1 (1926): Jin Tao, "The November Meeting of the Association," October 28, 1925, 7; 8, no. 12 (1927): Minutes of General Meeting, Beijing, November 22, 1927, 33–34; 11, no. 4 (1930): "Annual Meeting," 56; 19, no. 6 (1938): Samuel M. Dean, "Digest of Board of Directors' Meeting," 402.

5 JACAE, 10, no. 1 (1929): "The Journal's Reappearance," ii–iii.

6 JACAE, 8, no. 12 (1927): Minutes of General Meeting, Beijing, November 22, 1927, 33–34.

7 PJRF, 129: Yang Baoling to Freeman, February 27, 1929.

the 1920s and early 1930s. Although opinion about the Beijing government's condition was critical, the Nanjing government ultimately failed to ease their discontent.

Attempts at historicizing the development of engineering in China often conflate nationalization with professionalization. Following the Nationalist narrative of achieving political emancipation from foreign interference with the onset of the Nanjing decade in 1928, historians have claimed that Chinese civil engineers attained an accompanying form of professional independence during that era.[8] After all, they strategically required close relations with a strong state to realize their profession.[9] But the frustration engineers exhibited about the extent and outlook of their professional opportunities indicate a more nuanced attitude toward the Nanjing political environment.

After the Canal Board had shut down its operations in 1922, leading ACAE figure Yang Baoling (chapters 3, 5, Figure 25) had grown increasingly pessimistic about the outlook of Chinese politics. He had exclaimed that the political situation in China was "getting from bad to worse."[10] For him, the First World War had not done enough to make those in power renounce warfare, speculating that it needed "a still greater calamity to set human hearts right."[11] In this respect, his former colleague Oliver J. Todd summarized that the experiences with the Canal Board "taught caution in the field of government reliability,"[12] and made the involved engineers realize that the "greatest need of all was better government."[13]

After becoming involved in highway surveying, Todd traversed the Chinese regions more frequently than most of his engineering peers (chapter 5). Due to his high mobility, he became Yang's informant on the political situation in China. Todd at first welcomed the Nationalist revolution of 1927 as he expected

8 Köll, "The Making of the Civil Engineer in China," 150, 171; Le, "The Struggle for Autonomy from the State," 289–94; Lin-chun Wu, "Tuīdòng gōngchéng guójiā: Zhōng Měi gōngchéngshī xiéhuì zài Zhōngguó (1919–1941)," *Jìndàishǐ yánjiū* 5 (2018): 131; Lin-chun Wu, "China and the United States: Business, Technology, and Networks, 1914–1941," *Journal of American-East Asian Relations* 27, no. 2 (2020): 133–35; Wu, *Měiguórén wèijìng de Zhōngguó mèng*, 293.

9 Burrage, Jarausch and Siegrist, "An Actor-Based Framework for the Study of the Professions," 207–8; Le, "The Struggle for Autonomy from the State," 382–83.

10 PJRF, 129: Yang Baoling to Freeman, August 30, 1922.

11 PJRF, 130: Yang Baoling to Freeman, April 23, 1924.

12 Todd, *The China That I Knew*, 19.

13 Todd, *The China That I Knew*, 33. His assistant Joseph Henry Ehlers (1892–1988) shared his views: Ehlers, *Far Horizons*, 37.

it to consolidate politics and ease traveling.[14] His favorable depiction of the Southern regime made Yang, for a short time, see in the Cantonese government a potential "new dawn" that brought "sunshine and happiness."[15] After the success of the Northern Expedition, the optimism of their former supervisor John R. Freeman (chapter 2) temporarily returned as well. Freeman had been hoping for "strong and honest men" to restore "law and order" for years[16] and now envisioned an unselfish "patriot of the highest order" in the likes of Benito Mussolini (1883–1945) to lead China to prosperity.[17]

Yang agreed with Freeman that China needed "a man like Mussolini." A strong leader, he argued, could establish "peace and order," and help construction work to commence.[18] However, he did not see such a man in Generalissimo Jiang Jieshi. Already in the final months of the Northern Expedition, Yang lamented that poverty was spreading. Two years into the Nanjing decade, he felt that progress was continuously running backward. Yang expressed his frustration that government expenditure went to the army instead of river improvement.[19] In a change of mind not unusual at this time—Jiang temporarily stepped down in the light of public discontent in 1931—Yang's hopes in the Nationalist government had already faded. He anticipated that another political reset would eventually become necessary.[20]

ACAE honorary member Waddell had become a consultant of the Nationalist Ministry of Railways. In this position, he petitioned the US president-elect—and former engineer—Herbert Hoover to issue a $100 million gold loan for the Chinese "sister republic."[21] Yang despised the lobbying efforts. He believed that as a bridge engineer, Waddell should stick to his profession and "shut up" about politics. In Yang's eyes, the only foreign engineer that qualified to have an opinion about Chinese politics was Todd, as he had gained practical experience in the Chinese interior and was in contact with the locals (chapter 5).[22]

14 OJTP, 02.01: Todd to Ed. and Family, May 1, 1927; Fisher, "Service to China," 88–89.

15 PJRF, 129: Yang Baoling to Freeman, February 22, 1927.

16 PJRF, 130: Freeman to Yu Mingde, May 6, 1924.

17 PJRF, 129: Freeman to Yang Baoling, January 28, 1928.

18 PJRF, 129: Yang Baoling to Freeman, April 26, 1928.

19 PJRF, 129: Yang Baoling to Freeman, July 15, 1929.

20 PJRF, 129: Yang Baoling to Freeman, February 27, 1929; 129: Yang Baoling to Freeman, July 15, 1929.

21 PJRF, 129: J. A. L. Waddell to Freeman, February 7, 1929.

22 PJRF, 129: Yang Baoling to Freeman, April 4, 1931.

Like Yang, Todd soon became frustrated at the new government, as he found that militarists still ruled the country.[23] Furthermore, after experiences made during his highway surveying missions, he complained about the traditional "squeeze system," a form of institutionalized corruption, and about ongoing opium consumption. Despite his reliance on forced labor programs (chapter 5), Todd also voiced sympathy for the plight of the commoners: the real "bandits," he found, were not the "hungry peasants" but the land speculators who coerced survey engineers to tailor infrastructural projects to their favor against the common interest.[24] Similar to Yang, he believed the country needed an existential crisis that surpassed the 1931 Japanese invasion of Manchuria to establish an orderly government.[25]

Three years into the Nanjing decade, the consolidated Chinese state had failed to offer the new engineering opportunities that professionals had hoped for. Before the incremental return of the British Boxer Indemnity commenced in 1931 (chapter 7), the outlook for railroad construction looked especially bleak. The Siping–Taoyuan line was the only line conceived and constructed after the Nanjing government took over, and no new constructions were in sight after the foreseeable completion of the Gansu–Haizhou Railway. The Ministry of Railways was in debt and new railway loans were inconceivable.[26]

FIGURE 25
Yang Baoling, c. 1927 (JACAE 8.2, 1927).

23 OJTP, 30.06: Todd to J. B. Lippincott, February 23, 1934.

24 Todd, *The China That I Knew*, 33.

25 PJRF, 129: O. J. Todd to Freeman, October 23, 1931.

26 JACAE, 5, no. 10 (1924): John Earl Baker, "The Boxer Indemnity and Railway Construction," 1–3.

The situation for private enterprise was not promising, either. For instance, Chinese Engineering & Development, a company founded by Yang Baoling and other ACAE members (chapter 4), had specialized in the sale of American material for engineering projects. The renewed depreciation of silver after the Wall Street Crash of 1929 and the increased import tariff of the Smoot-Hawley Act of 1930 severely affected its business. The Great Depression itself was, in international engineering circles, thought to be the adverse effect of the profession and the technological unemployment it had enabled.[27] Adding to the Sino-American company's hardship, the return of the British Boxer Indemnity after 1931 diverted the focus to British material.[28] Therefore, despite his attempt to gain professional independence through the private market, Yang realized that the outlook of civil engineering always depended on politics and global developments.

Many engineers profited from Nationalist statism and served inside the potent new ministries and commissions.[29] For instance, Canal Board veteran and ACAE member Xue Zhuobin supported the Nationalist government because the stable regime allowed civil engineers to carry out "real constructive work."[30] However, Yang, who in Beiyang times had come to the Chinese bureaucrats' defense (chapter 2), was less apologetic toward the leadership of the Nationalist Party. Criticizing the new one-party rule, he complained that an engineer's career chances now depended on party membership, not solely on one's expertise.[31] Xue's old friend Tan Zhen likewise reflected two decades later that there was too much politics "entangled" with engineering work. As an engineer, he wanted "actual work done and not [be] playing with politics."[32]

Yang's interest in socialism, evident since his Purdue student days (chapter 1), had noticeably evolved through the years. He emphasized how Nationalist rule

---

27 JACAE, 15, no. 3 (1934): J. A. L. Waddell, "Engineering and the Depression," 37–45; Herbert Chatley, "The Relation between Engineering and the Great Economic Depression," *The China Reconstruction & Engineering Review* 5, no. 2 (1935): 62–63; Layton Jr., *The Revolt of the Engineers*, 228–29.

28 PJRF, 129: Yang Baoling to Freeman, April 4, 1931.

29 See Kirby, "Engineering China."

30 PJRF, 129: Xue Zhuobin to Freeman, August 1, 1928.

31 PJRF, 129: Yang Baoling to Freeman, February 27, 1929.

32 OJTP, 33.06: Tan Zhen to Todd, January 15, 1946. Similar comments from Todd's assistant Zhang Jichun: OJTP, 31.03: Zhang Jichun to Todd, August 3, 1937; 31.03: Zhang Jichun to Todd, September 10, 1946.

had, in his perception, made "people more poor and miserable."[33] In one of his last letters to Freeman before the latter's passing, Yang concluded:

> My personal opinion is that unless there is radical change of the present policy of selfishness and self-advertising, there will be an upheaval in China such as the world has never seen. I believe the most important right for the human being is the Right to Live, but we have now in China more than 200,000,000 people whose right to live has been taken away from them. They are called Bandits, Communists, etc., but in actual facts they are poor industrial people without the means of living. The world has seen the Russian Revolution, but if nothing is done to help the situation in China, there will be greater loss and worse complications.[34]

During the Beijing era, when Freeman had criticized that militarism impeded engineering projects, Yang had instead focused his blame on foreign meddling and the lack of international investment. In the Nationalist era, however, when Freeman saw Generalissimo Jiang as an able authoritarian leader, Yang became frustrated that the government diverted public money from infrastructure to the army.

In contrast to Yang's sympathies for the laboring class, the CIFRC—for which many of the ACAE core members worked—made use of compelled labor whenever it helped their construction projects (chapter 5). However, while most professionals, also for reasons of social distinction, did not identify with the conditions of the working class, they did not vehemently ignore them, either. For instance, Freeman also criticized industrial production, but from a solely Christian and often anti-Communist stance: his primary concern about the working conditions inside factories was that they were "crushing individualism." Nor did Freeman oppose socialist policies. When he suspected that landowners forestalled his scheme of straightening the Yellow River, he proposed the government expropriate the required land.[35] However, he, like Wang Jingchun and Ling Hongxun, opposed "the revolt against civilization," i.e., Bolshevism.[36] Yang Baoling and Tan Zhen, however, coming from the same

33 PJRF, 129: Yang Baoling to Freeman, April 4, 1931.

34 PJRF, 129: Yang Baoling to Freeman, April 4, 1931.

35 PJRF, 129: Freeman to W. W. Peter, June 29, 1922.

36 PJRF, 130: John Freeman, Memorandum for Address at Chinese Students Conference Brown University Providence, January 10, 1923, 4. For Wang, see chapter 3, for Ling, see chapter 7.

US civil engineering background and having shared the same disillusionments, came to different conclusions. To different degrees, they started to embrace the Communist movement.[37]

To set themselves apart as a profession that yields esoteric knowledge, engineers required "a world of not-knowers."[38] This self-aggrandizement generated frustration, as the engineers realized that the "not-knowers" who legitimized their professional claims remained gatekeepers who stymied their professional goals. In the US, technocratic tendencies among engineers emerged in response to the alleged incompetence of elected officials.[39] Similarly, in Weimar Germany, anti-democratic tendencies surged in reaction to a government not adequately responding to professional needs.[40] In both cases, historians nowadays conclude that the resulting calls for technocracy proved incompatible with democracy.[41] In contrast, the Anglo-American returned students were "liberal intellectuals" who saw no contradiction between democracy and nationalism.[42] Nevertheless, although it is true that the returned civil engineers reflected on their situation and the ideal state in a nuanced way void of agitation and anti-democratic dogma, they would not remain unreceptive to authoritarian ideas.

The turn from liberal, republican socialization to the desire for a stronger, authoritarian state may at first appear contradictory. However, it mirrors precisely the technocracy movement that took place inside the engineering profession in the US at that time. Democratic thought helped the new profession to argue against already established classes perceived as "selfish," be it monopolists in the US case or old-era bureaucrats in China. Eventually, the argument of one's superior expertise led to the call for strong, centralized planning. US historiography calls this conservationist engineering politics a "Janus-like

37 Tan eventually became a politician of the Communist Party in the 1950s. For Tan on industrialists and the working class, see Zhēn Tán, "Zhènxīng gōngyè zhī zhàng'ài jí qí sǎochú zhī fāngfǎ: Tán Quánfǔ zài gōngzhuān jīxiègōng xuéhuì zhī kēxué yǎnjiǎng cí," *Dàgōng bào*, May 15, 1929, 241.

38 Larson, "In the Matter of Experts and Professionals, or How Impossible It Is to Leave Nothing Unsaid," 43.

39 Akin, *Technocracy and the American Dream*, 1–2.

40 Jarausch, *The Unfree Professions*, 52.

41 Akin, *Technocracy and the American Dream*, 165–69; Layton Jr., *The Revolt of the Engineers*, 63–65; Renneberg and Walker, "Scientists, Engineers and National Socialism," 4.

42 Fung, *In Search of Chinese Democracy*, 10–12.

posture,"[43] a metaphor also used to describe the ideology of the fascistic wing of the Nationalist Chinese government.[44]

Regardless of whether the individual supported the Nationalists or communism, the Sino-American engineers had gradually moved away from political liberalism and arrived at supporting a form of command economy—a command economy that would enable infrastructural projects, not enlarge the military.[45] They were not alone in their turn of mind. Prominently, geologist Ding Wenjiang advocated a "new-style dictatorship" which based its rule on the guidance of technocratic elite to serve national interest.[46] The philosopher Li Zehou theorized that, in the first half of the 20th century, the two paradigms enlightenment (*qimeng* 啓蒙) and "saving the country" (*jiuwang* 救亡) characterized Chinese thought and reformist action, and that the latter dominated over the former.[47] Applying this conception to the field of civil engineering, practitioners may not have given up on their ideals but have given precedence to national reconstruction. As historian Fung makes clear, "no one favored a weak state"—not even democrats.[48]

3 The Resurrection of the ACAE

With the commencement of the Northern Expedition and the establishment of the Hankou government in 1926, the Guomindang and its Nationalist platform became the pivotal element of Chinese politics. The Beijing Tariff Conference of 1925–1926 remained the last occasion on which the US government actively tried to define a "new order" for the Far East.[49] While the rise of the GMD was in close connection to the anti-imperialist movement, which, by extension,

43 Layton Jr., *The Revolt of the Engineers*, 63–65, 116–18.

44 Maggie Clinton, *Revolutionary Nativism: Fascism and Culture in China, 1925–1937* (Durham: Duke University Press, 2017), 64.

45 While the Guomindang Leninist state economy, taken as a whole, is better described as a "planned economy," most civil engineers worked in a literal "command economy" which required them to swiftly react to the changing circumstances (chapter 7).

46 Charlotte Furth, *Ting Wen-chiang: Science and China's New Culture* (Cambridge: Harvard University Press, 1970), 217–20; Fung, *In Search of Chinese Democracy*, 104–7.

47 Zéhòu Lǐ, *Zhōngguó xiàndài sīxiǎng shǐ lùn,* Lǐ Zéhòu shí nián jí 3 (Hefei: Ānhuī wényì chūbǎnshè, [1987] 1994), 29–45. Discussed in Fung, *In Search of Chinese Democracy*, 6.

48 Fung, *In Search of Chinese Democracy*, 9.

49 Akria Iriye, *After Imperialism: The Search for a New Order in the Far East 1921–1931* (Chicago: Imprint Publications, [1965] 1990), 2–3.

built on anti-foreign sentiments, the nationalist rhetoric of the new government did not translate into a sudden expulsion of foreign experts from the Sino-American engineering bubble. On the contrary, the American engineers within the ACAE were quick to adapt.

Among the swiftest adapters were Todd and the ACAE's honorary member J. A. L. Waddell (chapters 2, 4). Despite Todd claiming to have "no sympathy with the attitude of foreigners who would interfere in Chinese politics,"[50] he networked with various Chinese politicians and offered his services to them. Todd's claim of noninterference was an opportunistic excuse to network with influential people regardless of their affiliation: from the Northern militarists to Southern nationalists to communist Zhou Enlai 周恩來 (1898–1976), Todd would rely on the support from a wide range of politicians throughout his career. With the rise of the Nationalists after 1926, he quickly saddled with the new powers, more precisely with its left wing in Hankou. The Song clan, including Kong Xiangxi 孔祥熙 (H. H. Kung, 1881–1967) and Sun Yat-sen's widow Song Qingling 宋慶齡 (1893–1981), invited Todd to their Hankou garden parties.[51] Todd proclaimed his "sympathies with the progressive movements" and invited the vice minister of foreign affairs, Tang Yueliang 唐悅良 (1888–1956), to dine at his home, recommending himself for new road constructions and river improvements.[52] Despite the US' nonrecognition of the Southern government and the ACAE's continuing allegiance to Beijing, Todd started accepting contracts from the Guomindang.

In early 1927, Sun Yat-sen's son Sun Ke and the "anti-imperialist" diplomat Eugene Chen (Chen Youren 陳友仁, 1878–1944) borrowed Todd's services from the China International Famine Relief Commission (CIFRC, chapter 5) to help with dike repairs in Hankou.[53] Sun Ke, himself a Columbia graduate, had previously promoted municipal reform in Guangzhou with the help of US-educated engineers.[54] Todd remained in close contact with Sun, writing to him directly and receiving responses from Sun with a penned signature. He did not need to go over secretaries or other bureaucrats but addressed him without intermediaries, even for minor, bureaucratic requests.[55]

50 OJTP, 02.01: Todd to H. B. Elliston, December 28, 1927.

51 OJTP, 02.01: Todd to Margaret Smith, June 5, 1927; Fisher, "Service to China," 81–89.

52 OJTP, 29.11: Todd to Tang Yueliang, April 22, 1928.

53 OJTP, 02.01: Todd to Margaret Smith, March 30, 1927. See also OJTP, 29.10: Sun Ke to Todd, February 5, 1927.

54 Jeffrey Cody, "American Planning in Republican China, 1911–1937," *Planning Perspectives* 11, no. 4 (1996): 343.

55 OJTP, 29.10: Todd to Sun Ke, March 11, 1927.

Todd's case shows that career upheavals depended more on political affiliation than nationality. This becomes clear when contrasted against the case of ACAE honorary member Wang Jingchun (chapters 1, 3, 4). In theory, the railway administrator had many connections to the new GMD regime. He had worked with the new Nationalist Minister of Finance, Wu Chaoshu 伍朝樞 (1887–1934), at the Versailles Peace Conference after he had studied under the scholarship of his father, Wu Tingfang. When the GMD "reunited" China through the Northern Expedition campaign, government ministries and other state institutions temporarily competed against the Beijing ministries. In the case of the Ministry of Communications, Wang Boqun 王伯群 (1885–1944) headed the new ministry in Nanjing, while at the same time Pan Fu, the former director of the Canal Board (chapter 2), led the old ministry in Beijing for another year before fleeing into exile. Like his former Chinese Students Alliance colleague Gu Weijun, against whom the new government issued an arrest warrant, Wang had remained loyal to the Northern government until the very end. As late as February 1928, the Northern Expedition in full swing, he had worked with a commission of telegraphy experts and linguists at the Beijing Ministry of Communications to create and publish a standardized version of alphabetized Chinese.[56] After the consolidation of the new government, the Nanjing Ministry of Railways, born out of the Nanjing Ministry of Communications in 1928, stocked its personnel with people from the GMD home province of Guangdong. In 1929, 37.5% of Railway Ministry employees came from that province.[57] There was no place for the loyal northerner Wang. For the rest of his life, he worked in remote positions in the US and London, never returning to influence at the ministry headquarters.

Meanwhile, Waddell had accepted an advisory position at Sun Ke's new Ministry of Railways. There, he petitioned Sun to install Todd as chief counselor of a new highway commission. While the highway commission never materialized due to a lack of funds and Sun's 1931 departure from the ministry, Todd used his direct access to Sun to collect accumulated dues of $5,000 (today approx. $75,000) of unpaid advertisement fees the railway bureaus owed to the ACAE.[58] Having secured the financial support of Sun's ministry, Todd and Waddell are to be credited with the "resurrection" of the ACAE in September 1929.

ACAE leaders used the 1929 World Engineering Congress in Tokyo as an opportunity to revitalize the organization. They hosted a reception in Beijing

---

56 Ching-chun Wang, "A Roman Alphabet for China," *Asia* 30, no. 6 (1930): 460.

57 Tyau, *Two Years of Nationalist China*, 493.

58 JACAE, 10, no. 1 (1929): "The Journal's Reappearance," ii–iii; PJRF, 129: Yang Baoling to Freeman, February 27, 1929; OJTP, 29.13: Todd to Sun Ke, April 2, 1929.

for engineers traveling through China[59] and printed advertisements for their association to distribute in Japan.[60] ACAE representation at the congress was disproportional. Attendants included John R. Freeman, Ling Hongxun, Shi Zhaoxiang, Oliver J. Todd, Yang Baoling, J. A. L. Waddell, and Wang Zhengfu 王正黼 (b. 1891). Although, in hindsight, the ACAE's gradual and ultimate decline in the 1930s commenced around the time of the Tokyo Engineering Congress,[61] it had nothing to do with it. On the contrary, the congress of 1929 coincided with, as Todd later put it, the ACAE's temporary recovery from its "nearly dead" status.[62]

After the revival of the ACAE, the association largely revolved around the same group of actors as during the twenties. Elected executives consisted of Grand Canal veterans such as Tan Zhen, Oliver J. Todd, Yang Baoling, and Xue Zhuobin, senior railway engineers such as Jin Tao 金濤 (T. King, b. 1888) and Wang Jinzhi, as well as a smaller group of foreign educators including Samuel M. Dean and John A. Ely (table 5). Although US-educated engineers from the post-1918 generation started entering the society, they rarely influenced the association in a visible way. For instance, when the acclaimed hydraulic engineer Zhang Hanying 張含英 (1900–2002) assumed his first influential posts in the 1930s, within the ACAE, a well-established circle already existed there to represent the professional field. Nonetheless, a small group of engineers from the younger generation, among them Li Shutian, Ling Hongxun (chapter 7), and Georg Gottlieb Stroebe (1900–1988), gradually assumed more prominent roles in the society leadership.

4 Crisis of Repute—The Salaqi Irrigation Project

The ACAE's decline started in the early 1930s, when a botched irrigation project severely damaged the association's reputation. From 1929 to late 1932, civil engineers of the CIFRC attempted to irrigate the Salaqi plains. The plains are part of today's Inner Mongolia Tumed Right banner, located between the cities of Baotou and Hohhot. The Beijing–Suiyuan Railway connected the region. The Yin Mountain range flanks the line to the north, while the Yellow River flows south of the plains. The idea was to use the water of the Yellow River

59 OJTP, 02.03: Todd to Yang Baoling, November 14, 1929.

60 OJTP, 02.03: Todd to Dwight W. Edwards, October 20, 1929.

61 Wu, "Tuīdòng gōngchéng guójiā," 131; Wu, "China and the United States," 133–35; Wu, *Měiguórén wèijìng de Zhōngguó mèng*, 293.

62 OJTP, 30.06: Todd to J. B. Lippincott, February 23, 1934.

TABLE 5 Timeline of ACAE executives after its "revival," 1929–1941

<table>
<tr><th></th><th>President</th><th>First Vice-President</th><th>Second Vice-President</th><th>Treasurer</th><th>Secretary</th></tr>
<tr><td>1929</td><td>Oliver J. Todd</td><td></td><td></td><td rowspan="3">Harry A. Raider</td><td rowspan="2">Xue Zhuobin</td></tr>
<tr><td>1930</td><td rowspan="2">Yang Baoling</td><td rowspan="2">T. N. Miller</td><td rowspan="2">Jin Tao</td></tr>
<tr><td>1931</td><td>Tan Zhen</td></tr>
<tr><td>1932</td><td>Jin Tao</td><td>Arthur M. Shaw</td><td>F. M. Titus</td><td>Wang Bi</td><td rowspan="8">Oliver J. Todd</td></tr>
<tr><td rowspan="2">1933</td><td>F. M. Titus</td><td rowspan="4">Samuel M. Dean</td><td>Wang Jinzhi</td><td rowspan="9">Jin Tao</td></tr>
<tr><td rowspan="2">Wang Jinzhi</td><td>Wang Bi</td></tr>
<tr><td>1934</td><td>Chen Bozhuang</td></tr>
<tr><td rowspan="2">1935</td><td>Wang Chongyou</td><td rowspan="2">Li Shutian</td></tr>
<tr><td>Samuel M. Dean</td><td>Chen Bozhuang</td></tr>
<tr><td>1936</td><td>Chen Bozhuang</td><td>Li Shutian</td><td>Georg G. Stroebe</td></tr>
<tr><td>1937</td><td>Li Shutian</td><td>Georg G. Stroebe</td><td>Ling Hongxun</td></tr>
<tr><td>1938
1939
1940</td><td>John A. Ely</td><td>Tan Zhen</td><td>Xue Zhuobin</td><td>Roy L. Creighton</td></tr>
<tr><td>1941</td><td>Tan Zhen</td><td>Xue Zhuobin</td><td></td><td>Samuel M. Dean</td></tr>
</table>

and have it flow through a new canal parallel to the railway (Figure 26). Gates would allow the water to enter laterals and irrigate the fields. Finally, the water would discharge in the east through what was then the Black River (nowadays the straightened Hasuhai Waste Canal) and flow back into the Yellow River. Initially, the costs figured at 750,000 to 800,000 dollars, of which the CIFRC was to contribute 500,000.[63] Many of the ACAE inner circle, including Samuel Dean, Sigurd Eliassen, Shao Fuwu, Tan Zhen, Oliver J. Todd, Xue Zhuobin, and Zhang Jichun, took part in the project.[64]

63 JACAE, 11, no. 11 (1930): Oliver J. Todd, "Getting Yellow River Water on the Saratsi Plain," 10.

64 Samuel M. Dean supported the Salaqi project when he sent his Beiping North China School of Engineering students to help the project. See JACAE, 11, no. 11 (1930): Oliver J. Todd, "Getting Yellow River Water on the Saratsi Plain," 10.

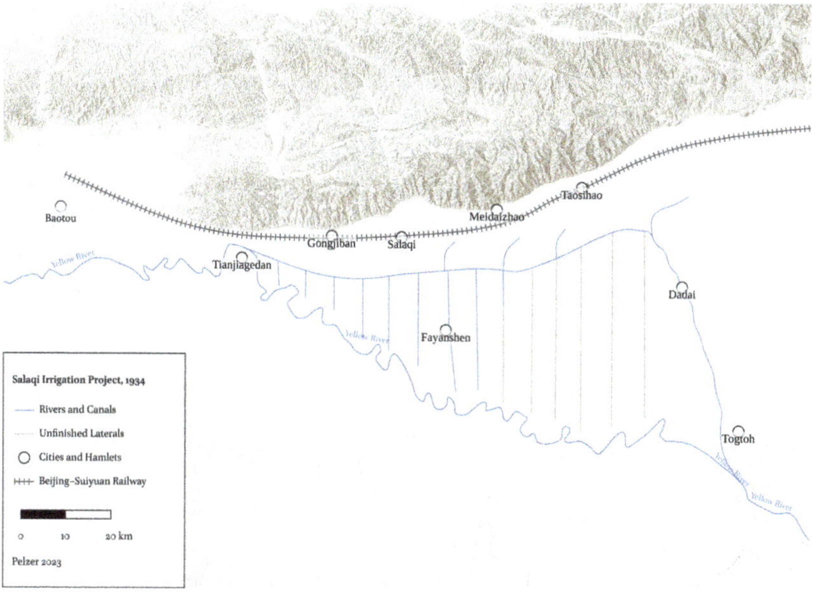

FIGURE 26 The Salaqi irrigation project's state in 1934.

Xue Zhuobin, one of the young engineers to aid the Grand Canal investigations in the late 1910s who had since ascended the ranks of the ACAE and had joined the CIFRC, led the Salaqi surveys that started in April 1929. Surveying time was just two months, not up to the professional standards the civil engineers had set for themselves and which they had frequently cited ten years earlier when explaining why construction at the Grand Canal could not begin before every measurement was taken and every blueprint drawn (chapter 2). This time, given the Salaqi project's designation as labor relief, work had to commence early before the famine was over, at which point the workers would no longer be in immediate need of work and food. The local Salaqi Construction Bureau (*Salaqi jianshe ju* 薩拉齊建設局) and the CIFRC co-managed the construction work that started in June of that year (Figure 27).[65]

The engineers did not have the time and budget for a proper survey, so the efforts ended in an unusable, silting canal system. Todd had known of the, as the laboratory results put it, "disturbing" quantities of alkali in the soil and

65 JACAE, 16, no. 3 (1935): Oliver J. Todd, "A Review of the Saratsi Irrigation Project," 139–140.

the nonoptimal silting since at least 1930.[66] Engineer Zhang Jichun had also warned of the system's potential decay.[67] Apart from design issues, the famine which led to the project had already passed. During the years of the construction, the local farmers enjoyed great harvests and saw no need to substitute their farm work with famine labor.[68] The engineers ended up short on workforce. The situation undermined the CIFRC's argument of constructing insufficiently surveyed projects to provide immediate work relief. The organization had to recruit outside labor and conscript soldiers to "save the situation."[69] Samuel M. Dean wrote that he and Zhang Jichun felt "rather lost" in the work.[70] In late 1932, in a move not unusual, the CIFRC, having spent its allocated funds,[71] handed the project over to the municipal construction office for completion.

In 1933, before the completion of the project, the Salaqi region saw abundant rainfall that led to mountain torrents, flooding, the breaking of the banks, and the sanding and silting of the canal.[72] In the eyes of the government, not only had the project not achieved its aim of immediate work relief, but it also had already failed to achieve its goal of long-term famine prevention before its initial opening. In response, the newly founded National Economic Commission sent its representatives to Salaqi in the spring of 1934 to investigate. In December of that year, the NEC's foreign advisors, European engineers from the League of Nations, arrived on site. Apart from the silting problem, the experts found that the even-leveled canals lacked an adequate pumping system. The League described the Salaqi project as having "very serious fundamental defects" that rendered it ultimately "useless." In remarks that moved between damaging the reputation and outright wrecking it, the report ended in proclaiming that the project was "a total failure almost unparalleled in the history of irrigation."[73] After receiving their report, the NEC decided to abort

66 OJTP, 37.02: Chas. F. Shaw to Todd, July 4, 1930; 37.02: Todd to David A. Brown, October 29, 1932. See also JACAE, 16, no. 3 (1935): Oliver J. Todd, "A Review of the Saratsi Irrigation Project," 144–146. Also Todd to Eliassen: "You are correct in stating that our own Commission should bear a great deal of blame," OJTP, 37.02: Todd to Siguard Eliassen, March 26, 1935.

67 OJTP, 37.02: William Kelly to Todd, November 26, 1932; 31.03: Zhang Jichun to Zhang Yuanshan, May 2, 1935.

68 JACAE, 11, no. 11 (1930): Oliver J. Todd, "Getting Yellow River Water on the Saratsi Plain," 2.

69 OJTP, 37.02: Todd to Zhang Yuanshan, August 12, 1930; 37.02: Sam Dean to Todd, August 21, 1930; JACAE, 16, no. 3 (1935): Oliver J. Todd, "A Review of the Saratsi Irrigation Project," 141.

70 OJTP, 37.02: Samuel M. Dean to Todd, September 15, 1930.

71 OJTP, 33.06: Todd to Tan Zhen, March 13, 1935.

72 JACAE, 16, no. 3 (1935): Sigurd Eliasson, "The Saratsi Project," 149–150; 16, no. 3 (1935): Oliver J. Todd, "A Review of the Saratsi Irrigation Project," 141.

73 A. Coode et al., *Report by the Committee of Experts on Hydraulic and Road Questions in China: Co-operation between the Organisation for Communications and Transit of the*

FIGURE 27 Digging the Salaqi main canal intake, May 1932 (CIFRC 1934, *Accomplishments*).

the project. CIFRC Secretary Zhou Yichun, who had broken off his friendship with Todd two years prior over his alleged lack of self-reflection, now prepared for Todd's contract termination.[74]

As chief engineers involved in the project for the CIFRC were also prominent executives of the ACAE, criticism of the project hit the Sino-American civil engineering segment at its core. The Salaqi issue was not a conflict between engineering and bureaucracy, politics, or the working class, as in previous cases (chapters 2, 3, 5). Rather, the criticism came from the technical experts of the League, engineers themselves. In reaction to their criticism, Todd aligned his fellow engineers to a public campaign aimed at saving their reputations.[75]

Simply put, the strategy was to blame the project's defects on the local administration that had taken over from the CIFRC in 1932. For ACAE member Li Shutian, the failure of the Salaqi irrigation project was rooted in a lack of time and authority. First, the engineers had sacrificed their principles of

---

League of Nations and the National Government of China, Series of League of Nations Publications 8.4 (Geneva: League of Nations, 1936), 110, 116.

74 OJTP, 02.04: Zhou Yichun to Todd, January 11, 1932; 30.06: Minutes of the 132nd Meeting of the Executive Committee of the CIFRC, Beijing, July 20, 1934; Fisher, "Service to China," 132–33.

75 OJTP, 33.06: Todd to Tan Zhen, March 13, 1935; 37.02: Todd to Zhang Jichun, October 6, 1936.

proper recording and surveying for a "rush work" to provide immediate "emergency famine relief." Then, they handed it over to the locals too early and could not guarantee its proper maintenance.[76] The involved engineers joined Li in shifting the blame to the locals. Their "unskilled irrigators," Zhang Jichun complained, applied too much water and were partly responsible for the area's flooding and silting.[77] Todd agreed that the "non-engineers" had shown "no intelligent leadership."[78] Befriended missionary William Kelly (1874–1957) likewise confirmed that the residents were "frontier people" who "do as they please."[79]

Oliver J. Todd and Tsinghua professor Zhang Ren 張任 (b. 1906) felt the NEC's criticism of the project's lacking profitability ignored the engineers' "humanitarian" duty and the original project aim of famine relief. While at other times, the civil engineers were eager to find fault with the government agencies' lack of efficiency, this time, Todd argued, "not all of [the civil engineer's] plans [were] to be based on the immediate profit motive."[80] The solutionist mindset of the civil engineers did not accept their project to have failed in principle. While they agreed current construction was insufficient, they were certain the project simply needed a fix. The solution was to build another weir at $500,000, which would lessen the danger of over-irrigation, increase the area of irrigated land from 533 km^2 to 800 km^2, and allow for the diversified harvest of oat, wheat, and potatoes.[81] Like Zhang Ren, Todd's right-hand Zhang Jichun argued that the NEC's decision was shortsighted, as future famines were certain to arrive and only a completed Salaqi canal would help against them.[82]

The ACAE invited the Shaanxi government and the NEC to a conference in Beijing, where the members attempted to save their reputations. However, the NEC declined to attend and to assist the ACAE and the CIFRC in their

76 JACAE, 16, no. 3 (1935): Shutian Li, "On the Saratsi Irrigation Project," 168.

77 JACAE, 16, no. 3 (1935): Jichun Zhang, "The Silt Problem and the Water Supply of the Saratsi Project," 160.

78 JACAE, 16, no. 3 (1935): Oliver J. Todd, "A Review of the Saratsi Irrigation Project," 139–140. See also Todd, *The China That I Knew*, 33.

79 OJTP, 37.02: William Kelly to Todd, November 26, 1932. See also JACAE, 16 (3): Oliver J. Todd, "A Review of the Saratsi Irrigation Project," 139–148: 143.

80 JACAE, 16, no. 3 (1935): Oliver J. Todd, "Further Discussion of Saratsi Irrigation Project," 165; 16, no. 3 (1935): Ren Zhang, The Feasibility of the Saratsi Project," 154.

81 JACAE, 16, no. 3 (1935): Oliver J. Todd, "A Review of the Saratsi Irrigation Project," 144–146.

82 JACAE, 16, no. 3 (1935): Jichun Zhang, "The Silt Problem and the Water Supply of the Saratsi Project," 163; 16, no. 3 (1935): Ren Zhang, The Feasibility of the Saratsi Project," 154–155. Also JACAE, 16, no. 3 (1935): Oliver J. Todd, "Further Discussion of Saratsi Irrigation Project," 164.

defense against the League's criticism.[83] While the *Journal* printed the conference papers in a special issue,[84] this did not sway public opinion, either. While framed as a platform of open discussion, the publication strategy rather followed a professional journal's function to bypass criticism and to reject influence and validation.[85] The affair, which dragged along from 1933 to 1936, demonstrates two sides of a coin: the spirit of solidarity within the ACAE but also its growing isolation as an echo chamber.

5 The Militarization of Civil Engineering

Apart from a crisis of repute, the Sino-American engineering segment also faced institutional changes. The reliance on international non-government organizations like the CIFRC was an embarrassment for a state that claimed national sovereignty, as famine relief had traditionally been a task of the government.[86] The Guomindang arrived in a position that enabled them to renationalize vital institutions. The Nanjing Republic was a reactionary restoration of a "bureaucratic-military administration" in which the bourgeois class, from which many of the US-educated civil engineers came (chapter 1), lost ground.[87]

Traditionally, professionalism, with its emphasis on autonomy, appears antithetical to state bureaucracy.[88] Accordingly, at the onset of the Nanjing era, civil engineers in reputable positions were few. A 1927 statistic about the educational background of 750 leading figures did not include engineering or technology as a category. Subsumed into the category of "others" which consisted of a few handfuls of individuals, they appeared irrelevant against the bulk of well-known political scientists, economists, and educators.[89]

---

83 OJTP, 37.02: Qin Fen to Todd, March 25, 1935; 33.06: Todd to Tan Zhen, March 30, 1935; 31.03: Zhang Jichun to Todd, May 8, 1935.

84 JACAE, 16 (3): 139–169, also see 17 (6): 333–357.

85 Crane, *Invisible Colleges*, 87–88.

86 Chi Kwong, *War and Geopolitics in Interwar Manchuria: Zhang Zuolin and the Fengtian Clique during the Northern Expedition* (Leiden, Boston: Brill, 2017), 130; Janku, "The Internationalization of Disaster Relief in Early Twentieth-century China," 7.

87 Bergère, "The Chinese Bourgeoisie, 1911–37," 722.

88 Celia Davies, "Professionals in Bureaucracies: The Conflict Thesis Revisited," in Dingwall; Lewis, *The Sociology of the Professions*, 177. See also Burrage, Jarausch and Siegrist, "An Actor-Based Framework for the Study of the Professions," 224; Layton Jr., *The Revolt of the Engineers*, 9–10, 74.

89 Y. S. Tsao, "A Challenge to Western Learning: The Chinese Student Trained Abroad—What He Has Accomplished—His Problems," *News Bulletin* (*Institute of Pacific Relations*) Dec (1927): 15.

However, after the establishment of the Nationalist regime in Nanjing, the situation changed gradually. In strong states, professionals depend on, profit from, and integrate into the bureaucracy.[90] Jiang Jieshi, while far from installing engineers in all-important positions, put the methods of scientific management on his broader agenda of economic reform.[91] Similar to how Communist thought eventually fathered the new status class of intelligentsias (*zhishi fenzi* 知識分子), new government institutions established a new social group of expert staff (*zhiyuan* 職員). These experts had often received a German or American education in engineering or metallurgy.[92] The group, similar to the Soviet intelligentsia, formed a relatively independent "technostructure" that collaborated, but not necessarily identified, with the political system.[93]

The concept of autonomy in engineering has always been ambivalent.[94] Adopted from the archetype of earlier Anglo-American professions, civil engineers upheld the ideal. However, especially in the case of civil engineers working on massive infrastructural projects requiring great amounts of capital,[95] there have been but few consultant engineers capable of living up to the professional ideal, while the majority worked in companies or state institutions. As was the case in other parts of the world, Chinese engineers mostly assumed salaried positions.[96] When pursuing large infrastructural projects, they depended on the goodwill of the state. Around the globe, from the US New Deal to the Soviet "new class," engineers turned to the state and became involved in state enterprises.[97]

---

90 Murphy, "Proletarianization or Bureaucratization," 81.

91 Chiang, "Development of the People's Economic Reconstruction Movement," 8. See also Margherita Zanasi, *Saving the Nation: Economic Modernity in Republican China* (Chicago, London: Chicago University Press, 2006), 177–178, 189–199.

92 Morris Bian, *The Making of the State Enterprise System in Modern China: The Dynamics of Institutional Change* (Cambridge, London: Harvard University Press, 2005), 94–99.

93 Bergère, *The Golden Age of Chinese Bourgeoisie*, 291. See also Galbraith, *The New Industrial State*, 74.

94 Burrage, Jarausch and Siegrist, "An Actor-Based Framework for the Study of the Professions," 219; Brante, "Professional Types as a Strategy of Analysis," 78; Davis, *Thinking Like an Engineer*, 165, 178; Thomas Hughes, *Human-built World: How to Think about Technology and Culture* (Chicago, London: University of Chicago Press, 2004), 73; Nye, "Medicine and Science as Masculine 'Fields of Honor'"; Torstendahl, "Introduction," 2–8.

95 Freidson, *Professionalism*, 165–69; Larson, "In the Matter of Experts and Professionals, or How Impossible It Is to Leave Nothing Unsaid," 43; Uekötter, "Techniker an der Macht," 397.

96 See Le, "The Struggle for Autonomy from the State," 7–11.

97 Siegrist, "Professionalization as a Process," 184; Larson, "In the Matter of Experts and Professionals, or How Impossible It Is to Leave Nothing Unsaid," 41–43; Murphy, "Proletarianization or Bureaucratization," 82–91.

In China, this development started with the Japanese invasion of Manchuria in 1931.[98] The Ministry of Industry swallowed the Ministry of Agriculture and Mining and grew into a new super ministry for geologists and mining engineers.[99] In addition, that year, the government formed new potent commissions on the same hierarchical level as ministries but working independently from them. Among these, the NEC—which played a role in the Salaqi investigations—concerned itself with infrastructural development and co-operated with the League of Nations' communications and transit organizations.[100] Meanwhile, the idea of the National Defense Planning Commission (*Guofang sheji weiyuanhui* 國防設計委員會, NDPC) was to investigate and plan economic construction to strengthen national defense.[101] Areas of responsibility were material and human resources, their investigation, surveying, and eventual exploitation.[102] Reporting to the Military Affairs Commission (*Guomin zhengfu junshi weiyuanhui* 國民政府軍事委員會, MAC), the new institution strengthened Jiang's position within the Nationalist government.[103]

The NDPC reorganized as the National Resources Commission (*ziyuan weiyuanhui* 資源委員會, NRC) in April 1935.[104] The NRC's accumulation of authority and responsibility led its chairperson to describe the commission as being its own "virtual cabinet" (*wuxing neige* 無形內閣) within the Nationalist government.[105] Its influence reached its peak in December of that year, when NRC commissioners took over the control of ministries in Jiang's "Cabinet of Action" (*xingdong neige* 行動內閣). Among others, they led the Ministries of Communications, Railroads, Industry, and Foreign Affairs. As such, the December of 1935 was a preliminary climax in the rise of a "technical elite."[106] Trained technicians assumed leading positions in the bureaucratic hierarchy. As the prime example, Weng Wenhao 翁文灝 (1889–1971), a Belgium-trained geologist, became head of the NRC and the Ministry of Economics in 1938 and

---

98 Línsūn Chéng, "Zhōngguó jìhuà jīngjì de qǐyuán yǔ zīyuán wěiyuánhuì," *Shí'èryī shìjì* 82 (2004): 89.

99 Yuan-li Wu, "Industrial Development and Economic Policy," in Sih, *The Strenuous Decade*, 243.

100 Chin, "National Economic Council," 364–65; Jürgen Osterhammel, "'Technical Co-Operation' between the League of Nations and China," *Modern Asian Studies* 13, no. 4 (1979).

101 Yǒukuí Zhèng, Línsūn Chéng and Chuánhóng Zhāng, *Jiù Zhōngguó de zīyuán wěiyuánhuì 1932–1949: Shǐshí yǔ píngjià* (Shanghai: Shànghǎi shèhuì kēxué yuàn chūbǎnshè, 1991), 1–11.

102 Zhèng, Chéng and Zhāng, *Jiù Zhōngguó de zīyuán wěiyuánhuì 1932–1949*, 18.

103 Zhèng, Chéng and Zhāng, *Jiù Zhōngguó de zīyuán wěiyuánhuì 1932–1949*, 1–9.

104 Zhèng, Chéng and Zhāng, *Jiù Zhōngguó de zīyuán wěiyuánhuì 1932–1949*, 1–9.

105 Zhèng, Chéng and Zhāng, *Jiù Zhōngguó de zīyuán wěiyuánhuì 1932–1949*, 22–23.

106 Zanasi, *Saving the Nation*, 190–92.

of the Executive Yuan in 1948. A civil engineer who climbed the ranks of the NRC was Shen Yi 沈怡 (Shen-yi, 1901–1980). However, the most prominent member of the committee, Qian Changzhao 錢昌照 (1899–1988), was an economist, and most other high-ranking members were non-technicians as well.[107] The NRC may have drawn from technocratic principles, but it was far from a fully formed rule of the technocrats,[108] as technocracy came to be understood in later decades.[109] While engineers disproportionately aligned with a strong state, they also rarely joined politics in the first place.[110]

Rather, the rise of the NRC marked the militarization of the engineering discipline.[111] Generalissimo Jiang was "a leader in the old Self-Strengthening mold who saw science as a means of fortifying the state."[112] The Nanjing Republic fought a military and an economic war, both of which the new institutions of the command economy helped to support.[113] NRC leader Weng Wenhao equated "scientific professionalism with both national achievement and good citizenship."[114] The CIE followed suit when it added national reconstruction and the "rescue of the nation" as professional aims to their preamble.[115]

At first, the militarization of engineering may not appear to relate to civil engineering. After all, the NRC became an employer for mechanical, metallurgical, and electrical engineers, but not civil engineers,[116] as it focused on the dismantling and relocation of China's heavy industry, on electricity, and on mining.[117] Accordingly, the NRC primarily adopted the culture of Europe-trained

---

107 Péngyuán Zhāng and Huáiyù Shěn, eds., *Guómín zhèngfǔ zhíguān niánbiǎo 1925–1949* (Taipei: Academia Sinica, 1987), 337–40.

108 Kirby, "Technocratic Organization and Technological Development in China," 29–33, 42; Kirby, "Engineering China," 148–152.

109 For modern Chinese history, see Andreas, *Rise of the Red Engineers*, 83, 257–8. For political theory, see McDonnell and Valbruzzi, "Defining and Classifying Technocrat-led and Technocratic Governments," 656–58.

110 See Jarausch, *The Unfree Professions*, 70–72.

111 See Zhèng, Chéng and Zhāng, *Jiù Zhōngguó de zīyuán wěiyuánhuì 1932–1949*, 49–52.

112 Shen, *Unearthing the Nation*, 112.

113 Zhèng, Chéng and Zhāng, *Jiù Zhōngguó de zīyuán wěiyuánhuì 1932–1949*, 49–52; Kirby, "Engineering China," 150–151.

114 Shen, *Unearthing the Nation*, 128.

115 Köll, "The Making of the Civil Engineer in China," 170–71.

116 Accordingly, Kirby's claim that the work of the Chinese Institute of Engineers' research groups and the work of the NRC became "indistinguishable" is exaggerated (Kirby, "Engineering China," 149–150).

117 Bian, *The Making of the State Enterprise System in Modern China*, 64, 169; Zhèng, Chéng and Zhāng, *Jiù Zhōngguó de zīyuán wěiyuánhuì 1932–1949*, 19–21, 24–27. For the office structure, see Zhèng, Chéng and Zhāng, *Jiù Zhōngguó de zīyuán wěiyuánhuì 1932–1949*, 19.

geologists, not of engineers.[118] Even if there was an ascendance of engineers, the large group of civil engineers that paved highways, constructed railroads, or conserved waterways were not part of it.[119] Nonetheless, the subordination of engineering under the efforts of war also reached the civil profession. After all, to exploit the resources of the Southwest and Northwest, the NRC required a better infrastructure to transport machinery and mobilize personnel. For instance, in order to connect the Yumen oil fields of Gansu with the wartime capital Chongqing, it had to cooperate with institutions such as the Northwestern Highway Bureau (*Xibei gonglu ju* 西北公路局),[120] which collaborated on the highway projects of Ling Hongxun (chapter 7). The GMD's militarization of the state institutions led to the gradual conceptualization of "wartime transport and communications" (*zhanshi jiaotong* 戰時交通).[121] The slogan concerned the question of how professional "civil" engineering expertise could help facilitate national interests of transportation and communication under the new priority of military survival.

Engineering in a military context was not a new development: arguably, in China, the modern engineering occupation, although not its professionalization, had started in the shipyards of Fuzhou and Tianjin in the 19th century and had been transmitted through foreign military instructors. However, military and civil engineering had traditions distinct from one another. The practices, ideals, and areas of responsibility differed between the two disciplines. The civil engineering ideal transmitted from the US was especially anti-militaristic. Freeman had portrayed military engineering as the antagonistic counterpoint to civil engineering, and Todd had publicly longed for a time in which "engineers instead of soldiers" oversaw infrastructure.[122]

In summary, militarization created problems for the Sino-American segment of civil engineers. The Chinese members of the Sino-American segment had to alter their professional ideals of autonomy and the civil use of their infrastructure. This development did not necessarily foster discontent, but it required reorientation. For the American members of the segment, the subordination

---

118 Shen, *Unearthing the Nation*, 7, 125–26.

119 Zhèng, Chéng and Zhāng, *Jiù Zhōngguó de zīyuán wěiyuánhuì 1932–1949*, 24–27, 47–48, 114, 169.

120 Zhèng, Chéng and Zhāng, *Jiù Zhōngguó de zīyuán wěiyuánhuì 1932–1949*, 89–98.

121 See, e.g., Jiā'áo Zhāng, *Zhànshí jiāotōng wèntí* (S. l.: Zhōngyāng xùnliàn tuán dǎng zhèng xùnliàn bān, 1940); Xuésuì Gōng, *Zhōngguó zhànshí jiāotōng shǐ* (Shanghai: Shāngwù yìnshūguǎn, 1947).

122 Oliver Todd and John Waddell, "Some Thoughts Concerning Economics in the Development of China: Memoirs Nos. 11 and 12," *Chinese Social and Political Science Review* 14, no. 2 (1930): 167.

of civil engineering under the wartime command economy was more problematic. They would hardly be allowed to, nor willing to, invest their energy in support of the Chinese war apparatus. While Sino-American engineering, as a professional culture, did not require the presence of US engineers, the profession, until that time, had always been a transnational, cosmopolitan practice. The impending war changed that.

6 The Plight of US Engineers

In Kuhn's classic study of the sociology of scientific knowledge, scientific communities compartmentalize when complex questions require specialization and disappear when methods are exhausted.[123] In contrast, the ACAE's demise occurred when neither was the case: its decomposition did not coincide with a process of differentiation, nor with the profession's exhaustion. At first sight, the ACAE's decline coincided with the rise of the CIE, the grand national umbrella organization, and strengthened confidence in domestic engineering. However, when looking at membership numbers and comparing its growth through the decades, the CIE grew only moderately in the 1930s, ten years in which the count rose from 2,169 to 3,290 members. The institute only really took off in the 1940s, through which numbers quintupled.[124] It is true that in the 1930s, the CIE's importance for the broader profession increased as it started the professional self-regulation of standards,[125] a process previously administered through the Ministry of Communications.

Nevertheless, there was no competition between the two organizations, as the two organizations had no claim to exclusivity. Renowned engineers were members of both organizations, which had friendly relations with one another (chapter 4). The CIE did not aim to inherit the function of the ACAE of representing the Sino-American segment. The US–Chinese exchange remained the uncontested characteristic of the latter, and as seen by the unbroken flow of articles from Chinese members in the ACAE's *Journal*, the unique character of the association remained in demand through the pre-war years (Figure 15 in chapter 4). Hence, rather than understanding the ACAE's fall in importance by looking at the rise of the CIE's national and international repute, understanding the association's demise means understanding the decline of the

123 Thomas Kuhn, *The Structure of Scientific Knowledge*, 2nd ed. (Chicago: University of Chicago Press, [1962] 1970), 92–173.

124 Fáng, *Jìndài gōngchéngshī qúntǐ de 'mínjiān lǐngxiù'*, 61.

125 Fáng, *Jìndài gōngchéngshī qúntǐ de 'mínjiān lǐngxiù'*, 149–58.

Sino-American segment and the connected exodus of American engineers from China.

Professional education was the earliest sector to see a decline in foreign participation vis-à-vis a trend of "domestication." This development started with the professors before it extended to the students. In the case of Jiaotong College in Shanghai, until 1920, half of its faculty was foreign, including US professors for civil and electrical engineering. The foreign quota dropped sharply in the early 1920s until, by 1927, the complete faculty was Chinese.[126] The prestige surrounding overseas education remained, as most of the new Chinese faculty had received their qualification abroad.[127] In the 1930s, the number of civil engineers graduating from domestic universities began to increase remarkably. The increase in engineering graduates was not an exclusively Chinese phenomenon. For instance, in Germany, graduates doubled between 1925 and 1933.[128] However, in China, the development was especially notable: while domestic colleges produced less than one hundred civil engineers per year in the early 1920s, over five hundred of them graduated annually around the pre-war peak in the mid-1930s (Figure 28). As these fresh graduates slowly commenced their careers in junior positions, in time, the sheer mass of domestic professionals gradually decreased the situations in which institutions saw the necessity to rely on foreign expertise. Furthermore, for the Chinese students, the increased quantity and quality of domestic colleges made it less urgent to seek education abroad. With domestic higher education institutions built up, it was no longer necessary to go abroad to receive a satisfactory engineering education.[129] CERD lists just 45 students who studied civil engineering in the US after 1930,[130] compared to the 196 who had completed their studies before that year.[131] In other words, while the total number of civil engineers rose steadily, the absolute number of newcomers into the Sino-American segment steadily decreased.

Additional factors added to the exodus of US experts from China, although not curtailing Sino-American engineering as a culture of Chinese professionals. Apart from public scrutiny in the wake of the Salaqi project, for American engineers, the cluster of nationalization efforts and a salary decrease resulted

126 Jesiek and Shen, "Educating Chinese Engineers," 132–34.

127 Rèn, Liáng and Lee, "Meritocracy and the Making of the Chinese Academe, 1912–1952," 945, 950.

128 Jarausch, *The Unfree Professions*, 82.

129 Köll, "The Making of the Civil Engineer in China," 163–66.

130 Pelzer et al., CERD, [{"t":"10"}, {"linked_to:940":[{"t":"58"}, {"f:9":">=1930"}, {"f:997":"9483"}, {"linked_to:964":[{"t":"57"}, {"linked_to:238":[{"t":"12"}, {"f:26":"119"}]}]}]}].

131 Pelzer et al., CERD, [{"t":"10"}, {"linked_to:940":[{"t":"58"}, {"f:9":"<=1930"}, {"f:997":"9483"}, {"linked_to:964":[{"t":"57"}, {"linked_to:238":[{"t":"12"}, {"f:26":"119"}]}]}]}].

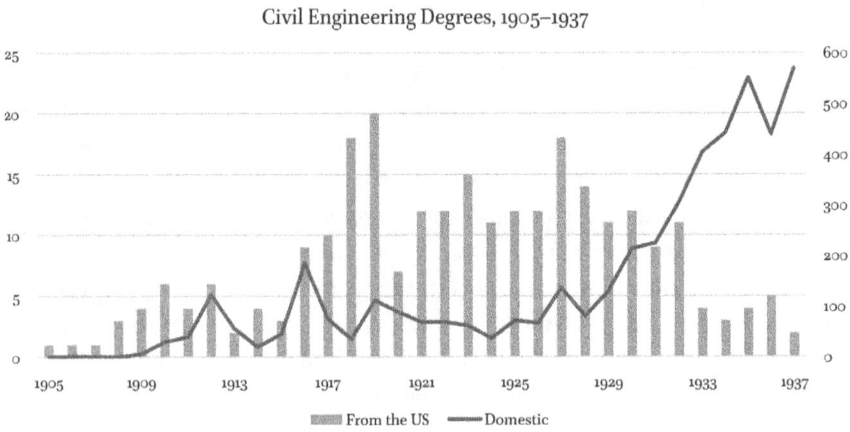

FIGURE 28 Civil engineering degrees, 1905–1937 (data from CERD).

in a "1930s turn."[132] The GMD regime aimed at incorporating big companies and non-governmental organizations into its statist economy. For instance, in 1929, it reorganized the Zhili River Commission, which no longer included foreigners.[133] Likewise, Sun Ke preliminarily planned to have the functions of the CIFRC taken over by his own department, and eventually, the NEC assumed the organization's construction work.[134] Furthermore, after 1930, the depreciation of silver—the standard of the Chinese currency in which the foreigners received their salary—resulted in less incentive for foreign members to stay in China.[135] Around the same time, the Manchurian Invasion of 1931 limited the government's leeway in paying grander salaries, as the state now had clear priorities for spending its budget.[136]

---

132 Will, "The Emergence of the Modern Civil Engineer in China, 1900–1940," 3:102–5.

133 OJTP, 02.03: Todd to E. W. Lane, February 19, 1929.

134 OJTP, 30.06: Minutes of the 132rd Meeting of the Executive Committee of the CIFRC, Beijing, July 20, 1934; JACAE, 19, no. 1 (1938): Oliver J. Todd, "Engineering Experience in China during the Last Two Decades," 33; Chin, "National Economic Council," 363–73. For other institutions competing with the CIFRC, especially in the field of rural reconstruction, see James Thomson Jr., *While China Faced West: American Reformers in Nationalist China, 1928–1937* (Cambridge: Harvard University Press, 1969), 122–23.

135 OJTP, 30.01: Todd to Wm. H. Gleysteen, February 18, 1930; 30.01: Zhang Yuanshan to Todd, April 15, 1930; 30.02: Hardy Jowett to the CIFRC Executive Secretary, June 20, 1930; 30.02: Zhang Yuanshan to Todd, January 28, 1931; JACAE, 11, no. 9 (1931): 59. For exchange ratio development, see Officer and Williamson, "The Price of Gold, 1257–Present," 1925–1935.

136 OJTP, 30.04: Todd to Arthur M. Shaw, April 11, 1932; 30.04: Arthur M. Shaw to Todd, April 22, 1932.

Adding to the lack of job opportunities and decrease in financial rewards, the country felt increasingly unsafe in the eyes of the Sino-American engineering community. While civil warfare had occasionally created situations that impeded surveys and fostered unease (chapters 2, 5), the civil engineers had mostly evaded physical harm. This changed in 1930, when bandits shot one of the CIFRC's foremen and beat up another.[137] In 1932, during an inspection trip outside the gates of Xi'an, rogue troops killed Huang Rongde, the engineer of the Guizhou highways (chapter 5), and tossed his corpse into a well. The walking distance between the scene of the deadly assault and the inner walls of the provincial capital led to the impression that, even in close proximity to state authority, things were not under control.[138] A year later, in the same province, communist insurgents took CIFRC engineer Sigurd Eliassen and his co-worker Henry Quan hostage.[139] The violent experiences of the early 1930s added to the impetus for foreign members to leave the country.

As to the CIFRC, donations from the US, one of the organization's three pillars for funds, dropped sharply with the beginning of the Great Depression in 1929, and the charity drive never recovered thereafter.[140] Todd was especially desperate to remain in China, applying to Sun Ke and recommending himself to the NEC, but to no avail.[141] He also unsuccessfully petitioned US President Franklin D. Roosevelt (1882–1945) for funds, proposing the establishment of new Sino-American commissions, primarily made up of engineers, that would draw up infrastructural development plans for China.[142] The outlook for foreigners in China was bleak. In the mid-1930s, Charles Crane, former US minister to Beijing and an acquaintance of Freeman from the time of the Grand Canal Board (chapter 2), paid for Todd's salary out of his private charity when the CIFRC was no longer capable of doing so.[143] American members gradually left China during the first years of the new decade. By 1934, Todd and the

137 OJTP, 37.02: Ernest W. Wampler to Todd, May 22, 1930.

138 OJTP, 30.04: Todd to Siufeng Huang, September 2, 1932; 30.04: Todd to Zhou Yichun, September 27, 1932; Sigurd Eliassen, *Dragon Wang's River* (London: Methuen, 1957), 206–7.

139 OJTP, 31.07: CIFRC Document ID 23499, June 22, 1933; Eliassen, *Dragon Wang's River*, 214–50; Todd, *The China That I Knew*, 82–83; Fisher, "Service to China," 137.

140 Huáng, "1921 zhì 1935 nián qíjiān fēizhèngfǔzǔzhī guójì císhàn mùkuǎn huódòng," 85.

141 OJTP, 02.02: Todd to Wang Zhengting, June 18, 1928; 29.13: Todd to Sun Ke, April 2, 1929; 29.13: Sun Ke to Todd, April 5, 1929; 02.03: Todd to Geo. Sellett, September 15, 1929; 13A.2: Todd to G. G. Stroebe, September 16, 1929; 30.05: Todd, "Proposed Program for Improving Living Condition in North China," Memorandum for 1933–1934.

142 OJTP, 02.05: Todd to Franklin D. Roosevelt, June 8, 1935.

143 JACAE, 20, no. 3 (1939): Oliver J. Todd, "A Friend of China–Hon. Chas. R. Crane," 101.

hydraulic engineer Georg Gottlieb Stroebe were the last practicing civil engineers from the US remaining in China.[144]

The lack of famines also hindered the CIFRC in commencing new labor relief programs. Its usual modus operandi was to recruit the famine-struck population to work on the constructions it supervised. However, without a new famine, farmers were busy caring for their farms. Usually, the CIFRC could rely on conscription labor, including soldiers, as an alternative, but by the summer of 1937, the military had other priorities.[145] With no new realizable projects in sight, the CIFRC engineering department formally closed its doors in April 1938.[146] With that, it survived a few months longer than the Chinese government's NEC, which had once aimed at replacing it. Meanwhile, the CIFRC's chief engineer Todd had ultimately lost his credit in the eyes of the government when, together with left-wing journalist Edgar Parks Snow (1905–1972), he edited the political journal *Democracy* promoting New Deal reforms for China.[147] Todd left the country in 1938, the same year in which the Nationalists, under Jiang's command, blasted the dikes of the Yellow River,[148] a river whose dikes Todd had once helped to repair.[149]

The exodus of US engineers diminished the ACAE's standout characteristics and its reason to exist alongside the CIE. Theoretically, the ACAE could have remained a forum of US-educated Chinese engineers, but clearly, without the interaction with American colleagues, the association lacked one of its original selling points. Overall membership no longer grew, and existing members stopped paying their fees.[150] During this time, grand names, such as the one of former ACAE president Sun Duoyu, quietly disappeared from the membership list. The *Journal* tried to appeal to the members' conscience, asking them to pay up or resign. Only because of the ACAE's "patriotic duty" would the association give troublemakers a second chance and not tell them "to go to hell."[151] The calls were unsuccessful: in the 1930s, membership numbers shrunk rapidly.

144 OJTP, 30.05: Todd to Murray Sullivan, November 7, 1933; 30.06: Todd to George G. Stroebe, January 25, 1934.

145 OJTP, 31.03: Zhang Jichun to Todd, August 3, 1937.

146 OJTP, 31.03: Todd to Zhang Jichun, April 14, 1938.

147 David Ekbladh, *The Great American Mission: Modernization & the Construction of an American World Order* (Princeton: Princeton University Press, 2010), 38; Fisher, "Service to China," 165–70.

148 Micah Muscolino, *The Ecology of War in China: Henan Province, the Yellow River, and Beyond, 1938–1950* (Cambridge: Cambridge University Press, 2014), 21–86.

149 Spence, *To Change China*, 210; Fisher, "Service to China," 158–60; OJTP, 27.08: Manuscript, "American Company Repairs Yellow River".

150 JACAE, 14, no. 2 (1933): "Notice to Members," n. p.

151 JACAE, 14, no. 6 (1933): Anonymous, "The Story of an Engineer (Who Was Reckless)," 44.

The last years of the ACAE concluded the ascent of Tan Zhen and Xue Zhuobin within the Sino-American segment: they had started out as interns for Freeman, worked as junior engineers at the Canal Board (chapter 2), assisted the CIFRC, and now became presidents of the ACAE. However, in their leading positions, they acted as little more than trustees of an estate which had passed its best days. The war against Japan eventually revitalized Sino-American cooperation in other realms of industrial engineering, but this support was mostly financial or consisted of short-time exchanges.[152] The days of long-term, on-site engagement between Chinese and American civil engineers were gone. After a few more years, during which the editors kept the *Journal* alive through an increasing number of reprints (Figure 15 in chapter 4), the magazine finally ceased publication in August 1941. With it, the association ceased as well. As the editors poetically described it, the Chinese wartime government had frozen the ACAE's assets "as a result of the general frost that ha[d] attacked the world."[153]

7 Conclusion

On the surface, the Nanjing government was the strong state that the engineers had been longing for and which they had previously only found on the provincial level in Guizhou (chapter 5). The Guomindang state received the Sino-American engineers' advance praise, and its support allowed for the resurrection of the ACAE. However, the Sino-American segment could not profit from the creation of potent new state institutions and the increased domestic college output that were important groundwork for a new era of civil engineering in China. Instead of promoting their segment by distinction from other groups in society, they came under attack from a different group of civil engineers, namely their colleagues from the League, who criticized the Famine Relief Commission's work in Salaqi. The involved practitioners were part of the vocal core of the ACAE, the pivot of the Sino-American engineering network, meaning the loss of face at Salaqi reflected on the larger segment.

For the Guomindang, nationalization of the discipline meant an increased politicization, which met the discontent of US-trained engineers like Yang Baoling and Tan Zhen. Furthermore, the emergence of the Nationalist technostructure resulted in the militarization of the civil engineering profession. This

152 William Kirby, "The Chinese War Economy," in Hsiung; Levine, *China's Bitter Victory*, 199–202.

153 JACAE, 22, no. 4 (1941): i.

forced civil engineers to reinterpret their service ideal, as they would continue to serve the nation, but in a modified form.

They would also need to pursue their discipline without the cosmopolitan element that had so far characterized their professional segment. The perceived increased danger of fieldwork, the depreciation of silver, the loss of face of the ACAE inner circle after the failure of the Salaqi project, and the lack of civil engagements during wartime all led to the remigration of US engineers, which in turn led to the demise of the ACAE. Starting from the mid-1930s, the association slowly lost its importance as an institution and gradually unraveled under the impression of world affairs. Nevertheless, the departure of US engineers and the ACAE's demise were not the end of Sino-American civil engineering in China, as the involved Chinese engineers did not disappear. However, left to themselves and operating in a command economy, they had to radically reinterpret the civil engineer's work.

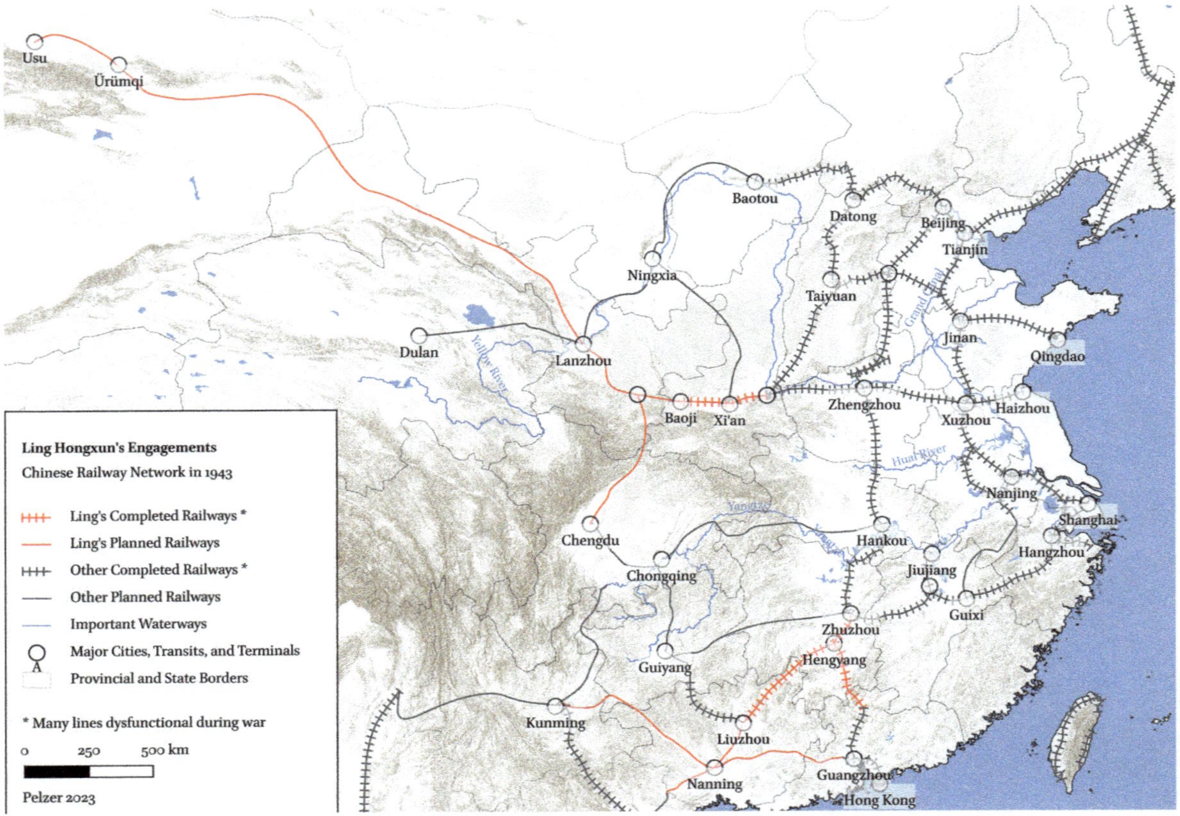

FIGURE 29 Map of Ling Hongxun's engagements and the Chinese railway network in 1943.

CHAPTER 7

Wartime Engineering

Ling Hongxun under Pressure, 1932–1945

The Nationalist era saw the emergence of a new technostructure, in which the central state integrated engineers into new potent institutions that managed the country's resources and infrastructures (chapter 6). The development took place institutionally, ideologically, and materially in the context of planned economy mechanisms and the impending war against Japan. Inside the new framework, some of the profession's earlier troubles were resolved while new discontents emerged. Civil engineers sacrificed several of their professional beliefs during this era: the claim of autonomy macerated in bureaucratic institutions, meticulous planning and the love for exactness gave way to swift reaction, and the essential divide between civil and military engineering became ambivalent in the wartime economy. Ling Hongxun was one of the most industrious and well-traveled railroad, highway, and bridge engineers of the period. The new command economy imposed psychological strains on him. Ling observed that his responsibilities grew while his authority remained subordinated.

1 A Career Start of Ups and Downs

Ling had come to the US as a self-paying student who paid for his evening engineering classes through internships and manual labor (chapter 1). On his return to China in 1918, Ling reported to the Ministry of Communications in Beijing for work. Despite not receiving a ministerial scholarship, therefore being ineligible for the guaranteed employment scheme of the Ministry, he later described how Minister Ye Gongchuo (chapter 3), a fellow Panyu native, happened to have family connections to his father. Apparently, the engineering profession's call for meritocracy over coterie fell silent when one's own career could profit. With Ye's support, the doors to the Ministry opened.[1]

---

1 Líng, *Qīshí zìshù*, 29–30.

 | DOI:10.1163/9789004549555_009

At the Ministry, the distribution of foreign-educated students followed the logic of the railroad concessions, and their resulting differences in working language, standards, and customs. For instance, students who studied in France or Belgium disproportionately went to the Francophone Beijing–Hankou, Gansu–Haizhou, and Zhengding–Taiyuan Railways. According to Ling, the US-returned engineering students primarily stayed inside the central Ministry, as the United States' "Open Door" policy and failed or withdrawn investments had seen the US left out of the race for local railway concessions. Consequently, the US-returned engineering students soon made up the new "middle class" at the Ministry headquarters and assumed posts in important offices such as the office for accounting and statistics (chapter 3).[2]

Accordingly, Ling joined the Ministry Commission on Railway Technics (*jishu weiyuanhui* 技術委員會), founded in January 1918. His idol, the senior railway engineer Zhan Tianyou, led the commission.[3] The commission asked the technical administrators of each railway line to produce national standards. At the time, while the railway gauges were already largely the same, there was no standard bridge type, no standardized track equipment, and the railway lines did not utilize a common rolling stock.[4] Under the new guideline, all new material had to follow national standards while the existing stock was to slowly be replaced. Ling met the different local railway administrators, meaning that despite his position at the central Ministry, his task allowed him to explore previously unvisited regions of the Republic, including Manchuria.[5] The combination of a firm position within the Ministry and frequent long-distance traversal of the country would characterize the career that followed.

Ling also took care of his US ties, as he joined the Association of Chinese and American Engineers (ACAE, chapters 4, 6). Paths of the members, no matter if railroad, bridge, or hydraulic engineers, frequently crossed. For instance, in 1921, Ling assisted government adviser J. A. L. Waddell, whom he knew through his internship at the American Bridge Company, in the international competition for the Beijing–Hankou Railroad Bridge, in which John R. Freeman, honorary member of the ACAE, participated (chapter 2). Eventually, after the consolidation of the Nanjing Republic in 1928, Ling's association with these

2 Líng, *Qīshí zìshù*, 30.

3 JACAE, 1, no. 1 (1920): "The Commission on Railway Technics," 6.

4 Wang, "China's Railway Ambitions," 412.

5 Líng, *Qīshí zìshù*, 32–35.

senior American engineers would enable his ascent to influence—ironic, as the new government rhetorically opposed foreign interference.

However, before that, Ling experienced his first of many professional frustrations linked to politics. In 1922, after a poor showing at the Washington Naval Conference, Zhili commander Wu Peifu had demanded Prime Minister Liang Shiyi's 梁士詒 (1869–1933) removal from office. The affair became linked to the personal network of the Communications Clique, of which Liang was the informal leader. Liang's close associate, Minister of Communications Ye, had asked Fengtian military leader Zhang Zuolin to intervene.[6] The Zhili–Fengtian coalition collapsed and Ye followed Liang's fate. By the time his successor Gao Enhong 高恩洪 (1875–1943) assumed office, the ruling Zhili faction's dismissal of Liang and Ye had translated into a general distrust of the "Cantonese Clique," as the Communications Clique was also known, and by extension of any Cantonese technician inside the ministry. Consequently, Gao purged the remaining Cantonese employees. Ling Hongxun, a Communications expert from Ye's native place Panyu, matched the criteria perfectly and had to leave.[7] Ling painted Gao as someone playing political and regional Shandong favors.[8] Disgusted by the lack of appreciation for his technical expertise, Ling became involved in private industrial construction companies in the South, which other returned US students had started. He also used the time to prepare textbooks on municipal and railway engineering.[9]

Tides turned after the Second Zhili–Fengtian War and the Beijing Coup in late 1924 had ended Wu Peifu's Zhili rule. The Ministry of Communications, back in Ye's hands, removed Zhili officer Chen Duheng 陳杜衡 (b. 1864) from his position as head of Shanghai Jiaotong University—or Nanyang University, as the school was temporarily called—and installed Ling as his replacement. This purge was more to Ling's liking and benefit.

Although Ling sympathized with the Nationalist idea,[10] he did not sympathize with engineers giving precedence to Nationalist politics over pursuing their work. Therefore, although in May 1925, Ling joined the jury

---

6 Jiǎ, "Sūn Zhōngshān jítuán yú jiāotōng xì de ēnyuàn," 52.

7 Like Wang Jingchun (chapter 3), Ling formally appears in Jia's list of individuals linked to the clique but plays no role for the clique's history. See Jiǎ, *Běiyáng jūnfá shíqī de jiāotōng xì*, 37.

8 Líng, *Qīshí zìshù*, 36.

9 Hóngxūn Líng, *Shìzhèng gōngchéng xué* (Shanghai: Shāngwù yìnshūguǎn, 1924); Hóngxūn Líng, *Tiělù gōngchéng xué* (Shanghai: Shāngwù yìnshūguǎn, [1925] 1931).

10 Also for reasons of efficient railroad locating: Hóngxūn Líng, "A Decade of Chinese Railroad Construction (1926–1936)," in Sih, *The Strenuous Decade*, 261.

that decided on the design of the Sun Yat-sen mausoleum,[11] when two weeks later, the May Thirtieth Movement erupted in Shanghai and sparked a series of anti-imperialist and anti-foreign demonstrations, Ling was not at the forefront. The strikes were a pivotal moment in the nationalist movement on which the Guomindang Republic would base its legitimacy.[12] Ling's attitude toward the movement, in his function as the administrative head of Shanghai's most prestigious government-run university, remained reserved. According to his own numbers, about one hundred of his students took part in the demonstrations, which would mean about 25% of the school was involved. Although Ling visited an injured student in the hospital, he was not sympathetic to the students' political goals. Ling objected to his students getting involved in politics at all, underlining the ideal of the efficient, seemingly non-ideological engineer who "saves the country" through his technical practice, not through his political action.[13] In this vein, he regretted how many of his "good students" became Communist "puppets" during these weeks. Ling was unable to calm his students, whose strike went over into examination week, and depended on the Beijing government to send an external mediator.[14]

When the Northern Expedition reached Shanghai in March 1927, Ling still received his salary from the Beijing government. It is unclear where his loyalties lay at that time. Decades later, after Ling had moved to Guomindang-ruled Taiwan, he had all interest to paint himself as a long-time follower of the party. However, in 1927, in his own retelling, he assumed an "unpolitical" middle ground for the safety of his students and to "save the tradition" of the university. According to his story, he had always planned to resign after the eventual Nationalist takeover. Regardless of whether he voluntarily resigned or the GMD forced him to hand over control, with some resemblance to Wang Jingchun's career setback after the Nationalist takeover (chapter 6), Ling departed from his university position in May of that year.[15]

Like during his first expulsion, Ling returned to the South and prepared textbooks, this time on bridge engineering and on factory design.[16] He departed for Wuzhou in Guangxi, where he became chief of the city's industrial labor

11 Delin Lai, "Searching for a Modern Chinese Monument: The Design of the Sun Yat-sen Mausoleum in Nanjing," *Journal of the Society of Architectural Historians* 64, no. 1 (2005).

12 Waldron, *From War to Nationalism*, 242.

13 Chéngdōng Lǚ, "Líng Hóngxūn de gāoděng jiàoyù sīxiǎng jí shíjiàn," *Guǎngzhōu dàxué xuébào (shèhuì kēxué bǎn)* 10, no. 5 (2011): 58–59.

14 Líng, *Qīshí zìshù*, 39–40.

15 Líng, *Qīshí zìshù*, 43.

16 Hóngxūn Líng, *Qiáoliáng* (Shanghai: Shāngwù yìnshūguǎn, 1929); Hóngxūn Líng, *Gōngchǎng shèjì* (Shanghai: Shāngwù yìnshūguǎn, [1933] 1939).

bureau. He managed the municipal motor roads, waterways, electricity, and running water.[17] Regardless of the technical gratification the participation in Guangxi's first provincial city government may have brought, from the level of prestige, the engagement was a substantial downgrade from being head of the most prestigious engineering school in China.

When the Ministry of Railways split from the Ministry of Communications in October 1928, its new minister Sun Ke invited Waddell, acclaimed bridge engineer and an honorary member of the ACAE, as technical advisor to the government. Waddell, who had worked with Ling through the years, requested Ling to join the new Ministry with him. It took a few months for Sun to locate Ling in his distant Wuzhou exile and to order him to Nanjing. After the call had reached him, Ling abandoned his uncompleted work in the south and departed for the new capital. He found himself lucky to have departed from Guangxi shortly before the Jiang–Guangxi War broke out in March 1929. The southwest provinces had joined forces in opposition to Jiang Jieshi's ruling faction within the Guomindang. Guizhou Governor Zhou Xicheng fell in one of the ensuing battles (chapter 5). Ling assumed that if he had still been in Guangxi by that time, the Nanjing Ministry would have assumed he belonged to the hostile New Guangxi Clique and would no longer have agreed to recruit him.[18]

In summary, Ling's integration into the Nationalist bureaucratic apparatus was an eventful journey of ups and downs: far from a straight meritocratic ascend, he sometimes profited from his personal connections and sometimes suffered because of them. Political events influenced his career more than his merits. Finally, the patronage of an American engineer advisor led to the opportunity to work at the central Ministry—a ministry run by a government that prided itself on regaining domestic control and independence from foreign experts.

2 Abortive Western Expansions

With the completion of the Northern Expedition in 1928, the Nationalist Party had established the strongest Chinese state in over a decade. The Nanjing regime marked a decisive political turn but not a clean slate. Living conditions did not improve instantly, nor did investments return *ad hoc*. While calmer, more consolidated times were on the horizon, civil wars did not

17 Líng, *Qīshí zìshù*, 43–46.
18 Líng, *Qīshí zìshù*, 47.

subside, and many of the previously despised "warlords" continued to rule over relatively autonomous provinces and regions. Nevertheless, despite the Nanjing Republic's limits, the political change brought new opportunities for civil engineers.

His new supervisor Sun Ke saw Ling as someone rooted in the south. As such, he asked him to start preliminary studies for a Guangdong–Yunnan Railway. The project never left the exploratory stage, as the semi-autonomous southern provinces remained outside the Ministry's control.[19] The way in which civil engineers explored, sometimes surveyed, but seldom completed their projects would become a repeating theme throughout the rest of Ling's career.

The government instead asked Ling to organize a surveying team for the Beijing–Hankou Railway Bridge. He had organized the abortive international tender competition for this bridge in 1921 (chapter 2). Connecting the lands south and north of the Yellow River, the bridge had long been known to be unstable, therefore limiting the speed of the trains running across it. As so often, Ling would commence but never complete the project. In the middle of the surveying mission, he returned to Nanjing for an engineering conference organized by the Ministry of Railways. During the conference, it became clear that the government needed Ling as the new chief engineer of the Gansu–Haizhou Railway instead (Figure 30).

The original plan of the Gansu–Haizhou Railway, as proposed by the modernizing tycoon Sheng Xuanhuai 盛宣懷 (1844–1916), saw the railway connect the Haizhou district of the eastern coastal city of Lianyungang, Jiangsu,[20] with Lanzhou, located on the Yellow River banks of Gansu province in the West. When Ling became chief engineer, the line spawned from the coast of Haizhou, crossed the Tianjin–Pukou line at Xuzhou, the Beijing–Hankou line in Zhengzhou, and then ended prematurely in Lingbao, a county close to the old capital Luoyang (Figure 29). If extended to Xi'an, Baoji, Tianshui, and ultimately to the Gansu capital of Lanzhou, its function of "opening up" the Chinese West and connecting it with the East would complement the function of the Guangzhou–Hankou–Beijing route, which was to enable traffic between north and south.

---

19 Hóngxūn Líng, *Shíliù nián zhùlù shēngyá* (Taipei: Zhuànjì wénxué chūbǎnshè, 1968), 6.

20 Originally terminating at Dapu near the Linhong River, the line later extended to the Lianyun harbor.

FIGURE 30 Ling locating a bridge site for the Gansu–Haizhou Railway, 1932 (Ling Hongxun papers at Columbia University).

Therefore, once completed, the railway would become the second main axis of the larger nationalization scheme that Sun Yat-sen had outlined in 1920,[21] and that his son Sun Ke, as Minister of Railways, wanted to realize. Sun Ke planned to construct the first 20,000 of the 100,000 miles his father had envisioned within the next decade—more than three times the length of what had been built to that date.[22] Apart from selling bonds, Sun's Ministry counted on the remittance of European Boxer indemnities to start the construction.[23] The Belgian government had recently decided to return their Boxer indemnity under the premise of buying Belgian material, such as rolling material and bridges, for the completion of the Gansu–Haizhou line.

However, the Gansu–Haizhou Railway turned out to be yet another interrupted assignment. After less than a year of engineering, the second phase of the Central Plains War erupted in 1930 and temporarily prevented a westward extension of the railroad. The Ministry of Railways used the suspension

21 Sun, *The International Development of China*.

22 Fo Sun, "National Scheme of Railway Construction," *The China Critic* 2, no. 8 (1929): 147–48.

23 Sun, "National Scheme of Railway Construction," 148; Fo Sun, "National Scheme of Railway Construction, cont." *The China Critic* 2, no. 9 (1929): 176.

to send Ling to Washington, D. C., where he was to participate in the Sixth International Road Congress. Ling used this opportunity for a yearlong voyage around the world, taking the Siberian Railway to Moscow, and traveling further to Poland, Germany, and Belgium.[24] On his trip, he inspected famous engineering sites such as the industrial Ruhr area and the ostentatious Leipzig train station, a state-of-the-art railway terminus that had been completed 14 years earlier.[25] In his role as chief engineer of the Gansu–Haizhou Railway, Ling investigated Belgian factories, spoke to Belgian managers and bankers, and met with the original constructors of the Beijing–Hankou Railway.

Upon Ling's return to China in 1931, the Central Plains War was over. In the stabilized environment, Ling was able to swiftly complete the Lingbao–Tongguan section of the Gansu–Haizhou Railway in August 1932. With it, the authority over the railway's construction went into Chinese hands, as the original contract with Belgium over the line ended west of Tongguan.[26] Now a sovereign project, chief engineer Ling received the opportunity to supervise the remaining Tongguan–Baoji section via Xi'an with comparably free rein. When the Japanese attacked Shanghai in 1932 and the Chinese government announced its plan to move their capital first to Luoyang and then to Xi'an, both stationed along the Gansu–Haizhou Railway, the political importance of the assignment increased. To crown his work at the Gansu–Haizhou Railway, Ling planned to present the new capital Xi'an with a "palace-like" train station, maybe inspired by the opulent station he saw in Leipzig.

Again, he could not see the project to an end. This time, the Ministry of Railways prematurely ordered him to Hunan, where the remittance of the British Boxer indemnity—which his senior Wang Jingchun (chapter 2, 4) had helped to broker in 1926—allowed for the completion of the Guangzhou–Hankou Railway.[27] In a speech he gave at Jiaotong, Ling lamented that the engineering world acted too hasty.[28] The interruptive and abortive projects he managed during his first four years of engineering under the Guomindang showed him that while a larger budget allowed new assignments to come in faster, they were not necessarily being completed more often than before.

---

24 Líng, *Qīshí zìshù*, 52–58.

25 Líng, *Qīshí zìshù*, 59–62.

26 Líng, *Shílìù nián zhùlù shēngyá*, 20.

27 Líng, *Qīshí zìshù*, 68; Líng, *Shílìù nián zhùlù shēngyá*, 23–24. For Wang, see Ching-chun Wang, "Report on the 'Construction Work,' to be Selected," in *Report of the Advisory Committee: Together with other Documents Respecting the China Indemnity*, ed. Austen Chamberlain (London: His Majesty's Stationery Office, 1926).

28 Hóngxūn Líng, "Yǎnjiǎng tǔmù gōngchéng shìyè zhī dàgài," *Jiāodà jìkān* 3 (1930): 13–14.

3 From Guangzhou to Hankou: Cutting Time

The engagement at the Guangzhou–Hankou Railway was where Ling first experienced how advancing war preparations tightened state expectations about the speed of operation, altered his perception of the impact of his work, and increased the extent of his responsibility.

The Guangzhou–Hankou Railway had a long and complex history. Running between the South and today's Wuhan, the line was to form a lengthy north–south axis in connection with the Beijing–Hankou Railway. However, in 1904, the American China Development Company, concessioners of the planned railway, came under Belgian control and had its right-of-way revoked. Regionalist merchants in the south opposed further foreign investments but rallied against railway nationalization efforts of the Qing government. The opposition to imperial expropriation efforts eventually led to the Wuchang Uprising and the Xinhai Revolution of 1911.[29] Years later, a privately financed line set off from Guangzhou and reached the Hunan border at Shaoguan in 1916. Coming from the north, a national line connected Hankou with Zhuzhou near Changsha in 1918. What remained was a 434 km gap between Shaoguan and Zhuzhou (Figure 29). The war-ridden global economy and the destabilized, decentralized domestic politics of the early 1920s kept international investors for this final stretch at bay. While ostensibly, the railway was close to completion, the Ministry of Communications had to shelve the nationalizing endeavor for almost a decade. The situation changed as the United Kingdom began to return its Boxer indemnity to China in the form of a fund, which a bilateral committee had decided to invest in closing this railroad gap of significant national importance.

The railroad had officially been transferred to the central Ministry of Railways, but the office in Guangzhou continued to exert partial authority.[30] To avoid the political meddling of provincial leaders, Ling moved his construction office to Hengyang, a politically insignificant city fit for river and land transport.[31] Ling claimed that, apart from obeying Boxer and labor regulations, the Ministry gave him free hand and responsibility of managing the construction. To him, having full administrative and technical rights was "an extremely

29 See Lee, *China's Quest for Railway Autonomy, 1904–1911*; Arthur Rosenbaum, "The Railway Controversy of 1911: The Bureaucratic Perspective," *Jìndàishǐ yánjiūsuǒ jíkān* 11 (1982); Yūzō Mizoguchi, "The Chaotic Late Qing and Early Republican Periods," *Inter-Asia Cultural Studies* 17, no. 4 ([2007] 2016): 583–84; Kaske, "Sichuan as a Pivot".

30 Líng, *Shílìù nián zhùlù shēngyá*, 29.

31 Líng, *Shílìù nián zhùlù shēngyá*, 33–34.

enlightened style of work."[32] Given that each of Ling's trips back and forth between Hengyang and the Ministry in Nanjing took about ten days, and that he only used telegram service for the most urgent matters,[33] his assertion of on-site autonomy appears plausible and partly the result of technical and infrastructural constraints.

In his original plan, Ling specified an ambitious construction time of four years. However, with war against Japan imminent, the government had an acute interest in connecting south and center as soon as possible. Accordingly, the government requested Ling to cut the construction time to 3.5 years. Ling felt the "great responsibility" lying upon him and, according to his memoirs, wrote the following slogans on a wooden board: "Each person works for two. Each day is spent like two. Each coin is used as two."[34] The efficiency paradigm (chapter 3) had been transformed into war preparation austerity.

Although Ling successfully undercut the original timetable, he lamented that he had been unable to complete the railway even earlier. He contemplated that were he to have finished the railway by 1933, the National Revolutionary Army (NRA) would have been able to utilize the railway to counter the Jiangxi Soviet forces in the Hunan–Jiangxi border zone during the Fourth Encirclement Campaign. With regret, Ling concluded that the Communist Party would never have been able to spread its "torment and menace" had the vital infrastructure existed just one or two years earlier.[35] The way in which Ling invoked a correlation between the railway construction time and the fortunes of Communist insurgency shows his awareness of the political and military dimensions of civil infrastructure. He realized the immense political power of civil engineers, even when rejecting the engineers' active participation in politics.

In April 1936, his team formally closed the gap between Guangzhou and Hankou—and, by extension, Beijing. The act became a publicity stunt that attracted media, politicians, and a puppy dog (Figure 31) to the small, rural Taipingli Station south of Chenzhou. For a short peaceful time in history, the closure of the gap enabled the longest connection of railway lines the world had seen, from Hong Kong via Beijing, Manchuria, and Siberia to Calais in France.[36] True to his frugal and humble image, Ling would later distance himself from the famous photograph that shows him driving the final spike into

32 Líng, *Shílìù nián zhùlù shēngyá*, 36.
33 Líng, *Qīshí zìshù*, 71–72.
34 Líng, *Shílìù nián zhùlù shēngyá*, 39–40.
35 Líng, *Qīshí zìshù*, 73; Líng, *Shílìù nián zhùlù shēngyá*, 44–45.
36 Dikötter, *The Age of Openness*, 96.

FIGURE 31 Ling (center in white) completing the Guangzhou–Hankou Railway, April 28, 1936 (Ling Hongxun papers at Columbia University).

the rail.[37] He claimed that he felt uncomfortable being dragged into the photo, as there was still more work to be done until traffic could truly commence.[38]

During the June 1 Rebellion of 1936, the provincial governments of Guangxi and Guangdong challenged Nanjing's policies against Japan, which they found lacked determination. When the Nanjing regime started its military intervention, the rebelling southern armies blasted two brand new steel bridges near Pingshi to prevent Jiang Jieshi's troops from advancing further. Seeing parts of the Guangzhou–Hankou Railway already destroyed before its opening for general traffic, Ling felt dejected. He had to rebuild the two bridges "at a time when just having relaxed [his] engineering efforts for one breath." With the NRA literally waiting behind his back, his team had to work "day and night" for seven days straight.[39] The swift repairs allowed the Nationalist Army to force the reintegration of the semi-independent southern provinces into the Nanjing Republic. While there are many descriptions of the wartime dismantling of railway tracks and of the catastrophic opening of the Yellow River dikes in 1938, the episode shows that the destruction of domestic infrastructures

37 Líng, *Shílìù nián zhùlù shēngyá*, 47.

38 Líng, *Shílìù nián zhùlù shēngyá*, 49–50.

39 Líng, *Shílìù nián zhùlù shēngyá*, 46–47.

already started before the Japanese invasion, due to domestic warfare during the "consolidated" Nanjing decade. Similar to the latter two-thirds of the Beijing era, which Nationalist historiography rejected as the time of the "warlords," domestic military conflicts kept upsetting civil engineering projects.

Ling regretted that the fast construction had led to administrative and technical mistakes.[40] The oversights damaged the reputation gained by having completed the project on time. Under Ling's supervision, more than thirty soldiers died in a fatal train accident. According to his autobiography, his superior at the Ministry scorned: "If this was the Soviet Union, you would be executed by the firing squad."[41] The professional ideals of planning and precision clashed with the state's expectation to subordinate engineering principles to military defense. Ling accepted both dimensions of responsibility but had difficulties reconciling them.

4 Lessons Learnt and Lessons Dealt

The swift completion of the Guangzhou–Hankou Railway was Ling's greatest engineering feat and remains what histories remember him most for to date. Ling, a frequent writer and speaker who often reflected upon his own work, framed the experience to fit his own engineering philosophy. This philosophy understood civil engineering as a public service, for which the professional had to live a life of frugality, discipline, and self-sacrifice. As the state would soon overstretch these demands, they require a closer examination.

For Ling, civil engineering was "a contribution to society," and serving the people was "the most joyful thing for humankind."[42] Similar to how his senior Wang Jingchun had described the railway's potential to change the space around it (chapter 3), Ling pointed out two dimensions of railway engineering. The first was "opening up mountains" (*kaishan* 開山), a metaphor for bringing the joys of civilization to greater parts of the interior. In this respect, he also called engineering the "sound of material civilization."[43] The second dimension was being "first at the mountains' peak," the Chinese expression for "vanguard" (*xianfeng* 先鋒). This was because, he explained, railway construction projects

40 Líng, *Qīshí zìshù*, 75. See also *The China Press*, "Buses, Troops Blamed for Rail Accidents," November 12, 1936, 2.

41 Líng, *Qīshí zìshù*, 129–30.

42 Líng, "Tiělù gōngchéng de duōbiān rènshí [7]," 6. See also Líng, "Yǎnjiǎng tǔmù gōngchéng shìyè zhī dàgài," 13–14.

43 Líng, "Yǎnjiǎng tǔmù gōngchéng shìyè zhī dàgài," 12.

acted as a kick-start for capital flow into new regions. The railway came first as the vanguard, and afterward, social, economic, and political change followed.[44] Engineers were to draw their satisfaction from enabling this development.[45]

Consequently, a serious engineer did not require satisfaction outside of work. The daily life could be basic. Ling liked to portray himself as living such a simple life. He explained that when working on the Guangzhou–Hankou Railway from his office in Hengyang, he earned $600 a month but had nothing to spend it on. He claimed that $25 for a single household or $45 for a family would have been enough.[46] During the war, when stationed in Guangxi, he took a salary cut of 40% and still earned more than the highest-ranked bureaucrats in the province. He described life in the hinterland as peaceful and happy.[47] In his opinion, instead of working for a high salary, the ideal engineer had to find content in improving the livelihood of the people and in the experience and status gained through it.[48]

In contrast, he expressed his disdain for Shanghai engineering students that dwelled in Western-style dormitories, ate Western food, drove cars, and frequented the cinema. To him, this lifestyle contradicted the experience engineering students had to be prepared for: weeks or months of straining fieldwork, with nights spent with an oil lamp.[49] He asked the students to give up their materialistic lifestyle and replace it with spiritual satisfaction. The interior was a site lacking in material benefits, culture, and food.[50] Accordingly, he asked his students not to overemphasize textbooks and to put more emphasis on practical fieldwork.[51] In order not to be seen as outsiders (chapter 6), engineers had to connect with the locals.[52] For instance, when working on the Guangzhou–Hankou Railway, he felt that even though the site of operation was geographically close to his native place of Panyu, having left Guangdong over twelve years earlier, he had grown unfamiliar with the people.[53] Like many, Ling had grown up in a big city, studied in the US, and accepted his

44 Líng, "Tiělù gōngchéng de duōbiān rènshí [7]," 6.

45 Líng, "Yǎnjiǎng tǔmù gōngchéng shìyè zhī dàgài," 12–13.

46 Líng, *Shíliù nián zhùlù shēngyá*, 38–39.

47 Líng, *Shíliù nián zhùlù shēngyá*, 65–66.

48 Líng, "Yǎnjiǎng tǔmù gōngchéng shìyè zhī dàgài," 15–16.

49 Líng, "Yǎnjiǎng tǔmù gōngchéng shìyè zhī dàgài," 15–16.

50 Líng, "Tiělù gōngchéng de duōbiān rènshí [7]," 6. See also Kirby, "The Chinese War Economy," 191; Joseph Needham, *Chinese Science* (London: Pilot Press, 1945), 29.

51 Líng, "Yǎnjiǎng tǔmù gōngchéng shìyè zhī dàgài," 14–15. See also James Reardon-Anderson, "Science in Wartime China," in Hsiung; Levine, *China's Bitter Victory*, 217.

52 Líng, *Shíliù nián zhùlù shēngyá*, 62.

53 Líng, *Shíliù nián zhùlù shēngyá*, 27–28.

first work assignments in developed areas of China: he and his students had grown accustomed to the abundance of the city, but only through experiencing the undeveloped interior, could they experience the actual state of affairs of the country.[54]

Ling's version of the ideal practicing engineer, with its emphasis on the self-restrained bodily regime, appeared in the tradition of the "care of the self" propagated in the ACAE (chapter 4). Ling painted the practice of civil engineering as a practice of constraint and sacrifice. After graduation, the (urban male) engineer would depart for fieldwork in the interior, where it was difficult to get married, and if married, difficult for children to receive a good education. As public servants, Ling found, it was the engineers' duty to adhere to central command and strive hard to speed up the work. This was reflected on one's body regime.[55] The body had to be physically trained and drilled, healthy and strong, swift and adaptive.[56] The mind, likewise, had to be cultivated and controlled, devoid of laziness and bad habits.[57] The control over the body did not exhaust itself in the compulsory physical exercises, which he had promoted as president of Jiaotong University: it was the professional's individual responsibility to live a disciplined life in service to the country.[58] Ling expected discipline not just from his assistants but also from his superiors. He lamented how uninvolved and ineffective the Ministry work in Beijing had been. According to his description, employees slowly started working after 11 am, while the minister and vice minister would not arrive at the office before 3 to 5 pm. Then, out of courtesy, the employees had to wait in the office until their superiors had finished too.[59] Ling urged his colleagues to abandon these old habits to protect the public reputation of the engineering profession, even if that meant sacrificing individual convenience, pleasure, and benefits.[60]

The urgency of Ling's lessons on austerity and discipline increased as the war against Japan escalated. In an article he published together with his Gansu–Haizhou colleague Qian Zongze 錢宗澤 (1891–1940), he called for the employees of his ministry to "sacrifice or be sacrificed by the enemy." The two asked their colleagues to take a cut in their salary to aid the air force. They

54 Líng, *Shíliù nián zhùlù shēngyá*, 12–13.

55 Hóngxūn Líng, "Duànliàn shēnxīn yǐ wèi jiā guó," *Lǒng Hǎi xúnkān* 15 (1931): 13.

56 Líng, "Duànliàn shēnxīn yǐ wèi jiā guó," 14; Hóngxūn Líng, "Tiělù gōngchéng de duōbiān rènshí [5]," *Xiāng Guì Qián xúnkān* 3, no. 28 (1948): 4.

57 Líng, "Duànliàn shēnxīn yǐ wèi jiā guó," 14.

58 Líng, *Shíliù nián zhùlù shēngyá*, 12–13; Líng, "Duànliàn shēnxīn yǐ wèi jiā guó," 13.

59 Líng, *Qīshí zìshù*, 31.

60 Hóngxūn Líng, "Yuè Hàn tiělù quánxiàn jiētōng hòu réng yīng jìxù nǔlì," *Tiělù zázhì* 2, no. 2 (1936): 108–9.

argued that because public servants depended on the nation's well-being for their income, saving the nation equaled saving one's own house and oneself.[61] The argument reveals the idea of an entanglement between the self and the nation, and further explains the duty that resulted from it. The engineers had argued that saving the nation was part of their obligation since their student days (chapter 1). The GMD state institutions had since employed a large portion of the profession (chapter 6). For many engineers, saving one's nation also meant saving one's employer. Different from his colleague Oliver J. Todd, who, despite calling himself a "sort of generalissimo," also prided himself in "the practice of non-interference in military affairs,"[62] Ling grew aware that despite its name and aims, civil engineers had a deep impact on the unfolding and outcome of the war.

5 Drawbacks of the Strong State

With the onset of the Sino-Japanese War, Ling started to handle a multitude of projects at once. While the potency and extent of central government authority had gradually evolved to one of the greatest in modern Chinese history, the problem of capital scarcity had not changed. What changed was that, in addition to the well-known constraints in financial capital, Ling now had to operate under an existential pressure of time, as any small acceleration or delay of the construction process could change the course of the war. Aware of the responsibility, he frequently pointed out that as the Japanese aggression had created a situation of "survival and perishing,"[63] railways had become a "matter of life and death."[64]

During the first years of the Second Sino-Japanese War, the enemy destroyed much of the progress civil engineers had made: Japanese planes bombed the Beijing–Suiyuan, Guangzhou–Hankou, Hunan–Guangxi, and Yunnan–Vietnam Railways.[65] More sobering, Ling witnessed his own government destroying critical infrastructure, which he had partly worked on himself, as

61 Zōngzé Qián, Hóngxūn Líng, and Shìbīn Yú, "Lǒng Hǎi tiělù chóu nǐ hángkōng mùjuān bànfǎ xuānyán," *Tiělù yuèkān Píng Hàn xiàn* 35 (1933): 8–9.

62 Todd, *The China That I Knew*, 7, 22, 66.

63 Hóngxūn Líng, "Wǒmen qìngzhù Guóqìng de shíhòu yīng xiān zhùyì dào zěnyàng qù jiěchú guónàn!?," *Yuè Hàn tiělù xúnkān* 7 (1936): 2.

64 Hóngxūn Líng, "Yóu Zuǒ Zōngtáng píngdìng Xīnjiāng shuō dào Gān Xīn tiělù zhī xīngzhú," *Bǎo Tiān lù kān* 2, 4–6 (1943): 133; Líng, "Yǎnjiǎng tǔmù gōngchéng shìyè zhī dàgài," 12–13.

65 Hóngxūn Líng, Hung-Hsun Ling Papers, Duì rì kàngzhàn bā nián jiāotōng dàshìjì chūgǎo. HHL, 1962, MS#0786, Columbia University Archival Collections, 37/8/1, 38/8/14, 39/8/23, 40/1/7.

a result of a scorched earth policy. The NRA destroyed the Qiantang, Yellow River, and Yunnan–Vietnam Railway Bridges, the Qingdao–Jinan Railway, the Haizhou harbor, and further ordered the dismantling of the Nanchang–Jiujiang, Gansu–Haizhou, Hunan–Guangxi, and Yunnan–Vietnam Railway tracks for use elsewhere.[66] By 1942, only 10% of railway tracks were still in operation, with twelve lines completely lost and many others partly destroyed or occupied.[67]

During his work at the Guangzhou–Hankou Railway, a line that had been planned for decades, Ling had argued that completing the construction three years earlier would have altered history. Instead of years, the priorities of new railway lines now changed within months. Cutting construction time by just a day could make a difference. Ling experienced this with the Hunan–Guangxi Railway, which he co-supervised between 1937 and 1939 (Figure 29).[68] In December 1939, on the very day the team connected Liuzhou to the line, the NRA used the railway to send its reinforcements from Guilin south to Liuzhou and then successfully fought back the Japanese troops in the Battle of Kunlun Pass.[69]

Highways grew in importance after the Japanese forces cut off the Hong Kong supply route in 1938, meaning the Chinese government had to rely on aid from the Soviet Union. The supplies came via trucks from Almaty that reached the Chinese interior via the old Hexi Corridor.[70] As part of executive reforms, the Ministry of Railways and the National Economic Commission Highway Department merged back into the Ministry of Communications.[71] In accordance with its extended area of authority, the Ministry of Communications started to assign Ling to highways, as well. In this respect, the bridge engineer Ling joined the tradition of Oliver J. Todd, who, despite also not being specialized in the field, had administered highways across the country (chapter 5).

Between 1938 and 1939 alone, the Northwest received new highways between Lanzhou and Hami, Tianshui and Fengyang, Tianshui and Lanzhou,

66 HHL, 37/12/23, 37/12/25, 38/1/31, 38/2/17, 38/4/29, 38/6/17, 39/2/21, 39/3/20, 40/8/11.

67 Geographical Section of the Naval Intelligence Division, *China Proper*, 495; Francis Pan, "Communications and Transportation," in *The Chinese Year Book: 1940–1941*, ed. The Council of International Affairs Chungking, 5th ed. (Shanghai: Commercial Press, 1941), 649, 652. See also Köll, *Railroads and the Transformation of China*, 193–202.

68 HHL, 38/2/3.

69 HHL, 39/12/16.

70 HHL, 37/8/31, 37/10/1; Geographical Section of the Naval Intelligence Division, *China Proper*, 452.

71 HHL, 38/1/1; Geographical Section of the Naval Intelligence Division, *China Proper*, 444.

Baoji and the Wei River, Tianshui and Shuangshipu, and Lanzhou and Suzhou (Jiuquan).[72] Highway construction was less centralized than the railways. Instead of reporting to a single railway office, Ling, as chief engineer, had to juggle between seven or eight partially responsible local bureaus situated along the Tianshui–Chengdu Road. When in 1938, the ACAE elected Ling as its vice-president, he was so occupied that he had to decline the poll. The withdrawal was so unusual that members had to assure themselves this was not a symbolic gesture of humbleness, but a serious objection due to the workload he faced.[73]

Through the spring and summer of 1940, the Nationalist government gradually restructured its transportation authorities and brought them under central supervision.[74] Instead of going through the Ministry, Ling now took orders from Jiang's administration directly. In January 1940, the government commanded Ling to become chief engineer of the Tianshui–Chengdu Railway, of which surveys immediately started.[75] The idea was to complement the general west–east orientation of the Chinese railroad network: in times of peace, the Beijing–Suiyuan, Gansu–Haizhou, Hunan–Guizhou, and Hunan–Guangxi Railways gave passengers means to travel west along different latitudes, but the traversal on the north–south axis was more limited. In the ACAE *Journal*, Todd further explained that the new railway would "bind" Sichuan "closer to Central China."[76] However, the Tianshui–Chengdu Railroad turned out to be yet another abortive undertaking: after less than a year of locating work, the government deemed it temporarily unaffordable.[77]

The strategic importance of the Northwest increased further after the Japanese cut off the Indochina route in 1941, and the British declared that they needed material the Chinese had requested for their own war at home.[78] The Ministry of Communications formed the Northwest Highway Administration Department (*Xibei gonglu guanli chu* 西北公路管理處) of which Ling became the chief engineer. Its area of responsibilities included the administration and improvement of highways through Shaanxi, Gansu, Ningxia, and Qinghai.[79] The scope of highways under Ling's command further expanded when in

---

72 HHL, 38/2/21, 38/3/1, 38/4/6, 38/6/17, 39/2/21, 39/3/15. See also Pan, "Communications and Transportation," 655–59.

73 JACAE, 19, no. 3 (1938): "Proceedings of the Annual Meeting of the Association," 212.

74 HHL, 40/3/13, 40/4/14, 40/7/1.

75 HHL, 40/1/1.

76 JACAE, 18, no. 5 (1937): Oliver J. Todd, "Future Work for the Engineer in China," 296.

77 Líng, *Shílìù nián zhùlù shēngyá*, 69–70.

78 Fāng Hé, "Luó Jiālún yǔ xīběi jiànshè kǎochá tuán" (M. A., Huázhōng shīfàn dàxué, 2012), 12–13.

79 HHL, 41/1/20.

November 1942, as general manager of the Military Affairs Commission's Northwest Highway Affairs Office (*Xibei gonglu gongwu ju* 西北公路工務局), he started overseeing the Gansu and Xinjiang highway network, as well.[80]

Although the Tianshui–Chengdu Railroad itself did not materialize, in 1941, the Ministry had added an additional segment to the line that survived the termination: the Tianshui–Lanzhou section.[81] Instead of acting as the northern portion of the new Chengdu line, the government reassigned the section to the Gansu–Haizhou Railroad with which Ling was well acquainted. Ling decided for an office in Tianshui to manage the railway and the various highways at the same time. The western expansion of the Gansu–Haizhou Railway started with the section from Baoji to Tianshui in 1942.[82] The track followed the mountain ridge along the Wei River. It was a difficult and time-taking endeavor. Over 120 tunnels would make up one-seventh of the track. Apart from the tunnels, more rocks awaited cutting above ground. There were no machines to use for this work. There were also no lime factories nearby, so the team had to use a kind of supplement cement. The enterprise lacked a feasible design, sufficient labor, and the infrastructure to supply a continuous flow of food. Ling later summarized, "Having to construct a railway during that time, in that place, especially a railway that particularly difficult to engineer, was an unimaginable undertaking."[83]

The war resulted in a thorough restructuring of the economy. Everything in state hands had to aid the war efforts. The Nationalists lost their productive coastal areas to the Japanese, and while they successfully dismantled and reconstructed many of the factories further west, these had become part of the war industry. The war economy challenged the management ideals of the civil engineers. Like his senior and colleague Wang Jingchun (chapter 3), Ling found that "all engineering project arrangements" had to be "rooted in economic benefit."[84] For him, the studies of economics and railway engineering stood in proximity and the latter was not practicable without the former.[85] Ideally, railway administrators needed to manage the lines like commercial

80 HHL, 42/11/20.

81 HHL, 41/6/12.

82 HHL, 42/1/21. See Geographical Section of the Naval Intelligence Division, *China Proper*, 503.

83 Líng, *Shílìù nián zhùlù shēngyá*, 76–78. Also Líng, "China's Epic Struggle in Developing Its Overland Transportation System During the Sino-Japanese War," 255.

84 Líng, *Shílìù nián zhùlù shēngyá*, 45.

85 Hóngxūn Líng, "Tiělù gōngchéng de duōbiān rènshí [1]," *Xiāng Guì Qián xúnkān* 3, no. 24 (1948): 5–6.

companies,[86] browsing the international market for the best offers in price and quality.[87] In contrast to these beliefs, the prioritization of the wartime railways was strategic, not economical. Against his professional ideals, surveying had to be short, not thorough, and construction fast and cheap, not safe and long-lasting. Accordingly, Ling cut the number of involved engineers, decreased the construction period, and reduced construction standards as much as possible.

Like his colleagues at the CIFRC (chapter 5), Ling was not shy of making use of the state's powers. Through the years, he had the government mobilize people to do the manual labor for his projects in Guangxi, Hunan, and Gansu.[88] In Guangxi alone, the state had conscripted 600,000 able-bodied men.[89] Apart from compulsory labor, Ling's railways also profited from the state's authority to expropriate lands and force resettlement. During wartime, he explained, the motto was "do first, deal with it later." He put pressure on local magistrates to force residents to resettle. He later admitted that constitutional and legal rights were ill observed.[90]

Civil engineering had always been an exhausting job—for instance, already in the 1920s, Henry Ehlers of the ACAE had lamented the "terrible physical and mental strain" of his eighteen-hour days[91]—but wartime engineering had stress and pressure reach new levels. Ling described the years of his highway engagements as the most difficult times of his career.[92] As he put it, "what appears as honor above shoulders is fear below it." In the time span of one and a half years, he unsuccessfully begged the government to accept his resignation eight times.[93] Retrospectively, he thought he could not "hold [his] strength." Evidently, even Ling, with his propagated ideal of the frugal and disciplined hinterland engineer, was close to succumbing to the demands of the military government. Twenty years later, Ling remembered that engineers like him "wandered about in a desperate plight."[94] In his autobiography, he wrote,

---

86 Líng, *Shílìù nián zhùlù shēngyá*, 60–61.

87 Hóngxūn Líng, "Tiělù gōngchéng de duōbiān rènshí [3]," *Xiāng Guì Qián xúnkān* 3, no. 26 (1948): 10; Hóngxūn Líng, "Tiělù gōngchéng de duōbiān rènshí [4]," *Xiāng Guì Qián xúnkān* 3, no. 27 (1948): 4.

88 Líng, "Tiělù gōngchéng de duōbiān rènshí [5]," 4.

89 Líng, "China's Epic Struggle in Developing Its Overland Transportation System During the Sino-Japanese War," 250–51. See also Diana Lary, "One Province's Experience of War: Guangxi, 1937–1945," in *China at War: Regions of China, 1937–1945*, ed. Stephen Mackinnon, Diana Lary and Ezra Vogel (Stanford: Stanford University Press, 2007), 319, 324.

90 Líng, "Tiělù gōngchéng de duōbiān rènshí [7]," 5.

91 Ehlers, *Far Horizons*, 50.

92 Líng, *Qīshí zìshù*, 127.

93 Líng, *Qīshí zìshù*, 128–29.

94 HHL, Foreword.

"Even as a superhuman, I would have found things difficult to endure. If in old times, people walked ten thousand miles out of honor, then I administered ten thousand miles out of fear."[95] Whereas ten years earlier, civil engineers had collaborated with strong governments to command massive workforces driven by fear (chapter 5), the engineers now realized that they, as well, had become impelled by a strong state.

6 The Gansu–Xinjiang Railway

In 1942, Ling worked as chief engineer of the Baoji–Tianshui section of the Gansu–Haizhou Railway. The tracks had not yet reached Tianshui when his office received a call to survey a further expansion: the Gansu–Xinjiang Railway. The railway was to start from Tianshui and follow the Hexi Corridor through the Gansu capital of Lanzhou and Xinjiang's capital Ürümqi, before terminating in Usu, located in the Ili Valley near the Soviet border (Figure 29).[96] Measuring almost 3,000 km, the line would be the vastest railway the government had ever conceived, about 1.5 times the length of the Beijing–Hankou–Guangzhou line that cut across China proper from north to south. While it was a project of unheard-of extent by itself, it was but one more item on Ling's agenda, who already had his hands full with the other highways and railways of the Northwest.

The foundations for a survey of the Gansu–Xinjiang line were laid in October 1940. That month, Jiang Jieshi established the Central Planning Board (*zhongyang shejiju* 中央設計局, CPB) to guide the ministries and commissions in implementing his "Triple System of Administration" (*xingzheng sanlianzhi* 行政三聯制, TSA). Revolving around the three steps of planning (*jihua* 計劃), implementing (*zhixing* 執行), and reviewing (*kaohe* 考核), TSA was essentially a call for a more efficient administrative management. It had its intellectual origins in the public administration reform movement in the US.[97] The system was conceived by the political economist Gan Naiguang 甘乃光 (1897–1956),[98] who had founded the Commission for Studying Administrative Efficiency (*xingzheng xiaolü weiyuanhui* 行政效率委員會) in 1934 under the impression of his meetings with the US public administration reformer

95 Líng, *Qīshí zìshù*, 128–29.
96 HHL, 42/6/2.
97 Bian, *The Making of the State Enterprise System in Modern China*, 153, 159.
98 Bian, *The Making of the State Enterprise System in Modern China*, 166.

Leonard D. White (1891–1958).[99] Between 1943 and 1944, the CPB supervised several plans for postwar reconstruction.[100] One of them was the Ten Years Plan for the Reconstruction of the Northwest (*xibei shinian jianshe jihua* 西北十年建設計劃). The reconstruction plan was to include areas of industrialization, infrastructure, irrigation and farming, and education. In accordance with TSA principles, the plan had to start with an exploration of the circumstances on the ground. For this, the CPB formed the Northwestern Expedition (*Xibei jianshe diaocha tuan* 西北建設調查團) in May 1943.[101] Ling's task within this expedition was to survey the route of the Gansu–Xinjiang Railway.

Although no serious explorations or surveys had been conducted, the idea of connecting the Han Chinese regions with Xinjiang and Central Asia was not new.[102] Already in 1912, Ye Gongchuo (chapter 3) had advocated connecting Europe and Asia through Chinese Turkestan to Omsk and Tashkent.[103] The Gansu–Xinjiang line appeared in Wang Jingchun's 1912 map of projected Mongolian railroads, called the "Imperial Chinese North-Western Trunk."[104] The projected line ran from Lanzhou to the Turkestani border. Wang saw similarities between the American frontier and the Western Chinese dependencies, as China would require a railway like the US Transcontinental Railway to facilitate mass Han emigration.[105] The railway had the potential to turn "deserts into productive lands" and bring the "wilderness within the reach of civilization."[106]

The Nationalist government's reasoning for the Gansu–Xinjiang Railway were based on these earlier thoughts. In the context of the war economy, the connection would act as a backdoor to augment deliveries through the Soviet Union. In addition, in the long term, it could support the economic and cultural integration of Xinjiang into China proper.[107] In theory, Ling agreed with this scenario.

---

99 Bian, *The Making of the State Enterprise System in Modern China*, 155–59.

100 Bian, *The Making of the State Enterprise System in Modern China*, 162; Chéng, "Zhōngguó jìhuà jīngjì de qǐyuán yǔ zīyuán wěiyuánhuì," 94; Hé, "Luó Jiālún yǔ xīběi jiànshè kǎochá tuán," 21.

101 HHL, 43/5/4; Línghé Mǎ, "Líng Hóngxūn yǔ xībù biānjiāng tiělù de guīhuà hé jiànshè," *Xīnjiāng shèhuìkēxué* 2 (2007): 102–3.

102 See my upcoming chapter in Elisabeth Kaske and Elisabeth Köll, *Age of Exploration* [tentative title].

103 Yè, *Jiāotōng jiùguó lùn*, 7; JACAE, 5 (2): Gongchuo Ye, "The Outlook of Communications in China," 5–20: 6.

104 Wang, "'The New China Will Be a New United States'."

105 Wang, "Economic Significance of Railway Location," 12–13.

106 Wang, "Economic Significance of Railway Location," 10.

107 See also Judd Kinzley, *Natural Resources and the New Frontier: Constructing Modern China's Borderlands* (Chicago: University of Chicago Press, 2018), 122–28.

In a review of Ye Gongchuo's development plan, he argued that in outer regions like Xinjiang, infrastructure could help to overcome ethnic troubles. These conflicts had become a hindrance to construction projects; for instance, a Kazakh insurgency stymied a highway survey between Dunhuang and Kashgaria.[108] As Ling explained, in order to propagate nationalist thought, the state had to support education and build up industry. Education and industry, in turn, required transport and communication infrastructures.[109]

The 1943 CPB expedition followed the northern silk road, passing through Wuwei, Jiuquan, the oil fields of Yumen, Anxi (Guazhou), traversing over the Xingxingxia Pass, reaching the Dzungarian towns of Ķumul (Hami) and Ürümqi. Its members joined local meetings, discussed with officials, and visited factories, schools, and hospitals. Eventually, they reached Ili Valley, where the railway was to terminate. In the valley's biggest town, the city of Ĝulja, Ling observed that the town was economically and culturally under influence of Soviet Russia. The Soviets had erected a transnational highway from Kazakhstan to Ürümqi via Khorgas, which led further from Ķumul to Xingxingxia, and connected to Lanzhou and Xi'an. Given the threat of the Japanese Empire, the anti-Communist engineer could not complain about the Soviet presence. Ling argued that Soviet infrastructure enabled imports that not only served the "forgotten" people of the Northwest but also served as a supply route when Japanese forces controlled other areas of entry.[110]

When the expedition team compiled their final report in Tianshui, Ling was the main editor of the chapter on railways. In September 1943, the fifth central assembly of the Republic convened for its eleventh and final sitting. Expedition leader Luo Jialun 羅家倫 (1887–1969), Xinjiang governor Sheng Shicai 盛世才 (1895–1970), and Gansu governor Gu Zhenglun 谷正倫 (1890–1953)—with whom Ling had co-led the Northwest Highway Affairs Office—agreed that construction of the various railway sections from Baoji to Ürümqi should be built in immediate succession. According to their plan, excluding the final stretch from Ürümqi to Usu, laborers were to place roughly 2,500 km of railway track by 1950.[111]

Meanwhile, Ling had grown skeptical about the project: the backdoor he had made out in the Ili Valley was too far away, as "water in a distant place"

108 Hé, "Luó Jiālún yǔ xīběi jiànshè kǎochá tuán," 35–36.

109 Líng, "Shù jiāotōng jiùguó lùn," 14.

110 Líng, *Qīshí zìshù*, 77; Hé, "Luó Jiālún yǔ xīběi jiànshè kǎochá tuán," 13.

111 Líng, *Qīshí zìshù*, 126; Hóngxūn Líng, "Tiělù piān," in *Xīběi jiànshè kǎochá bàogào*, ed. Jiālún Luō (Taipei: Academia Historica, 1968), 85.

was "of little use in putting out a fire" in the Chinese heartland.[112] Apart from questioning its appeal as a supply route, Ling also raised concerns about the ambitious schedule of construction. Even for Ling, who had undercut estimates when engineering the Guangzhou–Hankou Railway ten years earlier, the timeframe for constructing a railway from Baoji to Ürümqi appeared highly ambitious.[113] He wondered: given that for the first 150 km of the route between Baoji and Tianshui, he had already spent more than two years, how could he complete 2,500 km in the next six?[114] Different from how he had welcomed his earlier assignments as opportunities to materialize the potential of his profession, this time, he was to supervise an infrastructure he did not truly believe in. The plan would only be achievable in a pacified environment, both in terms of war with Japan and domestically.

Ling felt frustrated at the division between authority and responsibility. In his autobiography, he remarked, "It is unreasonable to have a person with responsibility but without authority."[115] When Ling argued that executive power over an engineering project should be united and centralized,[116] he did not mean a potent government commission overseeing the project from far away. He asked for the amalgamation of responsibility and power in not just one institution but one individual engineer to secure the best craft, best economy, and fastest handling possible for a project.[117] With the "technocratic" arrangement of the Guomindang era, Ling bore the responsibility for the swift completion of his projects, their safety, and their economic efficiency, but the ultimate authority over the projects remained in other places. Officially, he declared that willpower and decisiveness could fasten the construction of the Gansu–Xinjiang Railway. With enough effort and supported by the right national policies, he claimed, the railway could reach Ürümqi in seven years. However, this required the chief engineer to attain principal authority over the project.[118]

The limited extent of the civil engineers' authority in wartime China resembled that of Nazi Germany: where deemed useful, the regime offered professional experts to advance in the state hierarchy, but the bureaucratic

---

112 Líng, *Shílìù nián zhùlù shēngyá*, 57.

113 For an overview of the railways included in the first and second post-war ten year plans, see Geographical Section of the Naval Intelligence Division, *China Proper*, 508–9.

114 Líng, *Qīshí zìshù*, 126.

115 Líng, *Shílìù nián zhùlù shēngyá*, 48.

116 Líng, *Shílìù nián zhùlù shēngyá*, 73.

117 Líng, *Qīshí zìshù*, 75; Mǎ, "Líng Hóngxūn yǔ xībù biānjiāng tiělù de guīhuà hé jiànshè," 103–4.

118 Líng, "Yóu Zuǒ Zōngtáng píngdìng Xīnjiāng shuō dào Gān Xīn tiělù zhī xīngzhú," 134.

control remained.[119] What was a misunderstanding back then and remains a misconception to date: although engineers joined powerful state commissions that dealt with questions of resources and construction, the existence of these potent commissions did not place engineers as a collective, let alone every involved engineer, in charge of the process. There were a few powerful engineers in high-level positions (chapter 6), but they had performed what professionalization theorists have called an "exit": by accepting a bureaucratic and/or political position, they departed from their active professional work.[120] In this respect, engineers faced their unique dilemma of upward mobility. To gain authority over more than one's subordinated team members, e.g., the capital of one's firm or the control over one's ministry's direction, the engineer had to leave their occupation and become a manager, chief executive, government bureaucrat, or the like.[121] If the bureaucratic technostructure of the new powerful state commissions had brought any kind of rise of the "technocrats," they brought little advantage to practitioners in the field like Ling.

Given Ling's pressure and frustration, he may have been content that for the next ten years, the project did not commence as planned. Shortly after the expedition had completed their report, chaos ensued. In 1944, Uyghur activists in the North proclaimed an independent East Turkestan Republic. The NRA became occupied with their own civil war efforts against the Communist Party. With the government priorities no longer in the Northwest, Ling once again left his uncompleted work behind. He spent the last years of the Republic at the central Ministry as Vice Minister of Communications.

7 Conclusion

In theory, the conditions of Ling's assignments appeared destined to lead to professional pride. Reporting directly to Jiang's military government, without idle bureaucrats, reluctant international investors, or foreign senior engineers in between, Ling acted in a potent position and oversaw the construction of critical infrastructure. He received the public reputation that his senior Wang Jingchun, during his time at the Chinese Eastern, had felt cheated on (chapter 3). Apart from the appreciation for his achievements, the parameters of Ling's

119 Freidson, *Professionalism*, 140; Renneberg and Walker, "Scientists, Engineers and National Socialism," 5.

120 Burrage, Jarausch and Siegrist, "An Actor-Based Framework for the Study of the Professions," 214.

121 Freidson, *Professionalism*, 76, 168.

work also closely resembled his professional ideals: far away from the diversions of the city, Ling traveled across the northwestern peripheries, with all his contentment deriving from serving his nation and the state. Nonetheless, Ling experienced professional discontent. His discontent emerged less from the state ignoring or rejecting his professional ideals than from overstretching them. Commanded to oversee multiple railroads and highways at once, each of them critical to the war effort, Ling felt overburdened and asked to resign several times.

After the end of the Sino-Japanese War, Ling sat down and created an interrelated chronology of military and infrastructural events to demonstrate that modern war was inseparable from transport and communications. He knew that despite his education as a civil engineer, he had become deeply involved with the war. However, the wartime government did not share its power with the practitioners it relied on. Instead, Ling, while involved in more infrastructural projects than he could have dreamt of, worked at the mercy of the generalissimo, ceaselessly rushing from one assignment to the next. He shouldered the responsibility, and within his team, he also had the authority, but he never had the autonomy that his profession was longing for.

CHAPTER 8

Conclusion

The past seven chapters have described and analyzed the affairs of a short-lived professional segment that worked in an eventful period of Chinese history. Their experiences paint the picture of the challenges and frustrations of the civil engineering profession in Republican China and especially of those that received their training in the US and remained part of the Sino-American engineering community. Their cases lead to findings on how the young practitioners came to identify with their profession and the circumstances that led them to constantly modify and specify their understanding of it. The challenges entailed in the demands for more engineering expertise in leadership remain relevant today.

1 Summary

The young civil engineers began their careers in a time of optimism and revolutionary upheaval. Amid public calls for reform and reconstruction, they felt that the state needed graduates from their field of study. Enthusiastically, they absorbed professional knowledge and social practices from the United States and returned to China to offer their expertise. However, transport and communications required more than well-trained civil engineers. As the young generation soon realized, their skillset was of little use when the financial, economic, and political situation prohibited large construction projects from commencing.

One of the early experiences of failure was the Grand Canal improvement project. Having secured a loan from an American investment group, the new civil government of China appeared ready to revive the ancient waterway and restore its cross-regional economy. Although occasional warfare and insurgency hindered the surveys, the project's termination was ultimately not only a domestic outcome. The civil engineers realized that despite their theoretical power to reshape territories with their infrastructures, they were impotently subjected to unforeseeable world events. A concatenation of financial troubles led to the termination of the project. The engineers, material by profession, had been unable to transform their surveys into structures.

Wang Jingchun, a railway engineer assigned to work in the northeastern Fengtian domain, shared the professional discontent of the hydraulic engineers. The point of departure was different: While the American involvement with

 | DOI:10.1163/9789004549555_010

the Grand Canal attracted little suspicion of imperialist intent, the Chinese Eastern Railway, which Wang had to manage, was a dyed-in-the-wool colonial infrastructure. His experience of failure did not originate in financial straits. On the contrary, during the years of his presidency, the railway became a profitable enterprise. In Wang's case, it was the political quarrels in an unconsolidated, highly complex conflict situation that rendered the implementation of his professional ideals inconceivable. Instead of smoothly operating an efficient engineering enterprise under the premise of supporting the Chinese nation, Wang was caught in a fight between different Chinese states, which again struggled with different Russian and international agencies. Wang was forced to resign, and his project of technocratic management ended prematurely.

Despite the early discontent as demonstrated by these two case studies, the civil engineers with ties to the United States kept together and structured their professional exchange through a newly founded association. The Association of Chinese & American Engineers was an engineering society unique in that it promoted the Sino-American segment, achieved an executive balance between the nationalities, and co-existed with the broader engineering societies which it complemented. The association was not only a platform to network. Through the annual convention and the *Journal*, members exchanged and disseminated their version of the professional civil engineer, an ideal informed by shared experiences of discontent. Their common professional experiences united them more than their nationality could divide them, as apart from association business, the members of the segment also developed friendships.

Their past frustrations led to skepticism about non-technicians in leading positions. For the engineers working for the Chinese International Famine Relief Commission, an abortive road construction project in Yunnan appeared to reconfirm this distrust. In contrast, in Guizhou, the governor welcomed the engineers warmly. Through collaborating with the authoritarian provincial government, the engineers pronounced their distinction from the working class, who shared the profession's emphasis on physical labor and fieldwork. Finally equipped with the necessary capital to realize larger infrastructural projects, the involved engineers appeared to have lost track of their supposed service ideal that had the welfare of the people in mind.

The Nationalist transition brought changes and challenges to the Sino-American segment. The ACAE had clung to the Beijing government until the end and almost vanished along with it. The reputation of the segment suffered in the aftermath of another experience of discontent, this time the at least partly self-inflicted failure of the Salaqi irrigation project. Nevertheless, the reaction to the damaging public critique also demonstrated an enduring core within the segment. Despite the anti-imperialist rhetoric of Nanjing and its nationalization policies, the transnational engineering community managed to remain active and organized well into the 1930s. Not all engineers were eager to make use of and work for the new potent government institutions. As the state moved toward a war economy, the engineers had to rethink the civilian premise of their profession. They also had to reconsider the cosmopolitan element of their community, as their US colleagues left the country.

In the years leading to the Japanese invasion, the Nationalist government gradually established a command economy with broad authority and grand ambitions. The gravity of the situation led to increased awareness of responsibility and raised pressure for those in charge. Ling Hongxun oversaw the survey and construction of an extensive railway and highway system that was larger and more important in the national context than his contemporaries and seniors could have asked for. Instead of finding contentment in this assignment, his duties impelled him. He regularly begged for his dismissal and would have gladly exchanged some of his wide-reaching responsibilities for a hint of professional autonomy.

2 Findings

During the four decades of civil engineering covered in this book, the young professionals encountered many sources of discontent. These often resulted from contradictions between their objectives and what they found: They had believed in liberal economics (chapters 3, 7) and a political system oriented toward the United States (chapters 2, 4), but grew impatient with dispersed and reluctant governments and the dependency on unstable world finance (chapters 2–3, 5–6). They attempted to introduce efficient management but met political dissension (chapters 3, 5). They claimed that their work benefited all classes but defended the privileges of their own (chapters 4–6). They celebrated professional autonomy but relied on the state (chapters 6–7). They promoted the image of a restless, industrious engineer but had trouble living up to it (chapters 4, 7). They had high expectations and often achieved little (chapters 2–3, 6–7). These contradictions were not only a result of the meanings the practitioners had bestowed upon their profession, but they also invited the professionals to reimagine their craft. In this way, their professional discontent became a motor of their professionalization, as well.

Frustrated by seemingly endless surveys which they could not see to an end (chapter 2), the engineers desired a strong state that provided the capital they needed to materialize their plans. As engineers, although they followed a political goal in that they were nationalistically motivated, they nonetheless claimed to act as neutral technicians bound to supposedly rational logic and the paradigm of efficiency, which was celebrating its heyday in US engineering culture (chapter 3). Not a political power themselves, they remained dependent on the goodwill of others. The engineers were quick to form their ideals, voice their demands, and share their anger and frustration—but they were also quick to hide in the comfort of professional neutrality and collaborate with a remarkable spectrum of northern militarists, provincial despots, left-wing revolutionaries, and the Nationalist generalissimo (chapters 2–3, 5–7).

Their nonpartisan approach meant great adaptability. At the Chinese Eastern, Wang Jingchun and Ostroumov configured their arguments of reason to align the railway in the favor of their nation (chapter 3). For the famine labor highways, what was the benefit of some was the hardship of many others (chapter 5). For Ling Hongxun's highways and railways, efficiency came to mean military utility (chapter 7). The solutionist mindset of the engineers meant they tried to find a technical solution to every problem they encountered (chapters 3, 6), but with their own political alignment so loosely defined, it was not always clear to whose favor the problems were to be solved.

As members of a young profession, the civil engineers did not command a written code of ethics. For the longest time of the analyzed period, no Chinese engineering association had one. But the absence of a spelled-out code did not mean the absence of an ethical conception of their profession. The engineers promulgated their ideals through speeches and essays, but also internalized them through their practice. Honorary members of the ACAE set an example for other engineers to follow (chapter 4). The promulgated ideal of the self-restrained individual devoting oneself to the greater good did not always stand the test, as the limited benefit of the Guizhou Road suggests (chapter 5). But the service ideal was more than a strategy to elevate the profession in the eyes of the public. Ling Hongxun took the ideals seriously, tried to live up to them, and almost succumbed to the pressure it created (chapter 7). In this way, no matter how genuine or strategic in their origin, the elusiveness of the self-imposed work ethics led to further professional discontent.

While from Wang Jingchun to Yang Baoling, the engineers frequently lamented the lack of expertise in leadership, they did not always comply with their own meritocratic ideals. The networking aspect of their association, demonstrated by its many occupational links, at times resembled more a fraternity than a professional society (chapter 4). Likewise, Ling Hongxun had his patrons Ye Gongchuo and J. A. L. Waddell to thank for his career, not only his education (chapter 7). However, in the eyes of the professionals, their qualification was never in question, and so they understood their coterie as a countermeasure against a perceived lack of public appreciation and support.

The Sino-American engineering community was a transnational segment not only through educational upbringing but also through the close cooperation of Chinese and American engineers in China. The Chinese students did not just absorb the knowledge taught in US colleges and returned to their native place. Rather, their connection to the US engineering scene became a continuous performance, as engineers from the two nationalities worked together, networked through a professional society, and developed friendships that transgressed the professional realm (chapters 4, 6). Instead of joining the anti-foreign currents of the time, the US-educated engineers identified as part of a segment that distinguished itself through its cosmopolitan personal makeup. Their primary "others" were not the foreigners. As a group, the Chinese and American engineers distinguished themselves together from the bureaucrats (chapters 2–5), the working class (chapters 1, 5), and formally, although in a collegial way, from the broader civil engineering scene (chapters 4, 6). For the professional segment, expertise was more important than nationality.

In the instance that their expertise came under scrutiny, the segment reinforced its network for mutual support. Affairs such as the Yellow River Bridge controversy (chapter 2) or the vocal criticism against the Salaqi irrigation project (chapter 6) hurt the segment's professional pride. The involved engineers replied with organized opposition that had their reputation at least as much in mind as the state of the respective project.

As pointed out in the introductory chapter, "engineering trouble" has a two-fold meaning. It relates not only to trouble as a passive experience of the engineers but also to the active engineering of it. Instead of, or coincident to, bettering the country with their infrastructures, the work of the engineers had undesired consequences. The Grand Canal Board's failure reinforced the prejudice that engineers were more occupied with surveys than with construction (chapter 2). Wang Jingchun's failure to "solve" the problems of the Chinese Eastern saw the Soviet annexation of Vladivostok, the recognition of the Soviet Union, and the return to an old-style management (chapter 3). Tan Zhen created trouble by mismanaging his private company (chapter 6). The Famine Relief Commission's highways provoked protests from landowners and required forced labor. Huang Rongde's civilian highway was immediately used for military conquest (chapter 6). Ling Hongxun's race against the clock to finish the Guangzhou–Hankou Railway resulted in accidental deaths (chapter 7), and the Salaqi irrigation project, at least according to the League experts, was an engineered failure in itself (chapter 6). Engineering trouble was not only passively a result of political and economic conditions but also actively a product of engineering and its ideals.

3 Implications

It was the professional community itself that debated over the meaning of their craft and changed these meanings through performing it. The US-trained civil engineers may have shared much of their training and many of their ideals with other civil engineers. Nevertheless, they acted as a distinct group, as they identified with their education and continued this claim through their association with foreigners and with one another. The primary legitimization of professionalism was not the certification—in this study, the graduation from a renowned US college—but the professional organization and distinction process that followed and continued afterward. The returned engineers continued to share occupational links and decided to formalize their segment through a professional society. Instead of returning to China and associating solely with one's compatriots, they continued to spend time with their US counterparts and reaped their prestige from being part of a cosmopolitan community.

The findings of this study also add substance to emerging fields in the social sciences and connect them to the study of the professions. The study of socio-technical imaginaries is one of them. The conceptualization of the civil engineering profession relied on the meanings the civil engineers imposed on the infrastructures they planned to erect. Only by imagining railways, highways, and waterways to serve the national cause could the professionals arrive at the understanding that their work fulfilled a higher purpose, that the public had to appreciate it, and that the state had to support it. In this way, the socio-technical imagination was one of the prerequisites for the discontent that emerged after the infrastructures failed to materialize. Consequently, the findings also connect professionalization to the sociology of failure, pride, and discontent. Political meddling, economic crises, and institutional changes continuously questioned the high professional standing the engineers claimed, which originated from the imagined potential of their work. The engineers reacted to these experiences through frustration and reorientation. Accordingly, the findings of this study establish a reciprocal relationship between the professional ideals the engineers propagated and the frustrations they encountered along the way.

Apart from their significance for theory building, the findings are also relevant for scrutinizing the paradigms of engineering in the political field. The engineers employed a "cohesion frame" to legitimize their work on behalf of the greater good. The road constructions of the Famine Relief Commission, continued by Ling Hongxun's wartime efforts, demonstrate how cohesion, promoted as benefiting all, served to augment existing power structures. Finally presented with sufficient means to realize their projects, which they deemed necessary and desirable, the engineers followed suit, even if the people temporarily suffered in forced labor campaigns. In today's China, the government takes possession of a New Confucianist interpretation of social cohesion to enforce conformity. The cohesion frame, as a political logic, is not limited to the engineering profession, but observing it there can help in understanding and scrutinizing its rationale.

The technocratic trust in numbers has returned, as well. It serves as legitimization for the digitized authoritarianism in China, in which automatized systems rate, rank, discipline, and punish citizens on the basis of quantifiable parameters, seemingly objective algorithms, and deep learning technology. According to the rationale, technology keeps the state in order and leads it to prosperity. Sympathizers of this system argue with the paradigm of efficiency when they lobby for a wider adaptation of Beijing's technologies of control. While different in severity, Giorgio Agamben also identified a revival of technocratic sympathies in democratic states as a response to the Covid-19 pandemic. The propagated arguments are well known to historians of engineering: if only a government would listen to the experts, act according to the techno-scientific facts, and follow the paradigm of efficiency, the people would ultimately profit from it. However, historians of engineering also know that things are not that simple. The historic cases studied in this book demonstrate that rational parameters of efficiency are open to interpretation and political claim. The political adaptability of the engineers' solutionist mindset hints at the troubles supposed solutions can create.

4 Outlook

This study has focused on a specific specialization of engineers, namely civil engineers, working in a specific region, and further contained the case studies to a privileged group of individuals who had the means to study in the US in a relatively early period. On the one hand, the narrowly defined probe adds precision to a field of historical research that too often operates with vaguely defined concepts of the profession and, based on a few case studies, generalizes its findings to an overly large group of individuals. On the other hand, the narrowly specified group of actors also makes extrapolating the findings even more problematic than if the shared characteristics of the case studies were less clear.

Consequently, the study adds to the bodies of research on the sociologies of professionalization, failure, science and knowledge, and technical imaginaries, but the findings must be carefully put into relation to other complementing studies. To learn which observations can be confirmed for a greater group of individuals, and which conclusions are confined to the American segment, future research projects that could complement this study would look at civil engineers from other educational backgrounds, such as Germany, or study US graduates with a different specialization, such as mechanical engineers. The professional ideals promoted by the Association of Chinese & American Engineers should also be compared to those of larger bodies, such as the Chinese Institute of Engineers. Many engineering societies left behind journals waiting to be explored by future researchers. The restricted and impractical access to Chinese archives leaves open to imagination what other kinds of sources may be buried in dusted boxes and shelves.

Even more challenging than adding more domestic engineering voices is the task of complementing their impressions with the impressions of their counterparts, i.e., with the individuals who worked under them and were materially affected by their projects. While the observations of other privileged groups, such as intellectuals or political leaders with different educational backgrounds, have always been a hallmark of sinology due to the readily available texts, the experiences of the working classes are much more challenging to recover. They sometimes filter through indirectly, between the lines, in the form of complaints and conflict. It would require exceptional archival findings to bring their viewpoints back to the picture and offer a critical, but nonetheless subjective, counterweight to the biased accounts of the engineers.

The anthropological approach to the history of science and technology still has much untrodden ground to explore. Apart from hero narratives that celebrate the happy few, the focus of research has seldom been the engineer as an individual. Informed by legitimate but not exclusive research paradigms that range from international relations to anti-imperialist historiography, structural histories find a lot of conflict between nationalities. By not projecting this conflict on each individual, post-structuralist approaches that focus on the subjective experience can challenge existing histories and offer new grounds for comparison.

5 Coda: Loose Ends

The events described in this book end in the 1940s. There is a reason for it, as the decade marked a turning point for the Sino-American civil engineering segment.

Immediately after the unconditional surrender of the Japanese Empire in 1945, there was a brief outlook for a new consolidated era of reconstruction. The Chinese National Relief and Rehabilitation Administration, set up in cooperation between the United Nations and the Chinese Republic, looked poised to become the spiritual successor to the Chinese International Famine Relief Commission.[1] The latter's old employees, Oliver J. Todd and Zhang Jichun, started working for the new institution. Todd used his return to reconnect with the ACAE's last presidents, Tan Zhen and Xue Zhuobin. At dinner, they mused about resuming and finally completing the Grand Canal project and, unsuccessfully, tried to convince Zhou Enlai to voice his support.[2] However, after their reunion, the engineers would never hear from each other again.

The prospect of Chinese reconstruction with Western participation quickly waned as the civil war between the Nationalist government and the Communist insurgents recommenced. The proclamation of the People's Republic in 1949 and the expulsion of foreigners thwarted a lasting revival of the Sino-American engineering community. As the United Nations moved its famine relief program to Taiwan, the only area still under Nationalist control, Todd relocated to the island. Ling Hongxun, then appointed Vice Minister of Communications, likewise followed his government into Taiwanese exile. In retirement, he became a reader at the Academia Sinica and chronicled the modern history of transportation in many articles, monographs, and autobiographies.

Wang Jingchun did not follow his former associates to Taiwan. Since 1931, he had been summoned to the United Kingdom to oversee government purchases of railway material. Caught up abroad and experiencing the aerial bombings of London, he was informed of the destruction of his country through letter exchanges.[3] In 1949, he walked off from his assignment after he had stopped receiving orders from his silenced government. Never returning to China, he spent the remainder of his life in the US. Upholding his interest in techno-sciences, one of his sons, David Kehsin Wang (1919–2011), graduated from MIT and worked as an aeronautical engineer.[4]

Despite their socialization, not every US graduate ended up residing in Taiwan or in the United States. In accordance with their subtle socialist sympathies, Tan Zhen and Yang Baoling stayed in China, where they started working for the Tianjin municipal government. Tan became a delegate to the People's Congress and eventually, in 1959, indirectly succeeded Ling Hongxun as Vice

1 See Marilla Guptil, "The United States and Foreign Relief: UNRRA in China, 1942–1947" (PhD, University of Virginia, 1995); Chen, *Guilty of Indigence*, 176; Pietz, *The Yellow River*, 118–21.

2 OJTP, 33.06: Tan Zhen to Todd, January 15, 1946; 33.06: Todd to Tan Zhen, August 4, 1946.

3 Chinese Government Purchasing Commission Archive. Box 1, 1926–1951, GB 102 CGPC, School of Oriental and African Studies (SOAS) Archives, University of London, Wang Jingchun to Hang Liwu, September 9, 1940; Hang Liwu to Wang Jingchun, May 9, 1941; Hang Liwu to Wang Jingchun, July 26, 1941.

4 YARO, Box 723, Wong Ching-chun.

Minister of Communications. Yang never fully left his cross-Pacific connections behind him, as his children studied and worked in the US. When he was eighty years old, during the early phase of the Cultural Revolution, Red Guards beat him to death.[5] Today, a small plaque reminds visitors of the location of Yang's old residence inside the former Tianjin concession. Details about his work, let alone about his professional and personal association with foreign engineers, are not inscribed.

5 Luó, *Dōng chéng xī jiù*, 363.

Bibliography

ACCR Association of Cosmopolitan Clubs Records, 41/64/8, University of Illinois at Urbana-Champaign, Urbana

BOJP Benjamin O. Johnson Papers, 73070, Hoover Institution Archives, Stanford

CERD Pelzer et al., eds.: Chinese Engineers Relational Database, v. 1.6. Geneva: Zenodo, 2022. https://doi.org/10.5281/zenodo.4075601

CSM *The Chinese Students' Monthly*

DKP David Kinley Papers, 2/6/20, University of Illinois at Urbana-Champaign, Champaign

FRUS Foreign Relations of the United States, Office of the Historian, Washington D. C.

GCPC Chinese Government Purchasing Commission Archive, 1926–1951, GB 102 CGPC, School of Oriental and African Studies (SOAS) Archives, London

HCSC Archives of the Harvard Chinese Students' Club, 1908–1928. 008955637. Harvard Yenching Library Special Collections, Cambridge

HCSR Records of the Harvard Chinese Students' Club. HUD 3277.3000. Harvard University Archives, Cambridge

HHL Hung-hsun Ling Papers, Duì rì kàngzhàn bā nián jiāotōng dàshìjì chūgǎo. 對日抗戰八年交通大事記初稿. 1962, MS#0786. Columbia University Archival Collections

JACAE *Journal of the Association of Chinese & American Engineers* [*Zhong Mei gongcheng xiehui yuekan* 中美工程協會月刊], supertitled *The Oriental Engineer* between 1926 and 1927, Techn. 2055 s, Bavarian State Library, Munich, TA4.A84, Cornell University Library, Ithaca, TA4.J865, Smithsonian Libraries, Washington, D. C.

JRFP Papers of John Ripley Freeman, MC 0051, Massachusetts Institute of Technology, Cambridge

JSDG Modern History Archives [*Jinshisuo dang'anguan* 近史所檔案館], Academia Sinica, Taipei

NARA/CER Chinese Eastern Railway, RG 59/861.77, National Archives and Records Administration, Washington D. C.

MITR M. I. T. Student Records, AC-0058, Massachusetts Institute of Technology, Cambridge

OJTP Oliver Julian Todd Papers, 73087, Hoover Institution Archives, Stanford

UIUA University of Illinois Archives, 7/1/2, University of Illinois at Urbana-Champaign, Champaign

YARO Records of the Yale Alumni Records Office, RU 830, Yale University Archives, New Haven

Akin, William E. *Technocracy and the American Dream: The Technocrat Movement, 1900–1941*. Berkeley, Los Angeles, London: University of California Press, 1977.

Allen, Robert C., Jean-Pascal Bassino, Debin Ma, Christine Moll-Murata, and Jan van Zanden. "Wages, Prices, and Living Standards in China, 1738–1925: In Comparison with Europe, Japan, and India." *The Economic History Review* 64, S1 (2011): 8–38.

Amelung, Iwo. "Der Gelbe Fluß in Deutschland: Chinesisch-deutsche Beziehungen auf dem Gebiet des Wasserbaus in den 20er und 30er Jahren des 20. Jahrhunderts." [The Yellow River in Germany. Sino-German Relations on the Field of Hydraulic Engineering in the 20s and 30s of the 20th Century] *Oriens Extremus* 38 (1995): 151–82.

Amelung, Iwo. "Weights and Forces: The Introduction of Western Mechanics into Late Qing China." In *New Terms for New Ideas: Western Knowledge & Lexical Change in Late Imperial China*. Edited by Michael Lackner, Iwo Amelung and Joachim Kurtz, 197–232. Leiden, Boston, Cologne: Brill, 2001.

Amelung, Iwo. "Die ‚vier großen Erfindungen': Selbstzweifel und Selbstbestätigung in der chinesischen Wissenschafts- und Technikgeschichtsschreibung." [The 'Four Great Inventions': Self-Doubt and Self-Affirmation in Chinese Historiography of Science and Technology] In *Selbstbehauptungsdiskurse in Asien: China–Japan–Korea* [*Discourses of Self-Affirmation in Asia. China–Japan–Korea*]. Edited by Iwo Amelung, Matthias Koch and Joachim Kurtz, 243–74. Munich: Iudicium, 2003.

Amelung, Iwo. "Lokalität und Lokalisierung: Zur Entwicklung der Wissenschaften im China des späten 19. und frühen 20. Jahrhundert." [Locality and Localization. On the Development of the Sciences in China in the Late 19th and Early 20th Century] *Jahrbuch zur Europäischen Überseegeschichte* 14 (2014): 193–214.

Amelung, Iwo. "Science and National Salvation in Early Twentieth Century China." In *Discourses of Weakness in Modern China: Historical Diagnoses of the »Sick Man of East Asia«*. Edited by Iwo Amelung, 519–64. Frankfurt: Campus, 2020.

American Red Cross, and China International Famine Relief Commission. *Famine in China's Northwest: American Red Cross Commission's Findings and Rejoinders Thereto*. Beijing: China International Famine Relief Commission, 1930.

Andreas, Joel. *Rise of the Red Engineers: The Cultural Revolution and the Origins of China's New Class*. Stanford: Stanford University Press, 2009.

Arnold, Julean. *China: A Commercial and Industrial Handbook*. Washington, D. C.: Government Printing Office, 1926.

Association of Chinese & American Engineers. *Journal of the Association of Chinese & American Engineers* [*Zhōng Měi gōngchéng xiéhuì yuèkān*], supertitled *The Oriental Engineer* between 1926 and 1927. 中美工程協會月刊. JACAE, 1920–1941, Techn. 2055 s; TA4.A84; TA4.J865. Bavarian State Library; Cornell University Library; Smithsonian Libraries.

Association of Cosmopolitan Clubs Records. ACCR, 41/64/8. University of Illinois at Urbana-Champaign.

Atkison, Paul. "The Reproduction of the Professional Community." In Dingwall; Lewis, *The Sociology of the Professions*, 224–41.

Baark, Erik. *Lightning Wires: The Telegraph and China's Technological Modernization, 1860–1890*. Westport, London: Greenwood Press, 1997.

Bacino, Leo J. *Reconstructing Russia: U. S. Policy in Revolutionary Russia, 1917–1922*. Kent, Ohio, London: The Kent State University Press, 1999.

Baker, John Earl. *Explaining China*. London: A. M. Philpot, 1927.

Barton, Ruth. "'Men of Science': Language, Identity and Professionalization in the Mid-Victorian Scientific Community." *History of Science* 41, no. 1 (2003): 73–119. https://doi.org/10.1177/007327530304100103.

Bates, Lindell T., ed. *Class History 1910: Sheffield Scientific School*. New Haven: Yale University, 1910.

Bauman, Zygmunt. *Postmodernity and Its Discontents*. Cambridge: Polity, 1997.

Baur, Georg, and Elisabeth Kaske. *China um 1900: Aufzeichungen eines Krupp-Direktors* [China around 1900. Notes of a Krupp Director]. Weimar: Böhlau, 2005.

Beckman, Svante. "Professionalization: Borderline Authority and Autonomy in Work." In Burrage; Torstendahl, *Professions in Theory and History*, 115–38.

Benjamin O. Johnson Papers. BOJP, 73070. Hoover Institution Archives.

Bergère, Marie-Claire. "The Chinese Bourgeoisie, 1911–37." In *Republican China 1912–1949: Part 1*. Edited by John Fairbank, 721–825. Cambridge, New York, Port Chester: Cambridge University Press, [1983] 2008.

Bergère, Marie-Claire. *The Golden Age of Chinese Bourgeoisie: 1911–1937*. Cambridge: Cambridge University Press, [1986] 1989.

Bevan, Paul. "Zhou Xicheng's 'Guizhou Auto Dollar': Commemorating the Building of Roads for Famine Relief." *Journal of the Royal Asiatic Society* 29, no. 2 (2019): 345–59. https://doi.org/10.1017/S1356186318000561.

Bian, Morris L. *The Making of the State Enterprise System in Modern China: The Dynamics of Institutional Change*. Cambridge, London: Harvard University Press, 2005.

Bieler, Stacey. *"Patriots" or "Traitors"? A History of American Educated Chinese Students*. Armonk: M. E. Sharpe, 2004.

Black, Cyril E., Marion Levy Jr., and Gilbert Rozman. "Conclusions: China's Modernization in Historical Perspective." In *The Modernization of China*. Edited by Gilbert Rozman et al., 482–516. New York: The Free Press, 1981.

Boorman, Howard L., ed. *Biographical Dictionary of Republican China: Volume 3: Mao–Wu*. New York, London: Columbia University Press, 1970.

Boston Chamber of Commerce, ed. *Fifth International Congress of Chambers of Commerce and Commercial and Industrial Associations*. Boston: The University Press Boston, 1912.

Bourdieu, Pierre. *Distinction: A Social Critique of the Judgement of Taste*. Cambridge: Harvard University Press, [1979] 1984.

Brante, Thomas. "Professional Types as a Strategy of Analysis." In Burrage; Torstendahl, *Professions in Theory and History*, 75–114.

Bray, Francesca. "Technics and Civilization in Late Imperial China: An Essay in the Cultural History of Technology." *Osiris* 13 (1998): 11–33.

Bray, Francesca. "Technology." In *The Making of the Human Sciences in China: Historical and Conceptual Foundations*. Edited by Howard Chiang, 29–51. Leiden, Boston: Brill, 2019.

Brodie, Hugh. "Reconnaissance Trip through Honan and Shantung Provinces, China." *The Michigan Technic* 33, no. 4 (1921): 198–203.

Brooks, Charlotte. *American Exodus: Second-Generation Chinese Americans in China 1901–1949*. Oakland: University of California Press, 2019.

Bucher, Rue, and Anselm Strauss. "Professions in Process." *American Journal of Sociology* 66, no. 4 (1961): 325–34.

Buck, Peter. *American Science and Modern China: 1876–1936*. Cambridge, London, New York: Cambridge University Press, [1980] 2010.

Burrage, Michael, Konrad Jarausch, and Hannes Siegrist. "An Actor-Based Framework for the Study of the Professions." In Burrage; Torstendahl, *Professions in Theory and History*, 203–25.

Burrage, Michael, and Rolf Torstendahl, eds. *Professions in Theory and History: Rethinking the Study of the Professions*. London, Newbury Park, New Delhi: Sage, 1990.

Burton, Alfred E. "Report of the Dean." *Bulletin of the Massachusetts Institute of Technology Boston* 50, no. 2 (1914): 31–34.

"Buses, Troops Blamed for Rail Accidents." *The China Press*. November 12, 1936.

Bush, Michael L. *Servitude in Modern Times*. Cambridge: Polity, 2000.

Butler, Judith. *Gender Trouble*. New York, Oxon: Routledge, [1990] 2010.

"Calls China a New United States." *Lawrence Democrat*. December 18, 1912. 15.

Calvert, Monte A. *The Mechanical Engineer in America, 1830–1910: Professional Cultures in Conflict*. Baltimore: The John Hopkins University Press, 1967.

Campanella, Thomas J. "'The Civilising Road': American Influence on the Development of Highways and Motoring in China, 1900–1949." *The Journal of Transport History* 26, no. 1 (2005): 78–98. https://doi.org/10.7227/TJTH.26.1.5.

Camprubí, Lino. *Engineers and the Making of the Francoist Regime*. Cambridge: The MIT Press, 2014.

Cao, Nanyuan, Junbin Su, and Mingyan Hu. "Ethical Awareness in Chinese Professional Engineering Societies: Textual Research on Constitutions of Chinese Engineering Organizations." In *Philosophy and Engineering: Reflections on Practice, Principles and Process*. Edited by Diane Michelfelder, Natasha McCarthy and David Goldberg, 203–14. Dordrecht, Heidelberg, New York, London: Springer, 2013.

Carter, James. "Struggle for the Soul of a City: Nationalism, Imperialism, and Racial Tension in 1920s Harbin." *Modern China* 27, no. 1 (2001): 91–116.

Chang, Gordon H., and Shelley Fisher Fishkin, eds. *The Chinese and the Iron Road: Building the Transcontinental Railroad.* Stanford: Stanford University Press, 2019.

Chatley, Herbert. "The Relation between Engineering and the Great Economic Depression." *The China Reconstruction & Engineering Review* 5, no. 2 (1935): 62–64.

Chen, Hailian. "Technology for Re-Engineering the Qing Empire: The Concept of 'Arts' and the Emergence of Modern Technical Education in China, 1840–1895." *ICON: Journal of the International Committee for the History of Technology* 26, no. 1 (2021): 10–43.

Chen, Yanet Y. *Guilty of Indigence: The Urban Poor in China, 1900–1953.* New Jersey: Princeton University Press, 2012.

Chen, Zhihong. "Romance with the Periphery: Chinese Travel Writing about the 'Frontier' during the Nanjing Decade (1927–1937)." *Journal of Tourism History* 8, no. 3 (2016): 215–38. https://doi.org/10.1080/1755182X.2017.1293740.

Chéng, Línsūn 程麟蓀. "Zhōngguó jìhuà jīngjì de qǐyuán yǔ zīyuán wěiyuánhuì." 中國計劃經濟的起源與資源委員會 [The Origins of Economic Planning in China and the Resources Commission] *Shíèryī shìjì* 82 (2004): 88–100.

Chiang, Kai-shek [Jiang, Jieshi 蔣介石]. "Development of the People's Economic Reconstruction Movement." *The China Reconstruction & Engineering Review* 8, no. 1 (1935): 7–10.

Chiasson, Blaine R. *Administering the Colonizer: Manchuria's Russians under Chinese Rule, 1918–29.* Vancouver, Toronto: UBC Press, 2010.

Chin, Feng [Qín, Fén 秦汾]. "National Economic Council." In *The Chinese Year Book: 1936–37.* Edited by Chao-ying Shih and Chi-hsien Chang. 2nd ed., 363–88. Shanghai: Commercial Press, 1936.

"China Formally Takes Place on 1915 Exposition Map: New Republic's Flag Floats Over the Site for Building." *The San Francisco Call.* October 25, 1912. 147.

China International Famine Relief Commission, ed. *Custom Surtax of 1925 and Plan for Its Utilization.* B 15. Beijing: China International Famine Relief Commission, 1925.

China International Famine Relief Commission, ed. *Engineering Accomplishments: Famine Prevention and Relief Projects, 1932–33.* A 38. Beijing: s. n., 1934.

"China International Famine Relief Commission Urgently Appeals for Help." *The China Reconstruction & Engineering Review* 7, no. 3 (1935): 107.

Chinese Eastern Railway. NARA/CER, RG 59/861.77. National Archives and Records Administration.

Chinese Government Purchasing Commission Archive. Box 1, 1926–1951, GB 102 CGPC. School of Oriental and African Studies (SOAS) Archives, University of London.

Chu, Pao-chin. *V. K. Wellington Koo: A Case Study of China's Diplomat and Diplomacy of Nationalism, 1912–1966.* Hong Kong: The Chinese University Press, 1981.

Clements, J. Morgan. *China: Automobile Conditions and the Good Roads Movement.* Washington, D. C.: Government Printing Office, 1922.

Clinton, Maggie. *Revolutionary Nativism: Fascism and Culture in China, 1925–1937*. Durham: Duke University Press, 2017.

Cochran, Sherman. *Encountering Chinese Networks: Western, Japanese, and Chinese Corporations in China, 1880–1937*. Berkeley, Los Angeles, London: University of California Press, 2000.

Cody, Jeffrey W. "American Planning in Republican China, 1911–1937." *Planning Perspectives* 11, no. 4 (1996): 339–77. https://doi.org/10.1080/026654396364808.

Collaborative Research Center 1199. "Verräumlichungsprozesse unter Globalisierungsbedingungen: Finanzierungsantrag." [Processes of Spatialization under the Global Condition. Funding Application], Leipzig University, Leipzig, 2019.

Collins, Randall. "Changing Conceptions in the Sociology of the Professions." In Torstendahl; Burrage, *The Formation of Professions*, 11–23.

Coode, A. T., M. Coursin, M. Nijhoff, M. Omodeo, and M. Watier. *Report by the Committee of Experts on Hydraulic and Road Questions in China: Co-operation between the Organisation for Communications and Transit of the League of Nations and the National Government of China*. Series of League of Nations Publications 8.4. Geneva: League of Nations, 1936.

"Costs of Living in Peking: Wages." *Chinese Social and Political Science Review* 10, no. 3 (1926): 90–106.

Craft, Stephen G. *V. K. Wellington Koo and the Emergence of Modern China*. Lexington: The University Press of Kentucky, 2004.

Crane, Diana. *Invisible Colleges: Diffusion of Knowledge in Scientific Communities*. Chicago, London: The University of Chicago Press, 1972.

Culp, Robert Joseph. *Articulating Citizenship: Civic Education and Student Politics in Southeastern China, 1912–1940*. Cambridge, London: Harvard University Press, 2007.

Davies, Celia. "Professionals in Bureaucracies: The Conflict Thesis Revisited." In Dingwall; Lewis, *The Sociology of the Professions*, 177–94.

Davis, Michael. *Thinking Like an Engineer*. Oxford, New York: Oxford University Press, 1998.

Dayer, Roberta Allbert. *Finance and Empire: Sir Charles Addis, 1861–1945*. Houndmills, London: MacMillan, 1988.

"Dàzǒngtǒng lìng." 大總統令 [Presidential Orders]. *Dàgōng bào*. January 5, 1922.

Deng, Yunte 鄧雲特. *The History of Famine Relief in China*. Cambridge, New York: Cambridge University Press, [1937] 2020.

Dikötter, Frank. *The Age of Openness: China before Mao*. Hong Kong: Hong Kong University Press, 2008.

Dingman, Roger. "1917–1922." In *American–East Asian Relations: A Survey*. Edited by Ernest May and James Thomson Jr., 190–218. Cambridge: Harvard University Press, 1972.

Dingwall, Robert. "Accomplishing Profession." *The Sociological Review* 24, no. 2 (1976): 331–50. https://doi.org/10.1111/j.1467-954X.1976.tb00116.x.

Dingwall, Robert, and Philip Lewis, eds. *The Sociology of the Professions: Lawyers, Doctors and Others*. London, Basingstoke: MacMillan, 1983.

"Dr. C. C. Wang Makes Visit to Friends at Alma Mater." *The Daily Illini*. October 8, 1912. 19.

Dunne, Joseph, and James Kelly, eds. *Childhood and Its Discontents: The First Seamus Heaney Lectures*. Dublin: The Liffey Press, 2003.

Durkheim, Emile. *Professional Ethics and Civil Morals*. London, New York: Routledge, [1957] 2018.

Eastman, Lloyd E. *The Abortive Revolution: China under Nationalist Rule, 1927–1937*. Cambridge, London: Harvard University Press, 1990.

Economical Bureau, C. E. R. *The Chinese Eastern Railway and Its Zone*. Harbin: Economical Bureau, C. E. R., 1923.

Edgerton-Tarpley, Kathryn. *Tears from Iron: Cultural Responses to Famine in Nineteenth-century China*. Berkeley: University of California Press, 2008. https://doi.org/10.1525/9780520934221.

Ehlers, Joseph Henry. *Far Horizons: The Travel Diary of an Engineer*. New York: Carlton Press, 1966.

Ekbladh, David. *The Great American Mission: Modernization & the Construction of an American World Order*. Princeton: Princeton University Press, 2010.

Eliassen, Sigurd. *Dragon Wang's River*. London: Methuen, 1957.

Elleman, Bruce A. *Diplomacy and Deception: The Secret History of Sino–Soviet Diplomatic Relations, 1917–1927*. Armonk, London: M. E. Sharpe, 1997.

Elman, Benjamin A. *From Philosophy to Philology: Intellectual and Social Aspects of Change in Late Imperial China*. Cambridge, London: Harvard University Press, 1984.

Elman, Benjamin A. "Naval Warfare and the Refraction of China's Self-Strengthening Reforms into Scientific and Technological Failure, 1865–1895." *Modern Asian Studies* 38, no. 2 (1999): 283–326. https://doi.org/10.1017/S0026749X04001088.

Elman, Benjamin A. *On their Own Terms: Science in China, 1550–1900*. Cambridge, Massachusetts, London: Harvard University Press, 2005.

Elman, Benjamin A. "Toward a History of Modern Science in Republican China." In *Science and Technology in Modern China, 1880s–1940s*. Edited by Jing Tsu and Benjamin Elman, 15–38. Leiden, Boston: Brill, 2014.

Engel, Ulf. "Peace-building through Space-making: The Spatializing Effects of the African Union's Peace and Security Policies." *Journal of Intervention and Statebuilding* 14, no. 2 (2020): 221–36. https://doi.org/10.1080/17502977.2020.1717132.

Fáng, Zhèng 房正. "Yángwù yùndòng yǔ Zhōngguó jìndài gōngchéngxué de fāduān." 洋务运动与中国近代工程学的发端 [The Self-Strengthening Movement and the Origins of Modern Engineering in China] *Lánzhōu xuékān* 10 (2011): 165–68. https://doi.org/10.3969/j.issn.1005-3492.2011.10.031.

Fáng, Zhèng 房正. *Jìndài gōngchéngshī qúntǐ de 'mínjiān lǐngxiù': Zhōngguó gōngchéngshī xuéhuì yánjiū (1912–1950)*. 近代工程师群体的"民间领袖"：中国工程师学会研究 (1912–1950) ['Non-governmental Leadership' of Modern Engineering Societies. Research on China's Engineering Societies (1912–1950)]. Beijing: Jīngjì rìbào chūbǎnshè, 2014.

Feighan, Michael Aloysius. *81st Congress, First Session, Report No. 153: Ben Luke Pond, Shao Hung Pond, and David Yat Wei Pond*. Washington, D. C.: United States Government Printing Office, 1949.

Fisher, Michele Slavich. "Service to China: The Career of the American Engineer, O. J. Todd." PhD, Georgetown University, 1977.

"Ford Trade School Prepares Chinese Students for Career." *Ford News*. June 22, 1923.

Foreign Relations of the United States. FRUS. Office of the Historian.

Foucault, Michel. *The Care of the Self*. The History of Sexuality 3. New York: Pantheon Books, [1984] 1986.

Foucault, Michel. *The Use of Pleasure*. The History of Sexuality 2. New York: Vintage, [1984] 1990.

Freeman, John Ripley. "Grand Canal Investigations 1918–1921: American International Corporation Review and Conclusions." University of Michigan Copy, September 1921.

"Freeman Sees Great Future for China." *The Conference Daily*. September 11, 1923.

Freidson, Eliot. "The Theory of Professions: State of the Art." In Dingwall; Lewis, *The Sociology of the Professions*, 19–37.

Freidson, Eliot. *Professionalism: The Third Logic*. Cambridge, Malden: Polity, [2001] 2007.

Freud, Sigmund. *Das Unbehagen in der Kultur* [Civilization and Its Discontents]. Vienna: Internationaler Psychoanalytischer Verein, 1930.

Fu, Hong, and Calum Turvey. *The Evolution of Agricultural Credit during China's Republican Era, 1912–1949*. London: Palgrave Macmillan, 2018. https://doi.org/10.1007/978-3-319-76801-4.

Fu, Poshek. "Trans-Pacific Cultural Connections: The University of Illinois and China." *Journal of American-East Asian Relations* 24, no. 1 (2017): 27–40. https://doi.org/10.1163/18765610-02401008.

Fuller, Pierre. *Famine Relief in Warlord China*. Cambridge, London: Harvard University Press, 2019.

Fung, Edward S. K. *In Search of Chinese Democracy: Civil Opposition in Nationalist China, 1929–1949*. Cambridge, New York: Cambridge University Press, 2000.

Furth, Charlotte. *Ting Wen-chiang: Science and China's New Culture*. Cambridge: Harvard University Press, 1970.

Galbraith, John Kenneth. *The New Industrial State*. Boston: Houghton Mifflin, [1967] 1985.

Gamsa, Mark. *Harbin: A Cross-Cultural Biography*. Toronto, Buffalo, London: University of Toronto Press, 2020.

Gamson, William A. *Talking Politics*. Cambridge, New York: Cambridge University Press, [1992] 1996.

Gandhi, Mohandas Karamchand. *Hind Swaraj: And Other Writings*. Cambridge: Cambridge University Press, [1997] 2009.

Geographical Section of the Naval Intelligence Division, ed. *China Proper: Volume III. Economic Geography, Ports and Communications*. Geographical Handbook Series. Cambridge: Naval Intelligence Division, 1945.

Gilbert, James. *Whose Fair? Experience, Memory, and the History of the Great St. Louis Exposition*. Chicago: University of Chicago Press, 2009.

Gillin, Donald G. *Warlord: Yen Hsi-shan in Shansi Province 1911–1949*. Princeton: Princeton University Press, 1967.

Gōng, Xuésuì 龔學遂. *Zhōngguó zhànshí jiāotōng shǐ*. 中國戰時交通史 [History of China's Infrastructure during Wartime]. Shanghai: Shāngwù yìnshūguǎn, 1947.

Grubbs, Carolyn B. "American Railroaders in Siberia, 1917–1920." *Railroad History* 150 (1984): 107–14.

Guān, Gēnglín 關賡麟, ed. *Jiāotōng shǐ lùzhèng biān vol. 5*. 交通史路政編第五冊 [History of Transportation, Road Governance vol. 5]. Jìndài jiāotōng shǐ quánbiān 22. Beijing: Guójiā túshūguǎn, [1935] 2009.

Guptil, Marilla Bliss. "The United States and Foreign Relief: UNRRA in China, 1942–1947." PhD, University of Virginia, 1995.

"Hābù zhī qū Wò wèntí." 哈埠之驅沃問題 [The Harbin Issue of Dispelling Ostroumov]. *Dàgōng bào*. March 1, 1924.

Hadley, Arthur Twining. *Railroad Transportation: Its History and Its Laws*. New York, London: The Knickerbocker Press, [1885] 1886.

"Hǎiguān zhènzāi fùshuì zhī jīngguò." 海關賑災附稅之經過 [Process of the Customs Disaster Relief Surtax] *Jiùzāi yuèkān* 4, no. 1 (1926): 1–2.

Harrell, Paula. *Sowing the Seeds of Change: Chinese Students, Japanese Teachers, 1895–1905*. Stanford: Stanford University Press, 1992.

Harrison, Henrietta. *The Making of the Republican Citizen: Political Ceremonies and Symbols in China, 1911–1929*. Oxford, New York: Oxford University Press, 2000.

Harrison, John P. *Sources, Evils and Correctives of Professional Discontent: An Introductory Lecture Delivered Nov. 4th, 1845*. Cincinnati: Ben Franklin Printing House, 1845.

Harvard Chinese Students' Club. Archives of the Harvard Chinese Students' Club, 1908–1928. 1911 Revolution. HCSC, 008955637, Folder 31. Harvard Yenching Library Special Collections.

Harvard Chinese Students' Club. Records of the Harvard Chinese Students' Club. HCSR, HUD 3277.3000. Harvard University Archives.

Hawley, Ellis W., ed. *Herbert Hoover as Secretary of Commerce: Studies in New Era Thought and Practice*. Iowa City: University of Iowa Press, 1981.

Hayhoe, Ruth. *China's Universities 1895–1995: A Century of Cultural Conflict*. New York, London: Garland Publishing, 1996.

Hays, Samuel P. *Conservation and the Gospel of Efficiency: The Progressive Conservation Movement, 1890–1920*. Cambridge: Harvard University Press, 1959.

Hé, Fāng 何芳. "Luó Jiālún yǔ xīběi jiànshè kǎochá tuán." 罗家伦与西北建设考察团 [Luo Jialun and the Northwest Construction Investigating Group]. M. A., Huázhōng shīfàn dàxué, 2012.

Headrick, Daniel R. *The Tools of Empire: Technology and European Imperialism in the Nineteenth Century*. New York, Oxford: Oxford University Press, 1981.

Hsiung, James C., and Steven Levine, eds. *China's Bitter Victory: The War with Japan, 1937–1945*. Armonk, London: M. E. Sharpe, 1992.

Hsu, Madeline Y. *The Good Immigrants: How the Yellow Peril Became the Model Minority*. Princeton, Oxford: Princeton University Press, 2015.

Hsu, Madeline Y. "Transnationalism and the Emergence of the Modern Chinese State: National Rejuvenation and the Ascendance of Foreign-Educated Elites (*Liuxuesheng*)." In *A Century of Transnationalism: Immigrants and their Homeland Connections*. Edited by Nancy Green and Roger Waldinger, 132–58. Urbana, Chicago, Springfield: University of Illinois Press, 2016.

Huang, Carol. "The Soft Power of U.S. Education and the Formation of a Chinese American Intellectual Community in Urbana-Champaign 1905–1954." PhD, University of Illinois at Urbana-Champaign, 2001.

Huang, Fu-ch'ing. *Chinese Students in Japan in the Late Ch'ing Period.* East Asian Cultural Studies Series 22. Tokyo: The Centre for East Asian Cultural Studies, 1982. Translated by Katherine P. K. Whitaker.

Huáng, Wéndé 黃文德. "1921 zhì 1935 nián qíjiān fēizhèngfǔzǔzhī guójì císhàn mùkuǎn huódòng: Yǐ Zhōngguó Huá Yáng yìzhèn jiùzāi zǒnghuì wéi lì zhī yánjiū." 1921至1935年期間非政府組織國際慈善募款活動：以中國華洋義賑救災總會爲例之研究 [Fundraising Activities of INGO Charities, 1921–1935. A CIFRC Case Study] In *Jìndài Zhōngguó wàijiāo de xīn shìdài guāndiǎn.* 近代中國外交的新世代觀點 [*The New Generation's Viewpoints on Modern China's Foreign Affairs*]. Edited by Mǐnshū Liào, 71–96. Taipei: Zhèngdà chūbǎn shè, 2018.

Huenemann, Ralph William. "The Dragon and the Iron Horse: The Economics of Railroads in China, 1876–1937." PhD, Harvard University, 1982.

Hughes, Thomas P. *Human-built World: How to Think about Technology and Culture*. Chicago, London: University of Chicago Press, 2004.

Huò, Yǒuguāng 霍有光, and Lìmín Gù, eds. *Nányáng gōngxué: Jiāotōng dàxué niánpǔ (1896–1949)*. 南洋公學–交通大學年譜 (1896–1949) [Nanyang College. Jiaotong University Chronicles]. Xi'an: Shǎnxī rénmín chūbǎnshè, 2002.

Hyde, Theophilus R., ed. *Class History 1912: Sheffield Scientific School.* New Haven: Yale University, 1912.

Imperial Chinese Commission. *Catalogue of the Collection of Chinese Exhibits at the Louisiana Purchase Exhibition.* St. Louis: Shallcross Print, 1904.

Institution of Civil Engineers. *Transactions* 1. London: John Weale, 1836.

International Labour Organization. "C029: Forced Labour Convention, 1930 (No. 29)." Accessed June 25, 2021. https://www.ilo.org/dyn/normlex/en/f?p=1000:12100:0::NO::P12100_ILO_CODE:C029. 2017.

International Labour Organization. "Ratifications of C029: Forced Labour Convention, 1930 (No. 29)." Accessed June 25, 2021. https://www.ilo.org/dyn/normlex/en/f?p=NORMLEXPUB:11300:0::NO::P11300_INSTRUMENT_ID:312174. 2017.

Iriye, Akria. *After Imperialism: The Search for a New Order in the Far East 1921–1931.* Chicago: Imprint Publications, [1965] 1990.

Ishida, Fumihiko 石田 文彦. "Rigaku/kōgaku wo senkō shita chūgokujin no ryūnichi gakusei shi." 理学・工学を専攻した中国人の留日学生史 [The History of the Chinese Students Who Studied Science or Engineering in Japan] *Gijutsu-shi kyōiku gakkai shi* 6, no. 2 (2005): 22–27.

Ishida, Fumihiko 石田 文彦. "Rikōgaku wo senkō shita chūgokujin ryūnichi gakusei no shakaiteki katsudō." 理工学を専攻した中国人留日学生の社会的活動 [The Social Activities of the Chinese Students Who Specialized in Science or Engineering in Japan] *Gijutsu-shi kyōiku gakkai shi* 8, no. 2 (2006): 37–42.

Jackson, John A., ed. *Professions and Professionalization.* Sociological Studies 3. Cambridge: Cambridge University Press, 1970.

Jackson, John A., ed. "Professions and Professionalization: Editorial Introduction." In Jackson, *Professions and Professionalization*, 3–15.

Janku, Andrea. "The Internationalization of Disaster Relief in Early Twentieth-century China." *Berliner China-Hefte* 43 (2013): 6–28.

Jansen, Thijs, Jos Kole, and Gabriël van den Brink. "Introduction." In *Professional Pride: A Powerful Force.* Edited by Thijs Jansen, Gabriël van den Brink and Jos Kole, 15–21. Amsterdam: Boom, 2010.

Jarausch, Konrad H. *The Unfree Professions: German Lawyers, Teachers, and Engineers, 1900–1950.* New York, Oxford: Oxford University Press, 1990.

Jasanoff, Sheila. "Future Imperfect: Science, Technology, and the Imaginations of Modernity." In *Dreamscapes of Modernity: Sociotechnical Imaginaries and the Fabrication of Power.* Edited by Sheila Jasanoff and Sang-hyun Kim, 1–33. Chicago: University of Chicago Press, 2015.

Jesiek, Brent K., and Yi Shen. "Educating Chinese Engineers: The Case of Shanghai Jiao Tong University During 1896–1949." In *Engineering, Development and Philosophy: American, Chinese and European Perspectives.* Edited by Steen Christensen et al., 123–43. Dordrecht: Springer, 2012.

Jiǎ, Shúcūn 贾熟村. *Běiyáng jūnfá shíqī de jiāotōng xì*. 北洋軍閥時期的交通系 [The Communications Clique of the Warlord Era]. Zhengzhou: Hénán rénmín chūbǎnshè, 1993.

Jiǎ, Shúcūn 贾熟村. "Sūn Zhōngshān jítuán yú jiāotōng xì de ēnyuàn." 孙中山集团与交通系的恩怨 [Gratitude and Grudges between Sun Yat-sen's Group and the Communications Clique] *Yúnmèng xuékān* 31, no. 1 (2010): 51–54.

Jìnshǐsuǒ dàng'àn guǎn [Modern History Archives]. 近史所檔案館. JSDG. Academia Sinica.

Johnson, Benjamin Oliver. "New Light on Railroading in Sibiria." *The Journal of the Worcester Polytechnic Institute* 25, no. 2 (1922): 81–84.

Johnson, Malcolm. "Professional Careers and Biographies." In Dingwall; Lewis, *The Sociology of the Professions*, 242–62.

Kaske, Elisabeth. "Sichuan as a Pivot: Provincial Politics and Gentry Power in Late Qing Railway Projects in Southwestern China." In *Southwest China in a Regional and Global Perspective (c. 1600–1911)*. Edited by Ulrich Theobald and Cao Jin, 379–423. Leiden: Brill, 2018.

"Keen Rivals in Making Cloisonne Art Ware." *The St. Louis Republic*. August 25, 1904.

King, Helen. "Understanding Spatial Literacy: Cognitive and Curriculum Perspectives." *Planet* 17, no. 1 (2006): 26–28. https://doi.org/10.11120/plan.2006.00170026.

Kinzley, Judd C. *Natural Resources and the New Frontier: Constructing Modern China's Borderlands*. Chicago: University of Chicago Press, 2018.

Kirby, William C. "Technocratic Organization and Technological Development in China: The Nationalist Experience and Legacy, 1928–1953." In *Science and Technology in Post-Mao China*. Edited by Denis Simon and Merle Goldman, 21–43. Leiden: Brill, 1989.

Kirby, William C. "The Chinese War Economy." In Hsiung; Levine, *China's Bitter Victory*, 185–212.

Kirby, William C. "Engineering China: Birth of the Developmental State, 1928–1937." In *Becoming Chinese: Passages to Modernity and Beyond.* Edited by Wen-hsin Yeh, 137–160. Berkeley, Los Angeles, London: University of California Press, 2000.

Kobayashi, Michiko 小林 実智子, and Hideo Izumida. "Kindai chūgoku no gijutsusha kyōkai no kikanshi ni tsuite: Kindai chūgoku no gijutsusha kyōkai ni kan suru kenkyū sono 1." 近代中国の技術者協会の機関誌について—近代中国の技術者協会に関する研究 その1 [Bulletins of Architects and Engineers Association in Modern China. A Study of the Engineering and Architectural Society in Modern China, Part 1] *Journal of Architecture and Planning (Transactions of AIJ)* 76, no. 669 (2011): 2247–53. https://doi.org/10.3130/aija.76.2247.

Köll, Elisabeth. "Chinese Railroads, Local Society, and Foreign Presence: The Tianjin–Pukou Line in Pre-1949, Shandong." In *Manchurian Railways and the Opening of China: An International History*. Edited by Bruce Elleman and Stephen Kotkin, 123–48. New York: M. E. Sharpe, 2010.

Köll, Elisabeth. "The Making of the Civil Engineer in China: Knowledge Transfer, Institution Building, and the Rise of a Profession." In *Knowledge Acts in Modern China:*

Ideas, Institutions, and Identities. Edited by Robert Culp, Eddy U and Wen-hsin Yeh, 148–73. Berkeley: Institute of East Asian Studies, 2016.

Köll, Elisabeth. *Railroads and the Transformation of China*. Cambridge, London: Harvard University Press, 2019.

Kuhn, Thomas S. *The Structure of Scientific Knowledge*. 2nd ed. Chicago: University of Chicago Press, [1962] 1970.

Kwong, Chi Man. *War and Geopolitics in Interwar Manchuria: Zhang Zuolin and the Fengtian Clique during the Northern Expedition*. Leiden, Boston: Brill, 2017.

La Fargue, Thomas E. *China's First Hundred*. Pullman: The State College of Washington Press, 1942.

Lai, Delin. "Searching for a Modern Chinese Monument: The Design of the Sun Yat-sen Mausoleum in Nanjing." *Journal of the Society of Architectural Historians* 64, no. 1 (2005): 22–55. https://doi.org/10.2307/25068123.

Larson, Magali Sarfatti. "In the Matter of Experts and Professionals, or How Impossible It Is to Leave Nothing Unsaid." In Torstendahl; Burrage, *The Formation of Professions*, 24–50.

Lary, Diana. "One Province's Experience of War: Guangxi, 1937–1945." In *China at War: Regions of China, 1937–1945*. Edited by Stephen Mackinnon, Diana Lary and Ezra Vogel, 314–34. Stanford: Stanford University Press, 2007.

Lawn, Martin, and Romuald Normand. "Introduction." In *Shaping of European Education*. Edited by Martin Lawn and Romuald Normand, 1–13. Oxon, New York: Routledge, 2015.

Layton Jr., Edwin T. *The Revolt of the Engineers: Social Responsibility and the American Engineering Profession*. Baltimore, London: The John Hopkins University Press, [1971] 1986.

Le, Xiucheng. "The Struggle for Autonomy from the State: Professionalization of Engineering in China." PhD, Cornell University, 1989.

Lee, En-han. *China's Quest for Railway Autonomy, 1904–1911: A Study of the Chinese Railway-Rights Recovery Movement*. Singapore: Singapore University Press, 1977.

Lee, Hong Yung. *From Revolutionary Cadres to Party Technocrats in Socialist China*. Berkeley: University of California Press, 1991.

Leong, Sow-theng. *Sino-Soviet Diplomatic Relations: 1917–1926*. Canberra: Australian National University Press, 1976.

Lǐ, Bócōng 李伯聪. "Zhōngguó jìnxiàndài gōngchéng shǐ yánjiū de ruògān wèntí." 中国近现代工程史研究的若干问题 [Some Issues about the Modern History of Engineering in China] *Kēxuéjìshù zhéxué yánjiū* 30, no. 6 (2013): 61–67.

Lǐ, Bócōng 李伯聪. "Lüèlùn Zhōngguó jìnxiàndài gōngchéng shǐ de kāiduān." 略论中国近现代工程史的开端 [On the Beginning of Chinese Modern Engineering History] *Shāndōng kējì dàxué xuébào* 17, no. 2 (2015): 1–8. https://doi.org/10.16452/j.cnki.sdkjsk.2015.02.001.

Lǐ, Yìshùn 李益顺. "Jiǎnlùn *Xīnmín cóngbào* zhōng de kēxué rènzhī." 简论《新民丛报》中的科学认知 [Simple Discussion of the Understanding of Science in New Citizen] *Húnán kējìdàxué xuébào (Shèhuìkēxué bǎn)* 16, no. 6 (2013): 194–97.

Lǐ, Zéhòu 李泽厚. *Zhōngguó xiàndài sīxiǎng shǐ lùn.* 中国现代思想史论 [History of Contemporary Thought in China]. Lǐ Zéhòu shí nián jí 3. Hefei: Ānhuī wényì chūbǎnshè, [1987] 1994.

Liáng, Chén 梁晨, Yùnzhú Rèn, Yǔqián Wáng, and Lǐ Zhōngqīng 李中清. "Mínguó Shànghǎi dìqū gāoxiào shēngyuán liànghuà chúyì." 民国上海地区高校生源量化刍议 [Some Thoughts on the Quantitative Study of Student Backgrounds in Republican Shanghai] *Lìshǐ yánjiū* 3 (2017): 76–92.

Liáng, Qǐchāo 梁启超. "Lùn xuéxiào, 13: Xuéhuì." 論學校十三：學會 [On Schools, 13: Study Societies] *Shíwù bào* 10 (1896): 1–3.

Lichtenstein, Alex. "Good Roads and Chain Gangs in the Progressive South: 'The Negro Convict is a Slave'." *The Journal of Southern History* 59, no. 1 (1993): 85–110. https://doi.org/10.2307/2210349.

Lín, Wěi 林伟. "Xué zuò shìjiè gōngmín: Liú Měi Zhōngguó xuésheng yǔ shìjiè huì yùn dòng (1903–1914)." 学做世界公民：留美中国学生与世界会运动 (1903–1914) [Learn to be Cosmopolite: Chinese Students in the Cosmopolitan Club Movement (1903–1914)] *Gāoděngjiàoyù yánjiū* 37, no. 3 (2016): 87–93.

Líng, Hóngxūn 淩鴻勛. *Shìzhèng gōngchéng xué.* 市政工程學 [Municipal Engineering]. Shanghai: Shāngwù yìnshūguǎn, 1924.

Líng, Hóngxūn 淩鴻勛. "Shù jiāotōng jiùguó lùn." 述交通救國論 [Discussing the Doctrine of Saving the Country through Transport and Communications] *Guówén zhōubào* 2, no. 35 (1925): 13–15.

Líng, Hóngxūn 淩鴻勛. *Tiělù gōngchéng xué.* 鐵路工程學 [Railroad Engineering]. Shanghai: Shāngwù yìnshūguǎn, [1925] 1931.

Líng, Hóngxūn 淩鴻勛. *Qiáoliáng.* 橋梁 [Bridges]. Shanghai: Shāngwù yìnshūguǎn, 1929.

Líng, Hóngxūn 淩鴻勛. "Yǎnjiǎng tǔmù gōngchéng shìyè zhī dàgài." 演講土木工程事業之大概 [Speech Outlining Civil Engineering] *Jiāodà jìkān* 3 (1930): 12–16.

Líng, Hóngxūn 淩鴻勛. "Duànliàn shēnxīn yǐ wèi jiā guó." 鍛練身心以衛家國 [Training Body and Mind to Defend House and Country] *Lǒng Hǎi xúnkān* 15 (1931): 12–14.

Líng, Hóngxūn 淩鴻勛. *Gōngchǎng shèjì.* 工廠設計 [Factory Design]. Shanghai: Shāngwù yìnshūguǎn, [1933] 1939.

Líng, Hóngxūn 淩鴻勛. "Wǒmen qìngzhù Guóqìng de shíhòu yīng xiān zhùyì dào zěnyàng qù jiěchú guónàn!?." 我們慶祝國慶的時候應先注意到怎樣去解除國難！？ [When Celebrating National Day, Should We Not First Pay Attention to Lifting Our Nation's Distress!?] *Yuè Hàn tiělù xúnkān* 7 (1936): 1–2.

Líng, Hóngxūn 淩鴻勛. "Yuè Hàn tiělù quánxiàn jiētōng hòu réng yīng jìxù nǔlì." 粵漢鐵路全線接通後仍應繼續努力 [We Need to Proceed with Great Effort after Completing the Guangzhou–Hankou Railway] *Tiělù zázhì* 2, no. 2 (1936): 107–9.

Líng, Hóngxūn 淩鴻勛. "Yóu Zuǒ Zōngtáng píngdìng Xīnjiāng shuō dào Gān Xīn tiělù zhī xīngzhú." 由左宗棠平定新疆說到甘新鐵路之興筑 [From Zuǒ Zōngtáng's Pacification of Xinjiang to the Construction of the Gansu–Xinjiang Railway] *Bǎo Tiān lù kān* 2, 4–6 (1943): 131–34.

Líng, Hóngxūn 淩鴻勛. "Tiělù gōngchéng de duōbiān rènshí [1]." 鐵路工程的多邊認識 [The Versatile Knowledge of Railroad Engineering] *Xiāng Guì Qián xúnkān* 3, no. 24 (1948): 5–6.

Líng, Hóngxūn 淩鴻勛. "Tiělù gōngchéng de duōbiān rènshí [3]." 鐵路工程的多邊認識 [The Versatile Knowledge of Railroad Engineering] *Xiāng Guì Qián xúnkān* 3, no. 26 (1948): 10.

Líng, Hóngxūn 淩鴻勛. "Tiělù gōngchéng de duōbiān rènshí [4]." 鐵路工程的多邊認識 [The Versatile Knowledge of Railroad Engineering] *Xiāng Guì Qián xúnkān* 3, no. 27 (1948): 4–5.

Líng, Hóngxūn 淩鴻勛. "Tiělù gōngchéng de duōbiān rènshí [5]." 鐵路工程的多邊認識 [The Versatile Knowledge of Railroad Engineering] *Xiāng Guì Qián xúnkān* 3, no. 28 (1948): 4–6.

Líng, Hóngxūn 淩鴻勛. "Tiělù gōngchéng de duōbiān rènshí [7]." 鐵路工程的多邊認識 [The Versatile Knowledge of Railroad Engineering] *Xiāng Guì Qián xúnkān* 3, no. 30 (1948): 5–7.

Líng, Hóngxūn 淩鴻勛. Hung-Hsun Ling Papers, Duì rì kàngzhàn bā nián jiāotōng dàshìjì chūgǎo. 對日抗戰八年交通大事記初稿. HHL, 1962, MS#0786. Columbia University Archival Collections.

Líng, Hóngxūn 淩鴻勛. *Qīshí nián lái Dōng Qīng, Zhōng dōng, Zhōng Cháng tiělù zhī jīngguò: Zhōng piān.* 七十年來東清，中東，中長鐵路之經過：中篇 [Seventy Years of the Eastern Qing, Chinese Eastern, and Chinese Changchun Railway. Vol. 2]. Taipei: Jiāotōng bù jiāotōng yánjiū suǒ, 1965.

Líng, Hóngxūn 淩鴻勛. *Shíliù nián zhùlù shēngyá.* 十六年筑路生涯 [Sixteen Years Career of Road Construction]. Taipei: Zhuànjì wénxué chūbǎnshè, 1968.

Líng, Hóngxūn 淩鴻勛. "Tiělù piān." 鐵路篇 [Railroads] In *Xīběi jiànshè kǎochá bàogào.* 西北建設考察報告 [*Northwest Reconstruction Expedition Report*]. Edited by Jiālún Luō, 67–92. Taipei: Academia Historica, 1968.

Líng, Hóngxūn 淩鴻勛. *Qīshí zìshù.* 七十自述 [Autobiography at Seventy]. Taipei: Sānmín shūjú, [1969] 1988.

Líng, Hóngxūn 淩鴻勛. "A Decade of Chinese Railroad Construction (1926–1936)." In Sih, *The Strenuous Decade*, 255–88.

Líng, Hóngxūn 淩鴻勛. "China's Epic Struggle in Developing Its Overland Transportation System During the Sino-Japanese War." In *Nationalist China During the Sino-Japanese War, 1937–1945*. Edited by Paul Sih, 243–65. Hicksville: Exposition Press, 1977.

Líng, Hóngxūn 淩鴻勛, and Zǔkāng Zhào. "Gōngchéng jiàoyù diàochá tǒngjì zhī yánjiū." 工程教育調查統計之研究 [Research on Engineering Education Survey Statistics] *Gōngchéng: Zhōngguó gōngchéng xuéhuì huìkān* 2, no. 4 (1926): 242–51.

Lǚ, Chéngdōng 吕成冬. "Líng Hóngxūn de gāoděng jiàoyù sīxiǎng jí shíjiàn." 凌鸿勋的高等教育思想及实践 [On Ling Hongxun's Higher Education Thought and Practice] *Guǎngzhōu dàxué xuébào (shèhuì kēxué bǎn)* 10, no. 5 (2011): 57–61.

Luó, Yuánxù 羅元旭. *Dōng chéng xī jiù: Qī gè Huárén Jīdūjiào jiāzú yǔ Zhōng Xī jiāoliú bǎinián.* 東成西就: 七個華人基督教家族與中西交流百年 [Succeeding East and West. Seven Chinese Christian Families and One Hundred Years of Chinese–Western Exchange]. Hong Kong: Joint Publishing, 2012.

Mǎ, Línghé 马陵合. "Líng Hóngxūn yǔ xībù biānjiāng tiělù de guīhuà hé jiànshè." 凌鸿勋与西部边疆铁路的规划和建设 [Líng Hóngxūn and the Layout and Construction of the Railways in Western Frontier] *Xīnjiāng shèhuìkēxué* 2 (2007): 100–108.

Mǎ, Línghé 马陵合, and Fāng Wú. "Běiyáng shíqí jìshù guānliáo zài chǎnyè xíngzhèng zhōng de dìngwèi: Yǐ tiělù guǎnlǐ zhìdù biàngé wèi bèijǐng." 北洋时期技术官僚在产业行政中的定位—以铁路管理制度变革为背景 [The Status of Beiyang Era Technocrats within Industrial Administration—Reform in the Railway Administrative System as Background] *Shèhuì kēxué yánjiū* 6 (2019): 133–44.

Maclaurin, Richard C. "Report of the President." *Bulletin of the Massachusetts Institute of Technology Boston* 50, no. 2 (1915): 9–25.

Mallory, Walter H. *China: Land of Famine*. New York: American Geographical Society, 1926.

Marsden, Ben, and Crosbie Smith. *Engineering Empires: A Cultural History of Technology in Nineteenth-century Britain*. Houndmills, New York: Palgrave Macmillan, 2005.

Marvin, George. "Chinese Students in America: They Will Play Important Parts When They Return Home." *The Harvard Bulletin*, May 13, 1908.

Matsuda, Takeshi. "Woodrow Wilson's Dollar Diplomacy in the Far East: The New Chinese Consortium, 1917–1921." PhD, The University of Wisconsin-Madison, 1979.

Mazuzan, George T. "'Our New Gold Goes Adventuring': The American International Corporation in China." *Pacific Historical Review* 43, no. 2 (1974): 212–32.

McDonnell, Duncan, and Marco Valbruzzi. "Defining and Classifying Technocrat-led and Technocratic Governments." *European Journal of Political Research* 53, no. 4 (2014): 654–71. https://doi.org/10.1111/1475-6765.12054.

Mihailoff, I. A., ed. *North Manuchria and the Chinese Eastern Railway*. Harbin: C. E. R. Printing Office, [1922] 1924.

Milam, Erika Lorraine, and Robert Nye. "An Introduction to Scientific Masculinities." *Osiris* 30 (2015): 1–14.

Mizoguchi, Yūzō. "The Chaotic Late Qing and Early Republican Periods." *Inter-Asia Cultural Studies* 17, no. 4 ([2007] 2016): 574–605. https://doi.org/10.1080/14649373.2016.1244889.

Moon, Francis C. *Social Networks in the History of Innovation and Invention.* History of Mechanism and Machine Science 22. Dordrecht, Heidelberg, New York, London: Springer, 2014.

Morgan, Stephen L. "Transfer of Taylorist Ideas to China: 1910s–1930s." *Journal of Management History* 12, no. 4 (2006): 408–24.

Morozov, Evgeny. *To Save Everything, Click Here: The Folly of Technological Solutionism.* New York: PublicAffairs, 2013.

"Motor Road Penetrates Center of Kweichow." *Jiùzāi yuèkān* 5, no. 4 (1928): 25.

"'Motor-Car' Dollar Which Brought Bad Luck to Kweichow General." *The China Weekly Review.* November 11, 1933.

Murphy, Raymond. "Proletarianization or Bureaucratization: The Fall of the Professional?" In Torstendahl; Burrage, *The Formation of Professions*, 71–96.

Murray, Robert K. "Herbert Hoover and the Harding Cabinet." In Hawley, *Herbert Hoover as Secretary of Commerce*, 17–42.

Muscolino, Micah S. *The Ecology of War in China: Henan Province, the Yellow River, and Beyond, 1938–1950.* Cambridge: Cambridge University Press, 2014. https://doi.org/10.1017/CBO9781107785274.

Nathan, Andrew James. *A History of the China International Famine Relief Commission.* Cambridge: Harvard University Press, 1965.

Nathan, Andrew James. *Peking Politics, 1918–1923: Factionalism and the Failure of Constitutionalism.* Berkeley, Los Angeles, London: University of California Press, 1976.

"Nationwide Drive for Compulsory Labour to Facilitate Reconstruction, Dyke-Building and Afforestation." *The China Reconstruction & Engineering Review* 7, no. 3 (1935): 116.

Needham, Joseph. *Chinese Science.* London: Pilot Press, 1945.

Needham, Joseph, and Ling Wang. *Physics and Physical Technology: Part II: Mechanical Engineering.* Science and Civilisation in China 4.2. Cambridge: Cambridge University Press, 1965.

Needham, Joseph, Ling Wang, and Gwei-djen Lu. *Physics and Physical Technology: Part III: Civil Engineering and Nautics.* Science and Civilisation in China 4.3. Cambridge: Cambridge University Press, 1971.

"News Column (no. 6)." *The Chinese Students' Monthly* 2, no. 6 (1907): 125–33.

Noble, David F. *America by Design: Science, Technology, and the Rise of Corporate Capitalism.* Oxford, New York, Toronto: Oxford University Press, [1977] 1979.

Nye, Robert A. "Medicine and Science as Masculine 'Fields of Honor'." *Osiris* 12 (1997): 60–79.

O. J. (Oliver Julian) Todd papers. OJTP, 73087. Hoover Institution Archives.

Office of the Dean for Student Affairs. Student Record Cards. MIT, AC-0058. M.I.T. Archives.

Officer, Lawrence H., and Samuel Williamson. "The Price of Gold, 1257–Present." Accessed July 2, 2021. http://www.measuringworth.com/gold/. 2021.

"Officers and Committees of Chinese Student Christian Association in North America: 1918–1919." *The Chinese Students' Christian Journal* 5, no. 1 (1918): 71–74.

Osterhammel, Jürgen. "'Technical Co-Operation' between the League of Nations and China." *Modern Asian Studies* 13, no. 4 (1979): 661–80.

Pan, Francis K. [Pān, Guāngjiǒng 潘光迥]. "Communications and Transportation." In *The Chinese Year Book: 1940–1941*. Edited by The Council of International Affairs Chungking. 5th ed., 647–75. Shanghai: Commercial Press, 1941.

Pan, Liang. "National Internationalism in Japan and China." In *Internationalisms: A Twentieth-Century History*. Edited by Glenda Sluga and Patricia Clavin, 170–190. Cambridge: Cambridge University Press, 2017.

Papers of John Ripley Freeman. PJRF, MC.0051. MIT Libraries Distinctive Collections.

Parsons, Talcott. "The Professions and Social Structure." *Social Forces* 17, no. 4 (1939): 457–67.

Parsons, William Barclay. *An American Engineer in China*. Cambridge: Cambridge University Press, [1900] 2013.

Parsons, William Barclay. "The American Invasion of China: A Personal Survey of Commercial Conditions in the Interior of China." *McClure's Magazine* 14, no. 6 (1900): 499–510.

Parsons, William Barclay. "Engineering as a Profession: Founder's Day Address." Cornell University, January 11, 1901.

Parsons, William Barclay. *Jīqì dàyè piān*. 機器大業篇 [Engineering as a Profession]. Shanghai: American Presbyterian Mission Press, 1902.

Peking United International Famine Relief Committee. *The Famine of 1920–21 in North China*. Beijing: s. n., 1922.

Pelzer, Thorben, Elisabeth Kaske, and Hailian Chen. "Chinese Engineers Relational Database (CERD)." https://home.uni-leipzig.de/cerd/. August 7, 2022. https://doi.org/10.5281/zenodo.4075601.

"Personal, Local and Home News." *The Chinese Students' Christian Journal* 6, no. 6 (1920): 350–56.

Pietz, David Allen. *Engineering the State: The Huai River and Reconstruction in Nationalist China, 1927–37*. New York, London: Routledge, 2002.

Pietz, David Allen. *The Yellow River: The Problem of Water in Modern China*. Cambridge: Harvard University Press, 2015.

Pollock, Sheldon, Homi Bhabha, Carol Breckenridge, and Dipesh Chakrabarty. "Cosmopolitanisms." *Public Culture* 12, no. 3 (2000): 577–89.

Pomeranz, Kenneth. *The Making of a Hinterland: State, Society and Economy in Inland North China, 1853–1937*. Berkeley, Los Angeles, Oxford: University of California Press, 1993.

Porter, Robin. *Industrial Reformers in Republican China*. Armonk, London: M. E. Sharpe, 1994.

Powell, M. C., ed. *Who's Who in China*. Shanghai: The China Weekly Review, 1925.

Purdue Debris: Volume XXIII. Indianapolis: The Cheltenham-Aetna Press, 1911.

Qián, Zōngzé 錢宗澤, Hóngxūn Líng, and Shìbīn Yú. "Lǒng Hǎi tiělù chóu nǐ hángkōng mùjuān bànfǎ xuānyán." 隴海鐵路籌擬航空募捐辦法宣言 [Manifesto of Using Gansu–Haizhou Railway Funds to Contribute to Aviation] *Tiělù yuèkān Píng Hàn xiàn* 35 (1933): 7–9.

"Rah! Rah! Round the World." *Evening Star Sunday Magazine*. April 3, 1910.

Reardon-Anderson, James. *The Study of Change: Chemistry in China, 1840–1949*. Cambridge: Cambridge University Press, 1991.

Reardon-Anderson, James. "Science in Wartime China." In Hsiung; Levine, *China's Bitter Victory*, 213–34.

Records of the Yale Alumni Records Office. YARO, Accession 2001-A-066/RU 709/RU 830. Yale University Archives.

Reid, Carlton. *Roads Were Not Built for Cars: How Cyclists Were the First to Push for Good Roads & Became the Pioneers of Motoring*. Washington, D. C., Covelo, London: Island Press, 2015.

Reinsch, Paul S. *An American Diplomat in China*. Garden City, Toronto: Doubleday, Page & Company, 1922.

Rèn, Yùnzhú 任韵竹, Chén Liáng, and James Lee. "Meritocracy and the Making of the Chinese Academe, 1912–1952." *The China Quarterly* 244 (2020): 942–68.

Renneberg, Monika, and Mark Walker. "Scientists, Engineers and National Socialism." In *Science, Technology and National Socialism*. Edited by Monika Renneberg and Mark Walker, 1–29. Cambridge: Cambridge University Press, 1994.

Rives, John C., ed. *The Congressional Globe: Senate, 37th Congress, 2nd Session*. Washington, D. C.: Blair & Rives, 1862.

Robinson, Ronald E. "Introduction: Railway Imperialism." In *Railway Imperialism*. Edited by Clarence Davis and Kenneth Wilburn Jr., 1–6. New York: Greenwood Press, 1991.

Rogers, J. David. "The American Engineers That Built the Panama Canal." In *Engineering the Panama Canal*. Edited by Bernard Dennis Jr., 112–349., 2014.

Rosenbaum, Arthur. "The Railway Controversy of 1911: The Bureaucratic Perspective." *Jìndàishǐ yánjiūsuǒ jíkān* 11 (1982): 315–46.

Rueschemeyer, Dietrich. "Professional Autonomy and the Social Control of Expertise." In Dingwall; Lewis, *The Sociology of the Professions*, 38–58.

Salter, Arthur. "The Silver Problem." *Political Science Quarterly* 46, no. 3 (1931): 321–34.

Schillinger, Nicolas. *The Body and Military Masculinity in Late Qing and Early Republican China: The Art of Governing Soldiers*. Lanham, Boulder, New York, London: Lexington Books, 2016.

Schivelbusch, Wolfgang. *The Railway Journey: The Industrialization of Time and Space in the Nineteenth Century*. Oakland: University of California Press, [1977] 2014.

Schulte, Barbara. *'Zur Rettung des Landes': Bildung und Beruf im China der Republikzeit* ['On the Salvation of the Country.' Education and Occupation in Republican Era China]. Frankfurt, New York: Campus, 2008.

Schurman, Rachel A., and Dennis Takahashi Kelso, eds. *Engineering Trouble: Biotechnology and Its Discontents*. Berkeley, Los Angeles, London: University of California Press, 2003.

Selander, Staffan. "Associative Strategies in the Process of Professionalization: Professional Strategies and Scientification of Occupations." In Burrage; Torstendahl, *Professions in Theory and History*, 139–49.

Seow, Victor Kian Giap. "Carbon Technocracy: East Asian Energy Regimes and the Industrial Modern, 1900–1957." PhD, History and East Asian Languages, Harvard University, 2014.

Shao, Qin. *Culturing Modernity: The Nantong Model, 1890–1930*. Stanford: Stanford University Press, 2004.

Shen, Grace Yen. "Murky Waters: Thoughts on Desire, Utility, and the 'Sea of Modern Science'." *Isis* 98, no. 3 (2007): 584–96.

Shen, Grace Yen. *Unearthing the Nation: Modern Geology and Nationalism in Republican China*. Chicago: University of Chicago Press, 2014.

Sherwood, C. H., ed. *The 1920 Cornellian*. The Cornellian 52. Ithaca: The Cornell Annuals, 1920.

"Shih Lu Project Delayed by Local Problems." *Jiùzāi yuèkān* 4, no. 6 (1927): 31–32.

Siegrist, Hannes. "Professionalization as a Process: Patterns, Progression and Discontinuity." In Burrage; Torstendahl, *Professions in Theory and History*, 177–202.

Sih, Paul K.T., ed. *The Strenuous Decade: China's Nation-Building Efforts, 1927–1937*. Asia in the Modern World Series 9. New York: St. John's University Press, 1970.

Smith, Addie Viola, and Anselm Chuh. *Motor Roads in China*. Washington, D. C.: United States Government Printing Office, 1931.

Sörbom, Per. "The Reception of Western Technology in China and Japan." In *Technological Development in China, India and Japan: Cross-Cultural Perspectives*. Edited by Erik Baark and Andrew Jamison, 35–56. New York: Palgrave Macmillan, 1986.

South Manchuria Railway Co., 南満州鉄道株式会社, ed. *Tō Shi tetsudō wo chūshin to suru Ro-Shi seiryoku no shōchō*. 東支鉄道を中心とする露支勢力の消長 [Rise and Fall of Sino-Russian Power as Seen through the Chinese Eastern Railway]. Dalian: Minami Manshū tetsudō, 1928.

Spence, Jonathan D. *To Change China: Western Advisers in China*. New York: Penguin, [1969] 1980.

"State C. E. Convention Begins Session Thursday." *St. Louis Post-Dispatch*. June 22, 1904.

"Statistics of the Earnings of Graduate Engineers." *Engineering News* 74, no. 11 (1915): 505.

Storer, Norman W. *The Social System of Science*. New York: Holt, Rinehart and Winston, 1966.

Summary of Visit, 1908, Box 17, 2/5/3. President E. James General Correspondence, University of Illinois Archives.

Sun, Fo [Sūn, Kē 孫科]. "National Scheme of Railway Construction." *The China Critic* 2, no. 8 (1929): 147–49.

Sun, Fo [Sūn, Kē 孫科]. "National Scheme of Railway Construction, cont." *The China Critic* 2, no. 9 (1929): 172–77.

Sun, Yat-sen [Sūn, Yìxiān 孫逸仙]. *The International Development of China*. Shanghai: Commercial Press, 1920.

Tán, Zhēn 譚真. "Zhènxīng gōngyè zhī zhàng'ài jí qí sǎochú zhī fāngfǎ: Tán Quánfǔ zài gōngzhuān jīxiègōng xuéhuì zhī kēxué yǎnjiǎng cí." 振興工業之障礙及其掃除之方法 譚全甫在工專機械工學會之科學演講辭 [Obstacles to Reviving the Industry and How to Remove Them. Lecture by Tan Quanfu [Tan Zhen] at the Industrial Mechanics Association] *Dàgōng bào*, May 15, 1929.

Taylor, George E. "The Reconstruction Movement in China." In *Problems of the Pacific, 1936: Aims and Results of Social and Economic Policies in Pacific Countries*. Edited by W. Holland and Kate Mitchell, 376–408. Illinois: University of Chicago Press, 1936.

Teng, Emma Jinhua. "1925: The First Chinese Woman Student at MIT." Accessed May 4, 2021. https://earlychinesemit.mit.edu/1925-first-chinese-woman-student-mit.

Teng, Emma Jinhua. *Eurasian: Mixed Identities in the United States, China, and Hong Kong, 1842–1943*. Berkeley, Los Angeles, London: University of California Press, 2013.

The Chinese Students Alliance, U. S. A. The Chinese Students' Monthly. CSM, 1906–1931.

"The Compensation of Civil Engineers." *Engineering News* 73, no. 1 (1915): 34.

"The Harvard Chinese Club." *The Harvard Bulletin*. March 31, 1909.

"The Road Building Program in China." *The Oriental Motor* 3, no. 1 (1921): 7–9, 79.

Thomson Jr., James C. *While China Faced West: American Reformers in Nationalist China, 1928–1937*. Cambridge: Harvard University Press, 1969.

Todd, Oliver Julian. "The Open Road in China." OJTP 17.03. Unpublished manuscript, 1926.

Todd, Oliver Julian. "Construction Enterprises in South-west China." *The Far Eastern Review* 24, no. 9 (1928): 405–406, 419.

Todd, Oliver Julian. "Engineering and Altruism in China." *The Peking Leader*, November 7, 1929.

Todd, Oliver Julian. "The Si–Lan Road." *The China Critic* 7, no. 44 (1934): 1073–74.

Todd, Oliver Julian. *The China That I Knew*. Palo Alto: s. n., 1973.

Todd, Oliver Julian, and Sigurd Eliassen. "The Yellow River Problem." *Transactions of the American Society of Civil Engineers* 105 (1940): 346–416.

Todd, Oliver Julian, and John Waddell. "Some Thoughts Concerning Economics in the Development of China: Memoirs Nos. 11 and 12." *Chinese Social and Political Science Review* 14, no. 2 (1930): 153–71.

"Todd Returns from South-west." *Jiùzāi yuèkān* 5, no. 12 (1927): 11–13.

"Tóngrén xiāoxī." 同人消息 [Personal Notes] *Jiùzāi yuèkān* 4, no. 3 (1927): 14.

Tooze, Adam. *The Deluge: The Great War and the Remaking of Global Order, 1916–1931*. London, New York: Allen Lane, 2014.

Torstendahl, Rolf. "Essential Properties, Strategic Aims and Historical Development: Three Approaches to Theories of Professionalism." In Burrage; Torstendahl, *Professions in Theory and History*.

Torstendahl, Rolf. "Introduction: Promotion and Strategies of Knowledge-based Groups." In Torstendahl; Burrage, *The Formation of Professions*, 1–10.

Torstendahl, Rolf, and Michael Burrage, eds. *The Formation of Professions: Knowledge, State and Strategy*. London, Newbury Park, New Delhi: Sage, 1990.

Tsao, Y. S. [Cáo, Yúnxiáng 曹雲祥]. "The Relation of the Returned Students to the Chinese Revolution." *Journal of Race Development* 4, no. 1 (1913): 96–109.

Tsao, Y. S. [Cáo, Yúnxiáng 曹雲祥]. "A Challenge to Western Learning: The Chinese Student Trained Abroad—What He Has Accomplished—His Problems." *News Bulletin* (*Institute of Pacific Relations*) Dec (1927): 13–16.

Tseng, H. P., and Jean Lyon, eds. *China Handbook 1937–1944: A Comprehensive Survey of Major Developments in China in Seven Years of War*. Chongqing: Chinese Ministry of Information, 1944.

Turner, C., and M. Hodge. "Occupations and Professions." In Jackson, *Professions and Professionalization*, 16–50.

Tyau, Min-ch'ien Tuk Zung [Diao, Minqian 刁敏謙]. *Two Years of Nationalist China*. Shanghai: Kelly and Walsh, 1930.

Uekötter, Frank. "Techniker an der Macht: Der Ingenieur-Politiker im 20. Jahrhundert." [Technicians in Power. The Engineer-Politician in the Twentieth Century] *Historische Zeitschrift* 306, no. 2 (2018): 396–423. https://doi.org/10.1515/hzhz-2018-0010.

Urbansky, Sören. *Kolonialer Wettstreit: Russland, China, Japan und die Ostchinesische Eisenbahn* [Colonial Competition. Russia, China, Japan, and the Chinese Eastern Railroad]. Frankfurt, New York: Campus, 2008.

Urbansky, Sören. *Beyond the Steppe Frontier: A History of the Sino-Russian Border*. Princeton, Oxford: Princeton University Press, 2020.

van Laak, Dirk. *Weiße Elefanten: Anspruch und Scheitern technischer Großprojekte im 20. Jahrhundert* [White Elephants. Aspiration and Failure of Technical Grand Projects in the Twentieth Century]. Stuttgart: Deutsche Verlags-Anstalt Stuttgart, 1999.

Vermeer, Eduard B. "P'an Chi-hsün's Solutions for the Yellow River Problems of the Late 16th Century." *T'oung Pao* 73, 1/3 (1987): 33–67.

Waddell, John Alexander Low. "Dr. Waddell Flatly Repudiates Charges Made Against Him by Colleagues on Yellow River Bridge Board." *The Peking Leader*, September 1921.

"Waddell Proposes Concrete Program for China Regeneration." *The Conference Daily*. September 12, 1923.

Waldron, Arthur. *From War to Nationalism: China's Turning Point, 1924–1925*. Cambridge: Cambridge University Press, [1995] 2002.

Walker, Merle R. *Contemporary Chinese Leaders: Brief Sketches*. China Society Pamphlets 3. New York: The China Society of America, 1922.

Wang, Ching-chun [Wáng, Jǐngchūn 王景春]. "Economic Significance of Railway Location." A. M., University of Illinois, 1909.

Wang, Ching-chun [Wáng, Jǐngchūn 王景春]. "The Abolition of the Queue." *Atlantic Monthly* 107 (1911): 810–16.

Wang, Ching-chun [Wáng, Jǐngchūn 王景春]. "The New Chinese Currency." *The North American Review* 193, no. 667 (1911): 816–30.

Wang, Ching-chun [Wáng, Jǐngchūn 王景春]. "'The New China Will Be a New United States'." *The New York Times*, November 19, 1912.

Wang, Ching-chun [Wáng, Jǐngchūn 王景春]. "A Plea for Recognition of the Chinese Republic." *The Atlantic Monthly* 61 (1913): 42–45.

Wang, Ching-chun [Wáng, Jǐngchūn 王景春]. "China's Revolution and Its Effect." *The North American Review* 197, no. 687 (1913): 189–202.

Wang, Ching-chun [Wáng, Jǐngchūn 王景春]. "The Effect of the Revolution upon the Relations between China and the United States." *The Journal of Race Development* 3, no. 3 (1913): 268–85.

Wang, Ching-chun [Wáng, Jǐngchūn 王景春]. "Some Dangers of Railway Development in China and How to Prepare against Him [*sic!*], Part 1." *The Chinese Social and Political Science Review* 1, no. 4 (1916): 47–54.

Wang, Ching-chun [Wáng, Jǐngchūn 王景春]. "The Administration of Chinese Government Railways." *The Chinese Social and Political Science Review* 1, no. 1 (1916): 68–85.

Wang, Ching-chun [Wáng, Jǐngchūn 王景春]. "Some Dangers of Railway Development in China and How to Prepare against Them, Part 2." *The Chinese Social and Political Science Review* 2, no. 1 (1917): 83–105.

Wang, Ching-chun [Wáng, Jǐngchūn 王景春]. "Some Dangers of Railway Development in China and How to Prepare against Them, Part 4." *The Chinese Social and Political Science Review* 2, no. 3 (1917): 58–77.

Wang, Ching-chun [Wáng, Jǐngchūn 王景春]. "China's Railway Ambitions: Expansive Views of Mr. C. C. Wang." *The North-China Herald and Supreme Court & Consular Gazette*, August 16, 1919.

Wang, Ching-chun [Wáng, Jǐngchūn 王景春]. "Report on the 'Construction Work,' to be Selected." In *Report of the Advisory Committee: Together with other Documents Respecting the China Indemnity*. Edited by Austen Chamberlain, 77–89. London: His Majesty's Stationery Office, 1926.

Wang, Ching-chun [Wáng, Jǐngchūn 王景春]. "A Railroad Born in Sin." *World's Work* 58, no. 12 (1929): 62–65.

Wang, Ching-chun [Wáng, Jǐngchūn 王景春]. "A Roman Alphabet for China." *Asia* 30, no. 6 (1930): 437–439, 459–463.

Wang, Ching-chun [Wáng, Jǐngchūn 王景春]. "The Sale of the Chinese Eastern Railway." *Foreign Affairs* 12, no. 1 (1933): 57–70.

Wang, Ching-chun [Wáng, Jǐngchūn 王景春]. *Japan's Continental Adventure*. New York: Routledge, [1940] 2011.

Wang, Ching-chun [Wáng, Jǐngchūn 王景春], and L. Pu. "A Discussion with Mr. Gandhi Concerning His Opinions of the Chinese National Government." *The China Critic* 5, no. 2 (1932): 52–53.

Wang, Guanhua. *In Search of Justice: The 1905–1906 Chinese Anti-American Boycott*. Cambridge, London: Harvard University Press, 2001.

Wang, Zuoyue. "Saving China through Science: The Science Society of China, Scientific Nationalism, and Civil Society in Republican China." *Osiris, 2nd Series* 17 (2002): 291–322.

Wáng, Yǒngzé 王勇則. *Túshuō 1915 Bānámǎ sàihuì: Guāngyào shìbóshǐ de Zhōngguó piānzhāng*. 图说1915巴拿马赛会—光耀世博史的中国篇章 [The Panama Exposition in Pictures—China Sections of the Dazzling Expo History]. Shanghai: Shànghǎi yuǎndōng chūbǎnshè, 2010.

"Wáng Jǐngchūn dūbàn Dōng lù zhī yuányīn." 王景春督辦東路之原因 [Why Wang Jingchun Presides over the Chinese Eastern Railway]. *Dàgōng bào*. April 6, 1922.

"Wáng Nǎibīn cí Dōnglù dūbàn." 王迺斌辭東路督辦 [Wang Naibin Resigns from Chinese Eastern Railway Presidency]. *Dàgōng bào*. March 12, 1922.

Wèn, Shūfāng 问书芳, and Yǒuguāng Huò. "Yè Gōngchuò duì Jiāotōng dàxué de gǎigé jí yǐngxiǎng." 叶恭绰对交通大学的改革及影响 [Ye Gongchuo's Reform of and Influence on Jiaotong University] *Jiāotōng gāojiào yánjiū* 3 (2004): 21–24.

Will, Pierre-Étienne. *Bureaucratie et Famine en Chine au 18e Siècle* [Bureaucracy and Famine in Eighteenth-Century China]. Civilisations et Sociétés 66. Paris, The Hague, New York: De Gruyter Mouton, 1980.

Will, Pierre-Étienne. "The Emergence of the Modern Civil Engineer in China, 1900–1940." In *China and the World—The World and China: Transcultural Perspectives on Modern China*. Vol. 3. Edited by Barbara Mittler and Natascha Gentz, 91–110. Gossenberg: Ostasien Verlag, 2019.

Williams, William Appleman. *The Tragedy of American Diplomacy*. 2nd ed. New York: Dell, [1959] 1972.

Wilson, Woodrow. "Democracy and Efficiency." *The Atlantic Monthly* 87, no. 521 (1901): 289–299.

Wong, Wing Tak [Huáng, Róngdé 黃榮德]. "Kweiyang to Kwangsi by Motor Before Next Xmas." *Jiùzāi yuèkān* 5, no. 5 (1928): 34.

"Wong Ching Chun Visits Fair." *The St. Louis Republic*. May 3, 1904.

Wu, Lien Teh. "The Second Pneumonic Plague Epidemic in Manchuria, 1920–21: A General Survey of the Outbreak and Its Course." *The Journal of Hygiene* 21, no. 3 (1923): 262–88.

Wu, Lin-chun [Wú, Língjūn 吳翎君]. "Mínchū Měiguó Guǎngyì gōngsī yǔ Yùzhōng gōngsī duì Zhōngguó gōnggòng gōngchéng de tóuzī: Qǐyè, zhèngfǔ yǔ wàijiāo guānxì de kǎochá." 民初美國廣益公司與裕中公司對中國公共工程的投資—企業、政府與外交關係的考察 [Enterprise, Government and Diplomacy: American

International Corporation and Siems-Carey Company Invest upon the Engineering Works in Early Republic China] *Dōnghuá rénwén xuébào* 14 (2009): 161–93.

Wu, Lin-chun [Wú, Língjūn 吴翎君]. "Mínchū Měiguó qǐyè duì huánghé tiěqiáo hé nánfāng dàgǎng de tóuzī: Qǐyè, zhèngfǔ yǔ wàijiāo guānxì de kǎochá." 民初美國企業對黃河鐵橋和南方大港的投資—企業、政府與外交關係的考察 [Investment of American Corporations in the Yellow River Bridge and Canton Harbor Projects in Early Republic China: A Study of Enterprise, Government and Diplomacy] *Táidà lìshǐ xuébào* 47 (2011): 61–106.

Wu, Lin-chun [Wú, Língjūn 吴翎君]. "Foreign Engineers' Activities in China and the Process of China's Internationalization: The Case of the Engineering Society of China, 1901–1941." In *China's Development from a Global Perspective*. Edited by María Elizalde and Jianlang Wang, 375–403. Newcastle upon Tyne: Cambridge Scholars Publishing, 2017.

Wu, Lin-chun [Wú, Língjūn 吴翎君]. "Tuīdòng gōngchéng guójiā: Zhōng Měi gōngchéngshī xiéhuì zài Zhōngguó (1919–1941)." 推动工程国家—中美工程师协会在中国 (1919—1941) [Promoting the Engineering State: The Association of Chinese & American Engineers in China (1919–1941)] *Jìndàishǐ yánjiū* 5 (2018): 122–33.

Wu, Lin-chun [Wú, Língjūn 吴翎君]. "China and the United States: Business, Technology, and Networks, 1914–1941." *Journal of American-East Asian Relations* 27, no. 2 (2020): 119–41. https://doi.org/10.1163/18765610-02702002.

Wu, Lin-chun [Wú, Língjūn 吴翎君]. *Měiguórén wèijìng de Zhōngguó mèng: Qǐyè, jìshù yǔ guānxì wǎng.* 美國人未竟的中國夢—企業，技術與關係網 [The Americans' Unfulfilled China Dream. Enterprise, Technology, and Relational Network]. Xinbei: Linking Publishing, 2020.

Wu, Yuan-li. "Industrial Development and Economic Policy." In Sih, *The Strenuous Decade*, 237–49.

Wú, Qǐdí 吳启迪, ed. *Zhōngguó gōngchéngshī shǐ.* 中国工程师史 [History of China's Engineers]. Shanghai: Tongji University Press, 2017.

Wú, Shān 吴山. "Běn huì sì dà jìhuà zhī jìnxíng shū." 本會四大計畫之進行書 [Progress of this Society's Big Four Plans] *Dàolù yuèkān* 4, no. 2 (1923): 2–4.

Wǔ, Tíngfāng 伍廷芳. Why China and America Should be Friends, June 10, 1908, Box 17, 2/5/3. President E. James General Correspondence, University of Illinois Archives.

Xu, Xiaoqun. *Chinese Professionals and the Republican State: The Rise of Professional Associations in Shanghai, 1912–1937*. Cambridge: Cambridge University Press, 2001.

Yang, Pao-ling [Yáng, Bàolíng 楊豹靈]. "Improvement Begins on China's Famous Grand Canal." *Millard's Review of the Far East*, September 28, 1918.

Yáo, Yuǎn 姚远, Ruì Wáng, and Shùfēng Yáo, eds. *Zhōngguó jìndài kējì qīkān yuánliú 1792–1949*. 中國近代科技期刊源流 1792–1949. Ji'nan: Shāndōng jiàoyù chūbǎnshè, 2008.

Ye, Shirley. "The Grand Canal in Republican China." *Journal of the Economic and Social History of the Orient* 62, no. 4 (2019): 731–72. https://doi.org/10.1163/15685209-12341492.

Ye, Weili. *Seeking Modernity in China's Name: Chinese Students in the United States, 1900–1927*. Stanford: Stanford University Press, 2001.

Yè, Gōngchuò 葉恭綽. *Jiāotōng jiùguó lùn.* 交通救國論 [The Doctrine of Saving the Country through Transport and Communications]. Shanghai: Shāngwù yìnshūguǎn, [1924] 1926.

Yeh, Wen-hsin. *The Alienated Academy: Culture and Politics in Republican China.* Cambridge, London: Harvard University Press, 1990.

Yieh, Tsung-kao. "The Adjustment Problems of Chinese Graduate Students in American Universities." PhD Thesis, University of Chicago, 1934.

Zanasi, Margherita. *Saving the Nation: Economic Modernity in Republican China.* Chicago, London: Chicago University Press, 2006.

Zarrow, Peter. *After Empire: The Conceptual Transformation of the Chinese State, 1885–1924*. Stanford: Stanford University Press, 2012.

Zhāng, Jiā'áo 張嘉璈. *Zhànshí jiāotōng wèntí.* 戰時交通問題 [The Question of Infrastructure during Wartime]. S. l.: Zhōngyāng xùnliàn tuán dǎng zhèng xùnliàn bān, 1940.

Zhāng, Péngyuán 张朋园, and Huáiyù Shěn, eds. *Guómín zhèngfǔ zhíguān niánbiǎo 1925–1949*. 国民政府职官年表 (1925–1949) [Annals of the Nationalist Government Office Holders, 1925–1949]. Taipei: Academia Sinica, 1987.

Zhèng, Yǒukuí 鄭友揆, Línsūn Chéng, and Chuánhóng Zhāng. *Jiù Zhōngguó de zīyuán wěiyuánhuì 1932–1949: Shǐshí yǔ píngjià.* 旧中国的资源委员会 1932–1949. 史实与评价 [Old China's Natural Resources Commission 1932–1949. Historical Facts and Evaluation]. Shanghai: Shànghǎi shèhuì kēxué yuàn chūbǎnshè, 1991.

"Zhěngdùn Dōng tiě zhī jìhuà." 整頓東鐵之計畫 [Plan to Reorganize the Chinese Eastern Railway]. *Dàgōng bào*. May 24, 1922.

"Zhōngdōng lù zhī jìnxùn." 中東路之近訊 [Chinese Eastern Railway Latest News], *Dàgōng bào*. January 17, 1924.

Zhōngguó gōngchéngshī xuéhuì 中國工程師學會, ed. *Zhōngguó gōngchéngshī xuéhuì sānshízhōu jìniànkān: Sānshí niánlái zhī Zhōngguó gōngchéng.* 中國工程師學會三十周紀念刊—三十年來之中國工程 [China Institute of Engineers' Thirtieth Anniversary Commemoration Volume. Thirty Years of Chinese Engineering]. 2nd ed. Nanjing: Guó gōngchéng chūbǎn gōngsī, [1946] 1948.

Zhu, Qin. "Engineering Ethics Studies in China: Dialogue between Traditionalism and Modernism." *Engineering Studies* 2, no. 2 (2010): 85–107. https://doi.org/10.1080/19378629.2010.490271.

Zieger, Robert H. "Herbert Hoover, the Wage-Earner, and the 'New Economic System,' 1919–29." In Hawley, *Herbert Hoover as Secretary of Commerce*, 80–114.

Zīyuán wěiyuánhuì, 資源委員會. *Zhōngguó gōngchéng rénmínglù*. 中國工程人名錄 [Who's Who of Chinese Engineers]. Changsha: Shāngwù yìnshūguǎn, 1941.

Zuzindlak, Chelsea. "'Tell 'Em You're from Detroit': Chinese Americans in the Model City." In *Asian Americans in Michigan: Voices from the Midwest*. Edited by Sook Wilkinson and Victor Jew, 50–69. Detroit: Wayne State University Press, 2015.

Index

able persons government 102
Adams, Henry Carter (1851–1921) 90
American Bridge Company 35, 206
American China Development Company 213
American International Corporation (AIC) 55–7, 66–70, 74
American National Red Cross (ANRC) 54–5, 146–8, 150, 165–6, 168
American Society of Civil Engineers (ASCE) 3, 118, 124, 128, 131, 138
Americophilia 39, 43, 141
Andersen, Meyer & Co. 138
Association of Chinese & American Engineers (ACAE) 22–3, 31, 74, 112–42, 145–50, 165, 169, 171–91, 196, 200–2, 206, 209, 218, 221, 223, 231, 233, 238
authority 10, 23, 95, 121, 150–6, 175, 189, 193, 199, 205, 212–3, 219–20, 223, 227–9, 232
autonomy 19, 45, 78, 89, 115, 165, 191–2, 195, 205, 210, 214, 229, 232

Bai Lang Rebellion 92
Baker, John Earl (1880–1957) 94, 105, 147, 159, 167–8
banks
  Imperial Bank of Japan 56
  Russo-Asiatic Bank 94, 100, 105, 107
Bao, Guiqing 鮑貴卿 (1867–1934) 94–5, 110
Barrows, Harold Kilbrith (1873–1954) 32, 60
Beardsley, James Wallace (1860–1944) 59*n*44
Beijing Tariff Conference 182
Belgium 71–2, 85, 193, 206, 211–3
Bolshevism 93, 95, 180
Bonaparte, Napoleon (1769–1821) 105
Book of Mengzi 82
Boxer indemnities
  American 5, 24–5, 30–1, 43–4, 72
  Belgian 211
  British 82*n*4, 91*n*70, 178–9, 211–3
Brazil 55
bridges
  Beijing–Hankou *See* bridges, Yellow River
  Jiaozhou–Jinan 131
  Marco Polo 153
  near Pingshi 215
  Qiantang 220
  Tianjin–Pukou 65
  Yellow River 70–4, 206, 210, 220, 234
  Yunnan–Vietnam 220
Brodie, Hugh (1871–1932) 60*n*44
bureaucracy 11, 13, 15, 22, 44, 47, 85–91, 96, 114, 133, 137, 145, 154, 157, 174, 179–183, 189, 191–3, 205, 209, 217, 227–8, 234
Burton, Alfred E. (1857–1935) 32, 60

Cabinet of Action 193
Cai, Guangji 蔡光勣 (b. 1889) 47, 119
canals *See also* rivers, irrigation systems
  built by CIFRC 148
  Grand
    Jiangsu Grand Canal Improvement Board 68
    Shandong Grand Canal Improvement Board 8, 21–3, 52–79, 113, 117, 132, 139, 151, 153, 175–6, 179, 184–7, 199, 201, 230–1, 234, 238
  Hasuhai Waste 186
  Panama 1, 57, 60, 93
Cao, Rulin 曹汝霖 (1877–1966) 91
Carey, Clifton O'Neal (1875–1963) 60*n*44
Carey, William Francis (1878–1951) 139
centralization 45, 86, 181, 221, 227
Central Planning Board (CPB) 224–6
Changchun Freight Agreement 104
Charles River Dam 57
Chen, Duheng 陳杜衡 (b. 1864) 207
Chiang, Kai-shek *See* Jiang, Jieshi
China International Famine Relief Commission (CIFRC) *See* commissions
Chinatown 43
Chinese Education Mission (CEM) 4, 8, 58
Chinese Engineering & Development Company 77, 136
Chinese Engineers Relational Database (CERD) 9, 25, 27, 36, 49, 112, 116–7, 119, 134–6, 197–8
Chinese Exclusion Act (1882–1943) 28, 43
Chinese merchant boycott of 1905 5
Chineseness 39–41

Chinese Students' Monthly 46–50
Chinese Students' Quarterly 46
Christianity 26, 28, 29*n*14, 30, 39, 45, 83, 98, 114, 148, 151–2, 169, 180
cities
- Andover 35
- Baoding 147
- Baoji 210, 212, 220, 222, 224, 226–7
- Baotou 185
- Beidaihe 123
- Beijing 4–5, 7–8, 28, 38, 45, 53*n*1, 56, 59–60, 64–5, 69–71, 73–4, 77, 81, 83, 89–90, 92, 94, 96, 102, 104, 106–10, 114, 120, 123, 134, 137, 140, 146, 153–4, 157, 159, 165, 174–6, 180, 182–5, 190, 199, 205–8, 210, 212–4, 216, 218–9, 221, 224, 231, 236
- Boston 61, 90
- Changsha 213
- Chenzhou 214
- Chicago 14, 34, 83
- Chongqing 195, 204
- Detroit 24, 151, 157
- Dezhou 64
- Dunhuang 226
- Guangzhou 42, 45, 114, 134, 136, 158, 183, 210, 212–7, 219–20, 224, 227, 234
- Guilin 220
- Guiyang 156–7, 163–4
- Ĝulja 226
- Hami *See* Ķumul
- Handan 147
- Hangzhou 53–4, 55*n*14, 91*n*67, 149
- Hankou 45, 65, 71, 73–4, 82*n*4, 89–90, 94, 114, 120–1, 134, 136, 147, 182–3, 206, 210, 212–7, 219–20, 224, 227, 234
- Harbin 94, 98, 102, 103*n*143, 104–6
- Hengyang 213–4, 217
- Hohhot 55*n*14, 91*n*67, 185
- Hong Kong 45, 214, 220
- Huai'an 52–3
- Jinghua 147
- Jining 52, 63, 65
- Jiuquan 221, 226
- Khorgas 226
- Ķumul 226
- Kunming 151, 154, 204
- Lanzhou 55*n*14, 91*n*67, 168–9, 171, 173, 204, 210, 220–2, 224–6
- Leipzig 212
- Lianyungang 210
- Liuzhou 204, 220
- London 184, 238
- Luoyang 210, 212
- Manzhouli 105
- Moscow 92, 108–10, 212
- Mukden 89–90, 92*n*76, 120, 175
- Nanjing 22–3, 28, 76, 78, 81, 88, 134, 149, 171, 174–8, 184, 191–2, 194, 201, 206, 208–10, 214–6, 231
- Nantong 54, 68, 148
- New York City 1, 26, 35, 49, 66
- Omsk 225
- Panyu 205, 207, 217
- Paris 91
- Philadelphia 35
- Pingliang 169
- Providence 35, 60, 137, 139
- Qingdao 56, 120, 220
- San Francisco 1, 34, 64
- Shanghai 4–5, 32, 45, 59, 114, 120, 134, 138, 148–9, 159, 197, 206–8, 212, 217
- Shaoguan 213
- St. Louis 28–9, 83
- Tai'an 64
- Tangshan 28, 32
- Tashkent 225
- Tengzhou 63
- Tianjin 5, 53, 55–8, 65, 109, 134, 140, 147, 153, 195, 210, 238–9
- Tianshui 210, 220–2, 224, 226–7
- Tokyo 184–5
- Tongzhou 83, 146–7
- Ürümqi 204, 224, 226–7
- Usu 204, 224, 226
- Vladivostok 92, 99, 234
- Washington, D. C. 77, 95, 141, 207, 212
- Wuchang 57, 213
- Wuzhou 208–9
- Xi'an 168–73, 199, 204, 210, 212, 226
- Xuzhou 204, 210
- Yangzhou 65, 68
- Yumen 195, 226
- Zhengzhou 204, 210
- Zhenjiang 53
- Zhuzhou 69, 76, 204, 213
- Zunyi 156, 158, 164, 167

Cixi 慈禧 (1835–1908) 47
Chen, Eugene (1878–1944) 183
cliques
  Anhui 96, 98, 102
  Communications 96–7, 207
  Fengtian 96, 102–3, 109–10, 207, 230
  New Guangxi 209
  Zhili 96–7, 102, 207
code of ethics 14, 51, 125, 131, 142, 145, 233
colleges *See* universities
colonialism 6–7, 10, 30, 83, 92, 108, 111, 118, 231 *See also* imperialism
commissions
  China International Famine Relief C. (CIFRC) 112, 117, 138, 144, 147–73, 180, 183, 185–91, 198–201, 223
  Chinese Foreign Famine Relief Committee 147
  for Studying Administrative Efficiency 224
  for the Improvement of the River System of Zhili 57, 63, 69
  Military Affairs C. (MAC) 193, 222
  National Defense Planning C. (NDPC) 193
  National Economic C. (NEC) 171, 188, 190, 193, 198–9, 200, 220
  National Resources C. (NRC) 9, 193–5
  on Railway Technics 206
  on the International Régime of Ports, Waterways, and Railways 91
  on the Unification of Railway Accounts and Statistics 89
  on the Unification of Railway Terms 90
  Peking United International Relief Committee 147
  Wusong Port Development Administration C. 137
conservatism 47, 72, 76, 87
conservationism 58, 181
cosmopolitanism 10, 21, 36–8, 45, 51, 83, 110, 127, 140, 151, 153, 196, 202, 232–5
Crane, Charles Richard (1858–1939) 59, 98, 137, 199
Cromwell, Oliver (1599–1658) 105
customs service (*haiguan* 海關) 150
Czechoslovak Legion 93

Dean, Samuel M. 116, 130, 185–6, 188
Deng, Yunte 鄧雲特 (1912–1966) 145
dictatorship 104, 157, 175, 182
dikes 75, 146, 148, 153, 168, 183, 200, 215
Ding, Wenjiang 丁文江 (1887–1936) 91*n*70, 182
discipline 125, 127–8, 134, 142, 216, 218, 223, 236
dismantling 194, 215, 220, 222
Dong, Xianguang 董顯光 (1887–1971) 69, 107

efficiency 1, 3, 21, 39, 45, 47–8, 58, 81, 84–8, 90, 92, 94, 100–1, 107–8, 128, 131–2, 142, 146, 154–9, 162–3, 168, 173, 190, 208, 214, 224, 227, 231–3, 236
Ehlers, Joseph Henry (1892–1988) 119, 123, 138, 165, 223
Eliassen, Sigurd (1884–1960) 119, 138, 186, 188*n*66, 199

famine 22, 43, 58, 72, 75, 85, 145–8, 151, 156, 160, 163–4, 169–72, 187–8, 190, 200, 233, 238
famine relief *See* relief
famine-surtax 156
Far Eastern Republic 80, 95, 99
Far Eastern Times 104
Fengtian–Russian Agreement 110
Feng, Yuxiang 馮玉祥 (1882–1948) 148, 158, 166, 169
five-colored flag 70
floods 43, 53, 56–8, 61, 64, 69, 75–6, 138, 146–7, 165, 171, 188, 190
Ford Motor Company 24, 36, 150–1, 157, 166
Fosdick, Harry Emerson (1878–1969) 161
France 32, 71, 104*n*153, 160, 206, 214
Freeman, Evert Wendell (1898–1955) 60
Freeman, John Ripley (1855–1932) 35, 57–78, 93, 128–9, 137–40, 151, 177, 180, 185, 195, 199, 201, 206
Freud, Sigmund (1856–1939) 18
frontier 6, 17, 42, 190, 225
Fuzhou navy yard school 4

Gandhi, Mahatma (1869–1948) 111
Gan, Naiguang 甘乃光 (1897–1956) 224
Gao, Enhong 高恩洪 (1875–1943) 103*n*143, 207
Germany 32, 56, 77, 181, 192, 197, 212, 227, 237

gilded age 86
Goodrich, Ralph D. (1878–1961) 59*n*44
Great Britain 3, 15, 42, 73, 82*n*4, 89, 91*n*70, 118, 178–9, 212, 221
Great Depression 179, 199
Great Huabei Famine 75
Guomindang (GMD) 22–3, 96, 161, 171, 173, 175, 179, 182–4, 191, 195, 198, 200–1, 208–9, 212, 219, 222, 227
Guo, Taiqi 郭泰祺 (1888–1952) 59, 91*n*70
Gu, Weijun 顧維鈞 (1888–1985) 19, 39, 42, 91*n*70, 102*n*142, 108, 184
Gu, Zhen 顧振 (b. 1894) 119, 175
Gu, Zhenglun 谷正倫 (1890–1953) 226

Hadley, Arthur Twining (1856–1930) 86
Harbin Dawning 104
harbors
  Dairen 99
  Lianyun 210
  international 28
Harrison, John P. 18
Henry, Philip Walter (1864–1947) 68
Hetch Hetchy Aqueduct 57, 66
Hoover, Herbert Clark (1874–1964) 85, 130, 177
Horvat, Dmitry Leonidovich (1858–1937) 94
Huai River Conservancy 54–5, 76, 78
Huang, Rongde 黃榮德 (d. 1932) 138, 157–61, 164, 172, 199, 234
Hu, Shi 胡適 (1891–1962) 38
hygiene 45, 85, 127

imperialism 6, 29, 42, 45, 51, 89, 92, 108, 135, 141–2, 182–3, 208, 213, 225, 231, 237 *See also* colonialism
individualism 50, 180
intelligentsia 172, 192
Inter-Allied Railway Agreement 93
Inter-Allied Technical Board (IATB) 93, 95–9, 106
International Congress of Chambers of Commerce 90
International Highway Congress *See* International Road Congress
International Labour Organization 160
International Road Congress 141, 212
internationalism 10, 39, 83, 110
irrigation systems *See also* canals
  Dujiangyan 75
  Salaqi 22, 74, 169, 174, 185–91, 193, 197, 201–2, 231, 234
  Shishou 153

Japan 4–6, 10, 24–5, 39, 42, 44, 55–6, 69–71, 73, 93, 95, 97–9, 101, 104–5, 108, 116, 134, 160, 175, 178, 185, 193, 201, 205, 212, 214–6, 218–22, 226–7, 229, 232, 237
Jeme, Tien-yow *See* Zhan, Tianyou
Jesuits 2
Jiang, Jieshi 蔣介石 (1887–1975) 171–2, 177, 180, 192–4, 200, 209, 215, 221, 224, 228
Jiangnan shipyard 4
jiaotong jiuguo 交通救國 20 *See also kexue jiuguo*
Jin, Bangping 金邦平 (b. 1881) 55
Jin, Tao 金濤 (b. 1888) 119, 185–6
Joffe, Adolph Abramovich (1883–1927) 109
Johnson, Benjamin Oliver (b. 1878) 6
Johnson, Hewlett (1874–1966) 170
Journal of the Association of Chinese and American Engineers 122–30, 133, 138, 142, 149, 169, 174–5, 191, 196, 200–1, 221, 231

Kaiping Mines 28
Kang, Youwei 康有爲 (1858–1927) 38
Kappel, Vladimir (1883–1920) 95
Kawamura, Takeji 川村 竹治 (1871–1955) 108
Kazakh insurgency 226
Kelly, William (1874–1957) 190
kexue jiuguo 科學救國 21 *See also jiaotong jiuguo*
Knickerbocker Crisis 86
Kong, Xiangxi 孔祥熙 (1881–1967) 183
Koo, Vi Kyuin Wellington *See* Gu, Weijun
Kuang, Jingyang 鄺景揚 (1863–1929) 58, 119, 130, 138, 140

lakes
  Baoying 53
  Dongping 64
  Gaoyou 53
  Hongze 53
  Weishan 56, 63
Lam, Glines & Co. 140
League of Nations 188–9, 193, 201, 234
Li, Bing 李冰 (third century BCE) 75
Li, Jinlun 李錦綸 (1886–1956) 78

Li, Shutian 李書田 (1900–1988) 116, 131, 185–6, 189
Li, Yuanhong 黎元洪 (1864–1928) 57, 59, 117
Liang, Dunyan 梁敦彥 (1857–1924) 58–9, 90–1
Liang, Qichao 梁启超 (1873–1929) 3, 37
Liang, Ruhao 梁如浩 (1863–1941) 148
Liang, Shiyi 梁士詒 (1869–1933) 207
Library of Congress 61
Lin, Quancheng 林全誠 (b. 1891) 60–3, 68, 72, 77–8
Lincheng Outrage 109
Ling, Hongxun 凌鴻勛 (1894–1981) 1, 23–4, 28, 34–5, 71, 116, 137–8, 141, 172, 180, 185–6, 195, 204–29, 232–5, 238
Liu, Rusong 劉如松 (b. 1901) 137
Long March 170
Louisiana Purchase Exposition 28
Lowdermilk, Walter C. (1888–1974) 123
Lu, Xun 鲁迅 (1881–1936) 3, 85
Luo, Jialun 羅家倫 (1887–1969) 226

Mallory, Walter Hampton (1892–1980) 147
Manchuria 6, 42, 92, 95, 99, 102, 104, 108, 178, 193, 198, 206, 214
Mao, Yisheng 茅以升 (1896–1989) 18
Martin, W. A. P. (1827–1916) 45
Marxism 6 *See also* Communism
Meloy, Thomas 119, 137
Merrick, Howard Benjamin (1871–1926) 59*n*44
Merrill, Robert Hall (1881–1955) 59*n*44
Military Affairs Commission (MAC) *See* commissions
mining 28, 31, 37, 42, 193–4
Ministries
  Agriculture and Commerce 54–5, 91
  Communications 1, 58, 71, 73, 81, 89–91, 94, 96, 102, 103*n*143, 106, 121–2, 127, 132, 137–8, 149, 159, 175, 184, 193, 196, 205, 207, 209, 213, 220–1, 228, 238–9
  Foreign Affairs 44, 183, 193
  Industry 193
  Railways 137, 150, 175, 177–8, 184, 193, 209–13, 220
monarchic restoration (1915–1916) 92
Mongolia 42, 185, 225
mountains
  Liupan 169
  Yin 185
movements
  Good Roads 146–50, 158, 161, 166, 171
  May Thirtieth 208
  New Life 171
  People's Economic Reconstruction 171
  railway protection 42, 45
  Self-Strengthening 4, 194
Mussolini, Benito (1883–1945) 177

National Conservancy Bureau 57
National Defense Planning Commission (NDPC) *See* commissions
National Economic Commission (NEC) *See* commissions
National Good Roads Association (NGRA) *See* societies
nationalism 20–1, 24, 37, 39, 42, 50–1, 81–5, 87–8, 91, 101, 111, 141, 145, 172, 175, 181, 183, 208, 226, 232
Nationalist Party *See* Guomindang
nationalization 22–3, 39, 41, 82, 86, 101, 121, 145, 173–4, 176, 191, 197, 201, 211, 213, 232
National Resources Commission (NRC) *See* commissions
National Revolutionary Army (NRA) 214–5, 220, 228
national socialism 277
New Deal 192, 200
New York City Board of Water Supply 66
Northern Expedition 175, 177, 182, 184, 208–9
Northwestern Expedition 225
Northwest Highway Administration Department 221

Ono, Eijiro 小野 英二郎 (1864–1927) 56
Open Door 6, 25, 141, 206
Oriental Engineer, The See *Journal of the Association of Chinese and American Engineers*
Ostroumov, Boris Vasil'evich (1879–1944) 100–10, 233

Pan, Fu 潘復 (1883–1936) 54–6, 59, 62, 66, 69, 73, 139, 184
Pan, Jixun 潘季馴 (1521–1595) 76
Panama-Pacific International Exposition 1, 90

Parsons, William Barclay (1859–1832) 6, 45–6
Peng, Binglu 彭炳祿 (b. 1891) 138, 141, 151–7, 172
pneumonic plague 95
Poland 212
Proceedings 124, 138
provinces
  Fengtian 96, 102–3, 109–10, 207, 230
  Gansu 145, 150, 169, 195, 210, 221–7
  Guangdong 26, 34, 41–3, 45, 152, 177, 84, 207, 210, 215, 217
  Guangxi 156, 164, 208–9, 215, 217, 219–21, 223
  Guizhou 22, 145, 150, 156–68, 172–3, 199, 201, 209, 221, 231, 233
  Hunan 45, 147, 168, 212–4, 219–21, 223
  Jiangxi 153, 214
  Ningxia 221
  Qinghai 221
  Shaanxi 69, 145, 150, 169–70, 190, 221
  Shandong 52–70, 97, 147, 150, 165–6, 207
  Shanxi 138, 147–8, 150
  Sichuan 38, 75, 151, 156, 221
  Suiyuan 150, 185, 219, 221
  Tibet *See* Tibet
  Xinjiang 222, 224–8
  Yunnan 22, 42, 145, 150–8, 166, 172, 210, 219–20, 231
  Zhejiang 34, 42
  Zhili 28, 57, 63, 69, 77, 96–7, 102, 147, 198, 207

Qian, Changzhao 錢昌照 (1899–1988) 194
Qian, Zongze 錢宗澤 (1891–1940) 218
Qing government 4–5, 8, 30–2, 40, 44, 47, 213
Qi, Xieyuan 齊燮元 (1885–1946) 77
Quan, Henry 199

racialism 60
railways
  Beijing–Hankou 65, 71, 74, 89–90, 94, 206, 210, 212–4
  Beijing–Kalgan 154 *See also* railways, Beijing–Suiyuan
  Beijing–Mukden 89–90, 175
  Beijing–Suiyuan 185, 219, 221
  Bukhara 100
  Chinese Eastern 21, 23, 80–1, 90, 92–100, 111, 228, 231, 233–4
  Gansu–Haizhou 131, 169, 178, 206, 210–2, 218, 220–2, 224
  Guangdong–Yunnan 210
  Guangzhou–Kowloon 42
  Hunan–Guangxi 219–21
  Hunan–Guizhou 221
  Jiangsu–Zhejiang 42
  Kaiping Colliery Tramway 28
  Kolchuginsky 100
  Longhai *See* railways, Gansu–Haizhou
  Nanchang–Jiujiang 220
  Qingdao–Jinan 220
  Siping–Taoyuan 178
  South Manchurian 80, 99, 104, 108
  Tianjin–Pukou 58, 65, 109, 210
  Tianshui–Chengdu 221–2
  Transcontinental 1, 34, 82–3, 225
  Trans-Siberian 80, 92, 99
  Xinjiang–Gansu 224–8
  Yunnan–Sichuan 151
  Yunnan–Vietnam 219–20
  Yuzhsib 100
  Zhengding–Taiyuan 206
  Zhuzhou–Jinzhou 69, 76
Raindre, Jacques 105
Rautenstrauch, Walter (1880–1951) 162
regions
  Bohai Bay 28
  Central America 55
  Central Asia 225
  Chinese Northwest 17, 22, 169, 195, 220–9
  Dawenkou 64
  Far East 71, 92–3, 99, 182
  Hexi Corridor 220, 224
  Himalayas 42
  Ili Valley 224, 226
  Kashgaria 226
  Ruhr 212
  Siberia 92–3, 95, 98–9, 212, 214
  Tumed Right banner 185
  Turkestan 225, 228
  Xingxingxia Pass 226
  Yangzi Delta 149, 168
  Yumen oil fields 195
relief 22, 43, 56, 58, 69, 85, 145–8, 150–1, 158, 160, 163–4, 166, 168, 171–3, 187–8, 190–1, 200, 238
republicanism 21, 37, 39, 44–5, 50–1, 81, 83, 181

responsibility 18, 47, 51, 54, 72, 89, 110, 131–3, 141, 152, 171, 190, 193, 195, 205, 213–4, 216, 218–9, 221, 227, 229, 232
revolutions
  Russian (1917) 180 *See also* wars, Russian Civil
  Second (1913) 92
  Xinhai (1911) 42, 45, 88, 213
Ripley, Joseph (1854–1940) 60, 63, 65–6, 117
rivers *See also* canals
  Amur 93, 103*n*149, 105
  Bai 83
  Black 186–7
  Hai 56
  Huai 52, 54–5, 69, 74–8
  Linhong 210*n*20
  Qiantang 54, 220
  Si 56
  Wei 221–2
  Wen 56
  Yangzi 42, 52–3, 149, 168
  Yellow 42, 52–3, 55, 61, 63–5, 70–4, 138, 180, 185–7, 200, 210, 215, 220, 234
  Yongding 136
roads *See also* movements, Good Roads
  Baoji–Wei River 220–1
  Handan–Daming 147
  Jinghua–Qiyuan 147
  Kunming–Shuifu 151
  Lanzhou–Hami 220
  Lanzhou–Suzhou 221
  Libo–Guiyang–Chishui 156–7, 164
  Nanchang–Fuzhou 153
  Shanghai–Hangzhou–Nanjing 149
  Shanghai municipalities 149
  Tianjin–Baoding 147
  Tianjin–Tongzhou 146–7
  Tianshui–Chengdu 220
  Tianshui–Fengyang 220
  Tianshui–Lanzhou 220
  Tianshui–Shuangshipu 221
  Xi'an–Lanzhou 168–73
  Yunnan–Burma 172
Roosevelt, Franklin D. (1882–1945) 199
Roosevelt, Theodore (1858–1919) 5, 25, 43
Russia 4, 42, 85, 92–107, 110, 180, 226, 231

Sa, Fujun 薩福均 (1886–1955) 31, 116, 119, 136
scientific management 81, 85, 88, 132, 162, 192 *See also* Taylorism
Second Consortium 95
Semyonov, Grigory Mikhaylovich (1890–1946) 95
Shao, Fuwu 邵福旿 (b. 1897) 138, 169, 186
Sheng, Xuanhuai 盛宣懷 (1844–1916) 210
Siberian Intervention 93
Siems-Carey Company 55, 63, 68, 90–1
silting 53, 75, 187–8, 190
silver dollar 54, 62–3, 67, 69, 101, 179, 198, 202
Sinophilia 60
Sinophobia 106
Sino–Soviet Agreement 109–10
Sino-Soviet Conflict 111
Shen, Qi 沈琪 (1871–1930) 116
Shen, Yi 沈怡 (1901–1980) 194
Sheng, Shicai 盛世才 (1895–1970) 226
Shi, Zhaoxiang 施肇祥 (1881–1961) 119, 136, 185
Smoot-Hawley Act 179
Snow, Edgar Parks (1905–1972) 200
Social Darwinism 86
societies
  Ai-Kwoh-Hwai (*aiguo hui* 愛國會) 44
  British Institution of Civil Engineers 3, 73
  China Engineering S. (CES) 8, 115–6
  China Society of America 98
  Chinese Institute of Engineering (CIE) 114–6, 120, 123–4, 131, 194, 196, 200
  Chinese Students Alliance (CSA) 39–41, 46, 51, 184
  Chinese Student Christian Association 151
  Chinese Students' Club 38, 151
  Cosmopolitan Club 37–8, 110, 127, 151
  Earth Studies S. (*dixue hui* 地學會) 119
  National Good Roads Association (NGRA) 148–50
  North China International Society of Famine Relief 147
  Phi Beta Kappa 114
  Phi Tau Phi 114
  Sigma Xi 114
  YMCA 37, 59, 114
solutionism 84, 108, 190, 233, 236

Song, Qingling 宋慶齡 (1893–1981) 183
Song, Xiaolian 宋小濂 (1863–1926) 95–6
Soviet Union 92*n*76, 98–100, 108–11, 192, 214, 216, 220, 224–6, 234
spatial entrepreneur 20, 78, 84
spatial literacy 19–20, 84
standardization 3, 11, 17, 45, 82, 90, 91*n*70, 101, 184, 206
Stearns, Alfred E. (1871–1949) 32
Stevens, John Frank (1853–1943) 93, 95, 98–9, 106, 128, 133
Stone, Charles A. (1867–1941) 66
Stone & Webster 61
Stroebe, Georg Gottlieb (1900–1988) 185–6, 200
Sullivan, Murray (1881–1941) 116, 119, 136
Sun, Chuanfang 孫傳芳 (1885–1935) 78
Sun, Duoyu 孫多鈺 (1882–1951) 116, 119, 200
Sun, Ke 孫科 (1891–1973) 137, 141, 150, 169, 183–4, 198–9, 209–11
Sun, Yat-sen 孫逸仙 (1866–1925) 34, 83, 109, 183, 208, 211
Switzerland 89, 109

Taiwan 208, 238
Tangent Manufacturing Works 140
Tang, Jiyao 唐繼堯 (1883–1927) 151
Tang, Shaoyi 唐紹儀 (1862–1938) 59, 72, 75
Tang, Yueliang 唐悅良 (1888–1956) 183
Tan, Zhen 譚真 (1899–1976) 24, 32–3, 35, 60–2, 71, 76, 78, 119, 136–41, 169, 179–80, 181*n*37, 185–6, 201, 234, 238
Taylor, Frederick Winslow (1856–1915) 81*n*2
Taylorism 88 *See also* scientific management
technocracy 12, 48, 81, 84–5, 88, 90, 100, 107, 110–1, 132, 162, 165, 181–2, 194, 227–8, 231, 236
telegraphy 1, 46, 184
Ten Years Plan for the Reconstruction of the Northwest 225
Tibet 42, 172
Todd, Oliver Julian (1899–1973) 7, 66–7, 72–3, 76–8, 119, 124, 128, 133, 136–8, 140–1, 149–69, 176–8, 183–90, 195, 199–200, 219–21, 238
Tohoro, Duanfang 託濶羅端方 (1861–1911) 30
Tongwen guan 同文館 4
Tredgold, Thomas (1788–1829) 14
Triple System of Administration (TSA) 224–5
Twenty-One Demands 56

universities
- Beijing 5, 45
- Beiyang 5, 169
- Cambridge 57, 111
- Columbia 1, 26, 35, 183
- Cornell 25–6, 30, 47, 151
- Dongwu 30
- Harvard 26, 43, 114
- Illinois 1, 25–6, 30, 37, 81
- Jiaotong 5, 32, 34, 207, 218
- Massachusetts Institute of Technology (MIT) 32–3, 35, 38, 49, 57, 60, 238
- Michigan 25–6, 60*n*44, 150, 152
- Nanyang *See* Jiaotong
- Ohio Wesleyan 30
- Purdue 25–6, 30–1, 42, 57, 179
- Qinghua *See* universities, Tsinghua
- Tsinghua 32, 190
- Yale 1, 25–6, 30–1, 33, 35–6, 48, 58, 81, 86, 88, 90, 102*n*142, 114, 136–7

unfree labor 162, 172
US Bureau of Foreign and Domestic Commerce 166–7
Uyghur activists 228

Veblen, Thorstein (1857–1929) 81*n*3
Versailles Peace Conference 184

Waddell, John Alexander Low (1854–1938) 71, 73–4, 124, 127–9, 132, 137–8, 150, 177, 183–5, 206, 209, 233
Wall Street Crash 179
Wang, Anshi 王安石 (1021–1086) 42
Wang, Boqun 王伯群 (1885–1944) 184
Wang, Chongyou 王寵佑 (1881–1958) 116, 119, 186
Wang, Jingchun 王景春 (1882–1956) 1, 21–2, 24, 28–31, 35–6, 38–44, 49–50, 55, 81–112, 128, 134, 150, 180, 184, 208, 212, 216, 222, 225, 228, 230–1, 233–4, 238
Wang, Jinzhi 王金職 (1892–1939) 136, 185–6
Wang, Naibin 王迺斌 (b. 1870) 96
Wang, Zhengfu 王正黼 (b. 1891) 185
Wang, Zhengting 王正廷 (1882–1961) 59, 91*n*70, 106, 148, 150

wars
  Central Plains 211–2
  First World 8, 21, 54–5, 59, 61–2, 67, 131, 176
  Jiang–Guangxi 209
  Russian Civil 93–4, 103, 110
  Second Sino-Japanese 99*n*116, 117, 201, 205, 213–4, 217–9, 221–2, 225, 227–9, 232, 238
  Second World *See* wars, Second Sino-Japanese
  Second Zhili–Fengtian 207
Washington Naval Conference 77, 95, 207
Wei, Yi 魏易 (1880–1932) 66
Weng, Wenhao 翁文灝 (1889–1971) 193–4
West Point 100, 101*n*129
White, Leonard D. (1891–1958) 225
Wiggin, Thomas Hollis (1873–1964) 66
Wilson, Henry (1812–1875) 82
Wilson, Woodrow (1856–1924) 6, 93
World Engineering Congress 184
Wuchang Uprising *See* revolutions, Xinhai
Wu, Chaoshu 伍朝樞 (1887–1934) 184
Wu, Peifu 吳佩孚 (1874–1939) 97, 109, 207
Wu, Shan 吳山 (1884–1936) 148
Wu, Tingfang 吳廷芳 (1842–1922) 30, 59, 89*n*61, 184
Wu, Yulin 吳毓麟 (1871–1944) 122
Wu, Zhuopan 伍灼泮 31, 119

Xi, Debing 席德柄 (b. 1891) 38
Xiong, Xiling 熊希齡 (1870–1937) 56–7, 59, 61, 63, 66, 69, 73, 117, 129, 136
Xue, Zhuobin 薛卓斌 (1895–1991) 24, 32–3, 35, 60–1, 76, 78, 124*n*63, 137–8, 153, 179, 185–7, 201, 238
Xu, Shichang 徐世昌 (1855–1939) 8, 96, 147
Xu, Shiying 許世英 (1873–1964) 75
Xu, Wenjiong 徐文炯 114
Xu, Zhaonan 許肇南 (1886–1960) 38

Yan, Deqing 顏德慶 (1878–1942) 114, 116, 119
Yan, Huiqing 顏惠慶 (1877–1950) 30, 47, 148
Yan, Xishan 閻錫山 (1883–1960) 138, 166
Yang, Baoling 楊豹靈 (1887–1966) 24, 29–31, 38, 42, 57, 59, 63, 68–70, 73–8, 117, 119, 124*n*63, 128–9, 136, 138–9, 147, 168, 176–80, 185–6, 201, 233, 238–9
Yang, Huayan 楊華燕 (1886–1948) 48–50
Ye, Gongchuo 葉恭綽 (1881–1968) 84, 89, 94, 96–7, 103*n*143, 132, 149, 205, 207, 225–6, 233
You, Yunlong 由雲龍 (1877–1961) 154–5
Yu the Great 大禹 75
Yu, Rizhang 余日章 (1882–1936) 59
Yuan, Jinkai 袁金鎧 (1870–1947) 103*n*143
Yuan, Shikai 袁世凱 (1859–1916) 64

Zang, Zhiping 臧致平 (1869–1944) 77
Zhang, Hanying 張含英 (1900–2002) 185
Zhang, Jian 張謇 (1853–1926) 54, 57, 59, 68–9, 73, 76, 148
Zhang, Jichun 張季春 (1902–1946) 137–8, 141, 169–70, 173, 186, 188, 190, 238
Zhang, Ren 張任 (b. 1906) 190
Zhang, Xueliang 張學良 (1901–2001) 175
Zhang, Zhidong 張之洞 (1837–1909) 5, 49
Zhang, Zhitan 张志潭 (1883–1946) 127
Zhang, Zuolin 張作霖 (1875–1928) 96–7, 102, 105–6, 109, 140, 175, 207
Zhan, Tianyou 詹天佑 (1861–1919) 1, 18, 34, 91, 93, 114, 137, 154, 206
Zhong, Guidan 鍾桂丹 (1889–1956) 137
Zhong ti Xi yong 中體西用 5, 49
Zhong, Ziyuan 鍾紫垣 (1861–1945) 137
Zhou, Enlai 周恩來 (1898–1976) 183, 238
Zhou, Xiangxian 周象賢 (b. 1890) 33
Zhou, Xicheng 周西成 (1893–1929) 156, 158, 161–2, 166, 168, 209
Zhou, Yichun 周詒春 (1883–1958) 136, 140–1, 148, 189
Zhou, Ziqi 周自齊 (1870–1923) 42, 47, 97
Zhu, Qinglan 朱慶瀾 (1874–1941) 106

www.ingramcontent.com/pod-product-compliance
Lightning Source LLC
Chambersburg PA
CBHW071324230926
51849CB00002B/313
* 9 7 8 9 0 0 4 5 4 9 4 7 0 *